# LOCAL ORDER IN CONDENSED MATTER PHYSICS

# ORGANIZATION

Chairman: P. Jena (USA)
Co-Chairman: S. N. Behera
Co-Chairman: S. D. Mahanti

# INTERNATIONAL ADVISORY BOARD

H. Alloul (France)
M. Ciftan (USA)
R. H. Cotts (USA)
S. K. Joshi (India)
E. Karlsson (Sweden)
W. D. Knight (USA)
C. K. Majumdar (India)
V. S. Ramamurthy (India)
C. N. R. Rao (India)
E. Recknagel (Germany)
A. M. Stoneham (U.K.)
T. Yamazaki (Japan)

# LOCAL ORGANIZING COMMITTEE

S. N. Behera: Chairman
B. Dev
A. M. Jayannavar
N. C. Mohapatra
B. K. Rao

# Local Order in Condensed Matter Physics

*Edited by*

**S. D. Mahanti**
Physics Department
Michigan State University
East Lansing, MI, U.S.A.

*and*

**P. Jena**
Physics Department
Virginia Commonwealth University
Richmond, VA, U.S.A.

Art Director: Christopher Concannon
Graphics: Elenor Kallberg and Maria Ester Hawrys
Manuscript Coordinator: Roseann Pena
Book Production: Tammy Sauter, and Michelle Lalo
Circulation: Irene Kwartiroff and Annette Hellinger

*Library of Congress Cataloging-in-Publication Data*

Mahanti, S. D.
Local order in condensed matter physics / S. D. Mahanti, P. Jena
p. cm.
Papers from the international symposium held 6/14-17/93 in Jekyll Island, Ga.
Includes bibliographical references and index.
ISBN 1-56072-220-7
1. Condensed matter -- Structure -- Congresses. 2. Hyperfine interactions -- Congresses. 3. Electronic structure -- Congresses. I. Jena, P. II. Title.
QC173.458.S75M34 1995 94-48606
530.4'11--dc20 CIP


*6080 Jericho Turnpike, Suite 207*
*Commack, New York 11725*
*Tele. 516-499-3103 Fax 516-499-3146*
*E Mail Novasci1@aol.com*



*Printed in the United States of America*

*Dedicated to*

*Professor Tara Prasas Das*
*on the occasion of his*
*sixtieth birthday*

# CONTENTS

# PREFACE

Many of the observed electronic properties of condensed matter systems (such as clusters of atoms, solids with long or short range order, amorphous and liquid metals) are governed by the local atomic arrangements around the probe site. One of the techniques that is particularly sensitive to the local environment involves the hyperfine interaction between the electrons and the probe nucleus. Experimentally, the hyperfine techniques have included nuclear magnetic resonance, electron spin resonance, Mossbauer spectroscopy, perturbed angular correlation, muon spin rotation and interaction with ion beams. Through these techniques, studies of the electronic properties have not only provided precise information about the electronic structure, but equally important information on local atomic structure and coordination have also emerged. It has been possible to understand, for example, local order in glasses, order-disorder transition, role of correlations and defects in high temperature superconductors, metal-insulator transition in expanded fluids, atomic structure in small metal clusters, diffusion and clustering of atoms on surfaces and other restricted geometries.

These experiments measure quantities at precise positions in space and consequently provide a critical test for the calculated electron eigenfunctions, structural models, and low frequency dynamics. Development of new computational methods aided by high speed computers has helped theorists to meet this challenge, and fundamental understanding exists in many areas although much more needs to be done.

An international symposium was held in Jekyll Island, Georgia between June 14-17, 1993 to review and discuss the effects of local atomic structure on properties of condensed matter systems. There were eleven plenary sessions and thirty-three invited papers. The topics included:

* Molecules and clusters
* Point defects and defect complexes in solids
* Hydrogen and positive muons in metals and semi-conductors
* Disordered solids and liquid metals

* Diffusion and clustering in bulk, on surfaces, and other restricted geometry
* High temperatures superconductors including $C_{60}$ assembled materials

Each session emphasized the manner in which electronic properties serve as a probe of local atomic structures. In atomic clusters, their small size makes the local atomic environment the only environment, while in imperfect solids, glasses, and surfaces the local environment is the immediate vicinity of the probe atom. The electronic and magnetic properties of clusters highlighted the effect of reduced size and dimension, while the studies of high temperature superconductors and the interaction of hydrogen in metals and semiconductors concentrated on the role of defects. The properties of Fullerene and Fullerene based materials illustrated the novel role local atomic order can play if solids are assembled from nanostructures instead of atoms. Atomic correlations in disordered systems such as alloys, glasses, and porous media were illustrated through nuclear magnetic resonance, Mossbauer spectroscopy, and computer simulation.

The symposium was two years in the planning stage and would not have been possible without the help of many individuals. These include the members of the International Advisory Board and the local organizing committee. We also would like to extend special thanks to Professor B. K. Rao and Mrs. Barbara Martin, who spent innumerable hours on this symposium. The financial support of the Army Research Office and, in particular, the generous advice of Dr. Mike Ciftan are greatfully acknowledged.

RICHMOND, VIRGINIA
APRIL 15, 1994

**S. D. MAHANTI**
**P. JENA**

Local Order
in Condensed
Matter Physics


ISBN 1-56072-220-7

# PERSPECTIVES IN HYPERFINE INTERACTIONS IN ATOMS

Stanley S. Hanna

*Department of Physics, Stanford University, Stanford, CA 94305, USA*

## PROLOGUE

This brief account of that area of science where atomic, nuclear, and solid-state physics are brought together by the atomic hyperfine interaction is dedicated to Professor Tara Prasad Das on the occasion of his sixtieth birthday. It is entirely fitting that we do this. Anyone seriously interested in this region of physics will come to know and depend on the works of Professor Das: from his early indispensable treatise of the quadrupole interaction, through his beautiful account of the nuclear hyperfine field and the atomic configurations that give rise to specific hyperfine interactions, to his recent clarifying work on the transient field, Professor Das and the school of physicists that he has nurtured have been a guiding and stimulating presence in this field. My only regret is that, in addition to our mutual interests in the nuclear and the transient fields, I have not taken greater advantage of the other, fine contributions of Professor Das – but I still have time before his seventieth and other anniversaries!

## 1. INTRODUCTION

It is a privilege and a pleasure to present the opening talk at this *International Symposium* in honor of my good friend Tara Das. Over the years our paths have been drawn together by our mutual interest in that fascinating area of science where the physics of atoms, nuclei, and condensed matter intersect – a region that is defined and nurtured by the *hyperfine interaction* in atoms.

In a very real sense, however, our paths originated in the same region of the World – his on one side of the Bay of Bengal and mine on the other. I have always valued my early connection with the land that produced physicists like Bose, Raman, and Chandrasekhar. I recall nostalgically that in my first college course in astronomy I learned of the great work of Meghnad Saha, who interpreted the temperature dependence of the solar spectrum as arising from the formation and destruction of ionic species in the sun. It was with great interest that I learned that Saha was one of Tara's teachers in his early career. I had hoped to return to that part of the World for this Symposium, but Jekyll Island is indeed an interesting substitute.

## 2. THE ATOMIC HYPERFINE INTERACTION

The hyperfine interaction (HFI) was discovered in high resolution atomic spectroscopy. Its interpretation confirmed that all the degrees of freedom in an atom – nuclear as well as electronic – must be taken into account. This quantum system is now extended to include the solid-state environment as well.

Once the level splittings of the hyperfine interaction were observed in direct transitions between the nuclear levels – as in NMR and Mössbauer spectroscopy – relatively low-resolution techniques were adequate to observe a myriad of new phenomena. It is to this new and exciting field that Tara Das has made so many seminal and beautiful contributions.

The theory of the atomic hyperfine interaction was first given in coherent form in 1936 in the prize essay by Casimir [1]. For historical interest we show the title page of this work in Fig. 1. In the now standard treatment the Hamiltonian of the hyperfine interaction is expanded over the static electromagnetic moments of the nucleus [2]:

ON THE INTERACTION
BETWEEN ATOMIC NUCLEI
AND ELECTRONS

BY

H. B. G. CASIMIR

**Maar juist hierin is voor een deel het nut van elke uitbreiding onzer natuurkennis gelegen, dat zij ons duidelijker voor oogen stelt, wat er nog te doen overblijft....**
*(H. A. Lorentz, Acad. Proefschrift)*

**PRIZE ESSAY PUBLISHED BY „TEYLER'S TWEEDE GENOOTSCHAP"**

HAARLEM
DE ERVEN F. BOHN N.V.
1936

Fig. 1. Title page of the prize essay on the *atomic hyperfine* interaction by H. B. G. Casimir [1]. "But just here lies the benefit that comes since each extension of our knowledge of nature shows us clearly what still remains to be done . . . " (H. A. Lorentz, Academic Thesis).

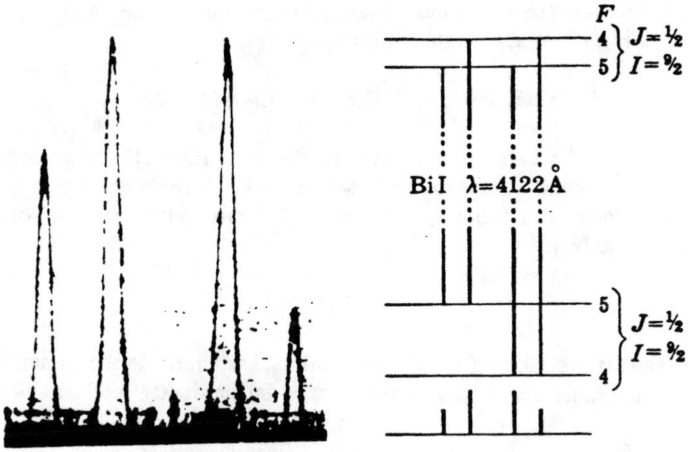


Fig. 2. Atomic hyperfine spectrum of the 4122 Å line of BiI [3].

$$\mathcal{H}_{HFI} = \mathcal{H}_{E0} + \mathcal{H}_{M1} + \mathcal{H}_{E2} + \ldots \tag{1}$$

Owing to its basic parity the nucleus does not possess the odd static electric moments, E1, E3, . . ., nor the even static magnetic moments, M0, M2, . . . The terms can be expressed in symbolic form as follows:

$$\mathcal{H}_{E0} = q_M V_0 \ , \tag{2}$$

$$\mathcal{H}_{M1} = -\mu \cdot B_0 \ , \tag{3}$$

$$\mathcal{H}_{E2} = Q \cdot \nabla E_0 \ . \tag{4}$$

where the energies are expressed in terms of the electronic potential $V_0$, magnetic field $B_0$, and electric field gradient $\nabla E_0$ (all at the nucleus) interacting with the nuclear monopole moment $q_M$, magnetic moment $\mu$, and quadrupole moment Q, respectively.

A classical atomic HFI spectrum, illustrated in Fig. 2, shows the spectrum for the $\lambda$ = 4122 Å line in BiI. Here we see the transitions between two different electronic states, both having J = 1/2, which are split into sublevels by the quantized orientation of the electron angular momentum relative to I = 9/2, the nuclear spin. While this spectrum provides a *textbook* example [3] of the HFI, it is the observation of transitions between the nuclear magnetic sublevels within the same electronic state, and the subsequent observation of transitions between two different levels within the same nucleus (still within the same electronic configuration) that provides the foundation of the exciting field we are discussing. The solid-state environment comes into play through the perturbations it produces on these levels, or in a major way, e.g., through phononic excitations. Indeed, it is necessary to include in this environment the influence of external (sometimes distant!) "solids" which produce electric, magnetic, or gravitational perturbations.

In our historical and sometimes nostalgic trip through the development of this field, we can only point out a few of the high points in each HFI category. However, we hope this brief account will stimulate readers to pursue some of the topics further and to learn more of the contributions made by Tara Das to this fascinating subject.

## 3. THE MONOPOLE INTERACTION

The first term in the expansion of the Hamiltonian Eq. (2), the monopole HFI, is perhaps best illustrated by the isomer shift as revealed in the Mössbauer spectrum [4]. For a nucleus of constant density and radius R, the energy shift due to the extension of the nuclear volume can be shown to be [5]

$$\delta E_{M0} = \frac{2\pi}{5} Ze^2|\psi(0)|^2 R^2 \quad . \tag{5}$$

Thus, an isomer shift is produced if the nuclear radius R changes in the transition between the ground and excited states and if the electron charge density $|\psi(0)|^2$ at the nucleus is differ-

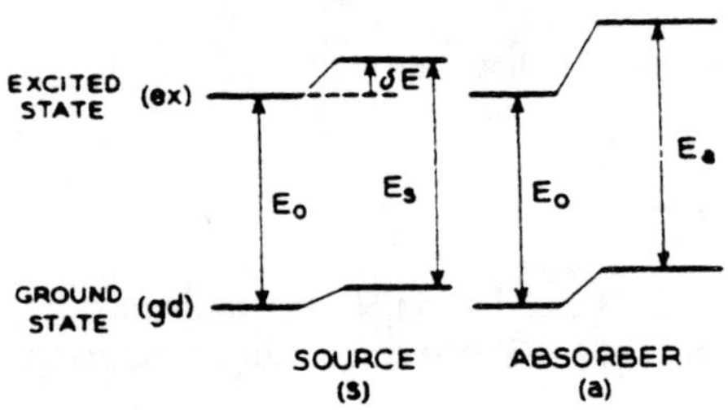


Fig. 3. The isomer shift in a Mössbauer experiment is $(2\pi/5)\,Ze^2[|\psi_a(0)|^2 - |\psi_s(0)|^2]\,[R_{ex}^2 - R_{gd}^2]$ [5].

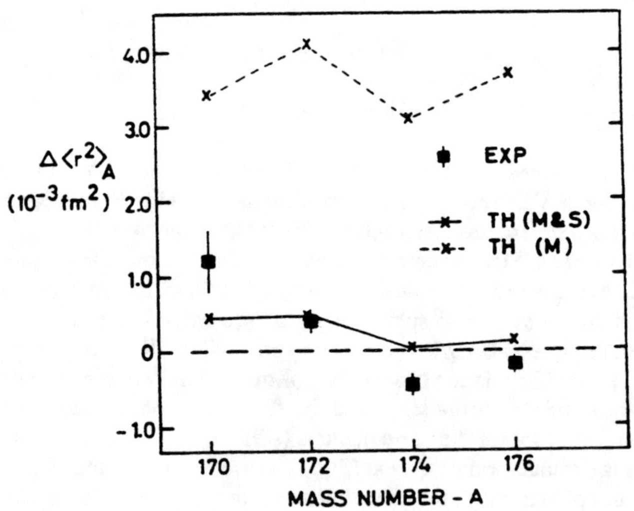


Fig. 4. Comparison of experimental [6] and theoretical values of $\delta\langle R^2\rangle$ for the $0^+ \rightarrow 2^+$ transitions in the even-even ytterbium isotopes, where $\delta\langle R^2\rangle$ is derived from the isomer shift; see Fig. 3.

ent in source and absorber, as illustrated in Fig. 3. In particular, if the electronic configurations at the nuclei in the source and absorber are known or calibrated, one can tell whether a nucleus expands or contracts on excitation and by how much.

A very interesting example is afforded by the nuclear isotope series $^{170,172,174,176}$Yb in which, as we go from left to right, two neutrons are added to the nucleus each time. These are rotational nuclei in which the Mössbauer transition is made between the ground state with angular momentum equal to zero, I = 0, and the first rotational level with I = 2. The result of the experiment [6], given in Fig. 4, shows that for $^{170}$Yb the nucleus expands when it rotates (as expected, classically speaking) but as neutrons are added, the expansion decreases, becoming quite negative at $^{174}$Yb and remaining negative for $^{176}$Yb. This astonishing result was explained beautifully by Meyer and Speth [7] who showed that it is due to the shell-model structure of deformed nuclei, wherein the nuclear rotation promotes nucleons into orbits with smaller effective radii. In Fig. 4 we see that the theory, labeled M+S, falls a little short in predicting the "negative" experimental result, but there is no doubt about the correctness of the explanation. A "classical" theory, labeled M, is shown by the dashed line.

Measurements of nuclear charge radii can also be carried out with high precision by means of electronic excitations with laser light [8]. An amazing series of measurements in the sequence of isotopes $^{190-214}$Pb is displayed in Fig. 5.

Reference to Fig. 3 shows that the isomer shift can also be used to determine relative changes in the electronic environment at the nucleus, if the nuclear transition is kept the same in a Mössbauer experiment. This powerful tool was used [9] to construct a chart, Fig. 6, in which the electronic charge densities at the Fe nuclei in Fe compounds are sorted out according to the electronic configuration in Fe, $3d^n4s^x$, where n = 4, ..., 8 gives the basic d-wave configuration and x the s-wave mixing. Similar charts have been determined for other Mössbauer nuclei [10] (see Fig. 7) to provide detailed and precise information on atomic configurations heretofore unavailable.

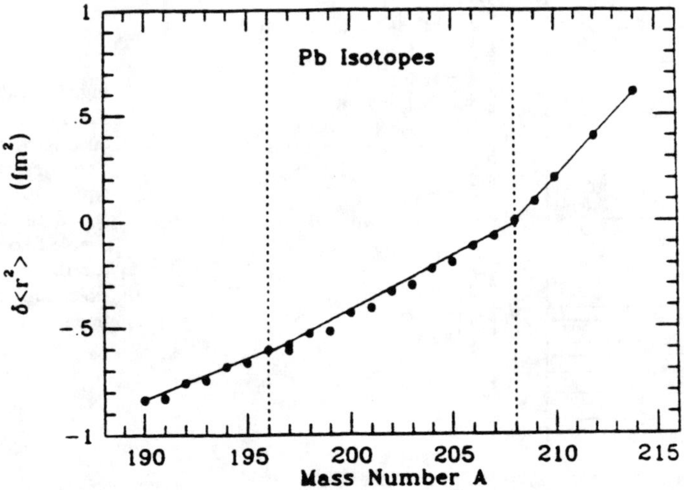


Fig. 5. Mean-square charge radii of the isotopes of Pb measured by laser spectroscopy [8]. Note the staircase effect produced by even and odd values.

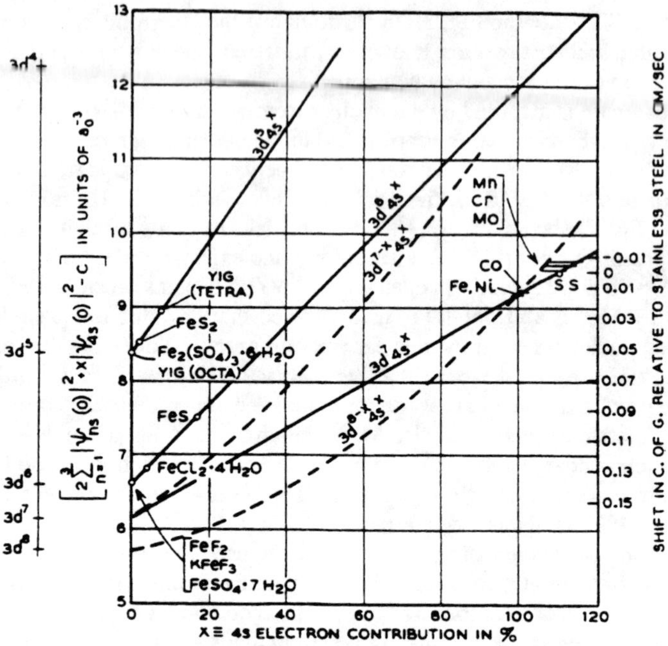


Fig. 6. Interpretation of the $^{57}$Fe Mössbauer isomer shifts in various solids containing Fe [9]. For various d-electron configurations the total s-electron density at the nucleus is plotted as a function of the percentage of 4s character. The data are located on the various lines according to their isomer shift given on the right ordinate.

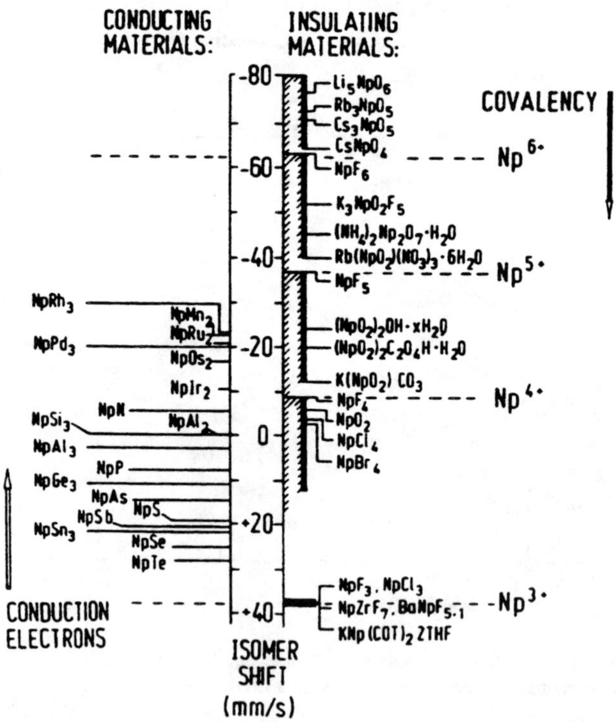


Fig. 7. Mössbauer isomer shifts for insulating and conducting Np compounds [10]. The designations $Np^{3+}$, $Np^{4+}$, $Np^{5+}$ and $Np^{6+}$ correspond to the shifts of the simple fluorides $NpF_n$. Only for the two lowest charge states are they equal to the free ion values.

## 4. THE DIPOLE INTERACTION

As with the monopole HFI, the magnetic dipole HFI can be used to provide both nuclear and atomic information and by way of the latter, solid-state information as well. The most striking example of the dipole HFI is the nuclear Zeeman effect which was first observed and explained in the Mössbauer effect in $^{57}Fe$ [11]. Here, instead of the electronic potential that gives rise to the isomer shift, it is the effective magnetic field produced at the nucleus by the electronic spin and orbital motion that produce the Zeeman splitting of the nuclear levels.

Figure 8 shows the first nuclear Zeeman spectrum [11] along with the Zeeman level diagram deduced from it. Since the nuclear ground-state magnetic moment of $^{57}Fe$ was known precisely [12], this spectrum measured the excited-state moment with essentially the same precision. The six-line Zeeman spectrum of $^{57}Fe$ was subsequently confirmed in many cases [5]. Many excited state nuclear moments have since been measured with high precision by the nuclear Zeeman effect. Figure 9 shows the early result for $^{119}Sn$ [13,14].

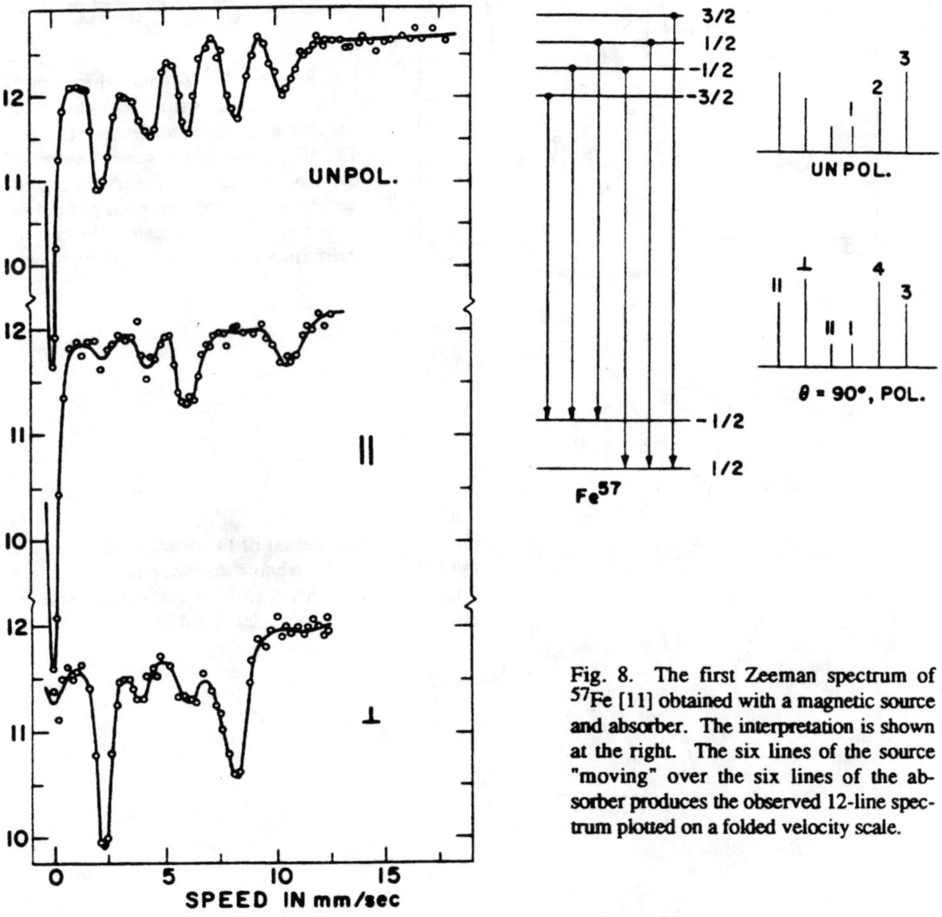


Fig. 8. The first Zeeman spectrum of $^{57}Fe$ [11] obtained with a magnetic source and absorber. The interpretation is shown at the right. The six lines of the source "moving" over the six lines of the absorber produces the observed 12-line spectrum plotted on a folded velocity scale.

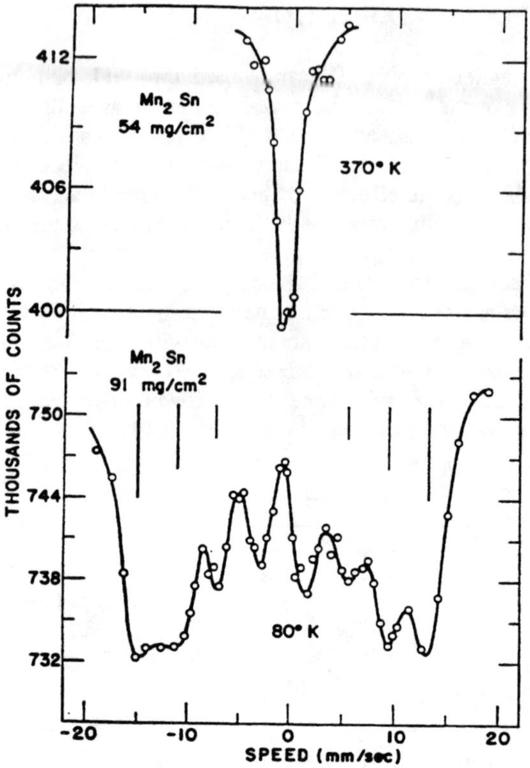


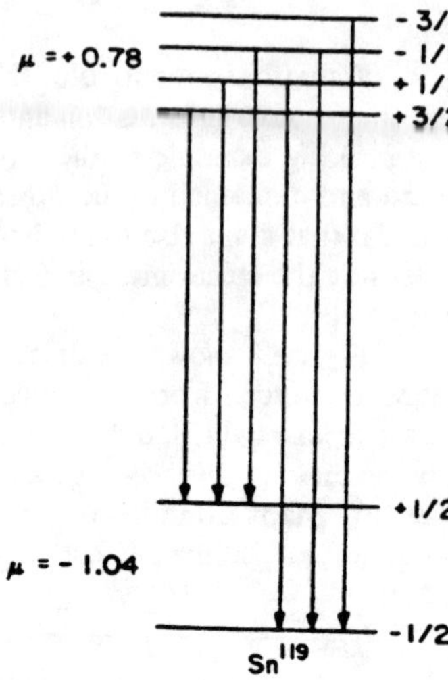


Fig. 9. The first Zeeman spectrum of $^{119}$Sn obtained with a non-magnetic source and a magnetic absorber [13,14]. The single line of the source "moving" over the six lines of the absorber produces the six line spectrum plotted on an unfolded velocity scale. The two interior lines are from an impurity.

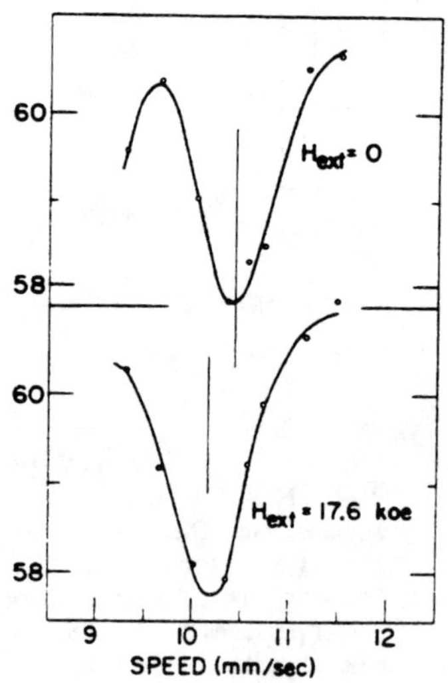


Fig. 10. Determination of the direction of the nuclear field of $^{57}$Fe [15]. When a magnetic field is applied to the source, the inward shift of the outermost line of Fig. 8 shows that the nuclear field in Fe is negative.

Additional striking phenomena were revealed by the determination of the magnetic field at the nucleus. (Knowledge of the nuclear ground-state moment also determines the magnitude of the effective magnetic field at the nucleus in the Zeeman spectrum.) In the case of metallic Fe this field turned out to be $H_{eff} = -333$ kG, the minus sign [15] indicating that the field was opposite to the direction of magnetization in the ferromagnetic environment (see Fig. 10). An early prediction [16] indicated a positive value for the field; although this value turned out to be wrong, the work indicated the mechanisms whereby a negative field could arise. These developments led Tara Das to his great series of papers in which he obtained the experimental result precisely and elegantly [17,18]. The heading of Ref. 18 is shown in Fig. 11 and the results of his calculations are reproduced in Table I of Ref. 18. See Fig. 12.

The magnetic field at the nuclei of magnetic media has become a testing ground for theories of magnetic configurations and other properties. We cite one further example of the Heussler alloy ($Pd_2MnSb$) [19] shown in Fig. 13; in this case the internal field is positive, showing that indeed the value of an internal field at a nucleus is the result of a delicate balance between positive and negative terms, providing thereby a stringent test of theoretical calculations. The systematics of the nuclear fields in Heussler alloys has been studied by one of our chairmen, Puru Jena [20].

HYPERFINE FIELDS AND NEUTRON SCATTERING IN FERROMAGNETIC METALS*

T. P. Das

Laboratorium für Festkörperphysik, ETH, Hönggerberg, 8093 Zürich, Switzerland

and

Department of Physics, State University of New York, Albany, New York, 12222, USA**

* Supported by U. S. National Science Foundation

** Permanent Address

ABSTRACT

The origin of ferromagnetism in the transition metal ferromagnets, iron, cobalt, and nickel is discussed, from an ab initio band structure point of view, with proper attention to the explicit roles of exchange, correlation and hybridization effects. The influence of these effects and all the mechanisms such as direct, exchange core polarization and many-body effects that have been found important for the hyperfine properties of atomic systems are included in attempting to understand the experimentally observed hyperfine fields at the nuclei in these metals. Spin-density distributions using calculated spin polarized band wavefunctions are used to make comparisons with experimental neutron scattering data. The impact of the results of analyses of hyperfine fields at the nuclei and spin density distributions on the origin of hyperfine fields at muon sites is discussed.

Fig. 11. Heading of paper [18] in which T. P. Das summarizes his work on nuclear fields in ferromagnetic metals and applies the results to neutron scattering.

Table III. Contributions[a] to Hyperfine Fields in Iron, Cobalt, Nickel (in KOe)

| Mechanism | | Iron | Cobalt | Nickel |
|---|---|---|---|---|
| | | Metal | | |
| ECP | 1s | -1 | 2 | -8 |
| | 2s | -622 | -278 | -71 |
| | 3s | 243 | -79 | 14 |
| | Total | -380 | -355 | -65 |
| Band Contributions | (A) d-like | | 13 | 40 |
| | (A) free-like | | -56 | -60 |
| | (B) unpaired electrons | 177 | 77 | |
| | (B) paired electrons | -144 | -120 | |
| | Total (A) or (B) | 33 | -43 | -20 |
| | Total Field | -347 | -398 | -85 |
| | Experiment[b] | -339 | -223 | -75 |

[a]References 21a,b,c. [b]References 19a-d.

Fig. 12. Table given by T. P. Das [18] showing calculations of nuclear fields in ferromagnetic metals. For Refs. a and b see Ref. 18.

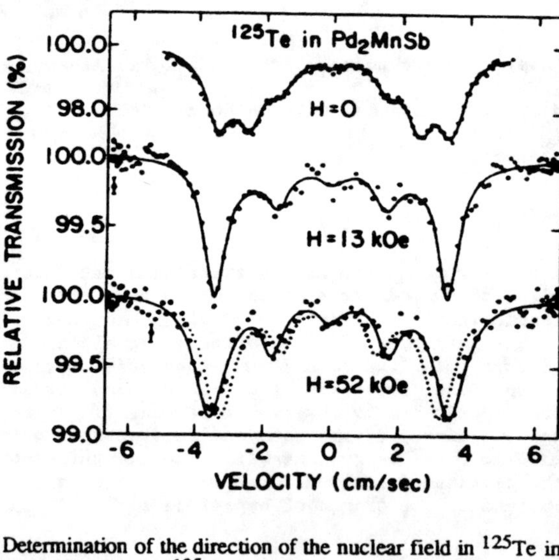


Fig. 13. Determination of the direction of the nuclear field in $^{125}Te$ implanted in the Heussler alloy $Pd_2Mn^{125}Sb$ [19]. When a longitudinal field of 32 kOe is applied to the $Pd_2Mn^{125}Sb$ source a polarized four-line spectrum is produced. When the 52 kOe field is applied the spectrum expands indicating a positive nuclear field. The Mössbauer absorber was ZnTe.

## 5. THE NUCLEAR HYPERFINE FIELD

The picture that has emerged for the nuclear fields is one in which the bulk magnetism is produced by the polarization of the d-electrons (in the transition elements) which, however, makes only a minor direct contribution to the nuclear field. It is the core polarization of the s-electrons produced by the d-electrons that makes a decisive contribution to the nuclear field – in particular by providing strong "negative" components.

The above picture is supported by experiments with polarized positrons [21,22], in which the positrons come to rest in magnetized iron and annihilate with the atomic electrons of Fe. The angular correlation between the two quanta of the (two-quantum) annihilation is measured when the direction of magnetization and that of the positron spin are parallel and then antiparallel. Annihilation with the low momentum s electrons produces a $\gamma$-$\gamma$ angular correlation tightly distributed about 180°. In contrast to this, annihilation with the high momentum d electrons produces a wider distribution about 180°. If the d electrons are selected geometrically in the angular correlation it is found that the two-quantum decay from annihilation with d electrons increases when the positron spin and magnetization are parallel and decreases when they are antiparallel. Since the electron spin is opposite the magnetization and the $J = 0$ (spins opposite) state produces the two-quantum annihilation, the experimental result in Fig. 14 shows conclusively that it is the spins of the d electrons that produce the bulk of the magnetization. (If it were the s electrons, just the opposite result would be obtained in Fig. 14.)

In summary, the d electrons provide the bulk magnetization but very little field at the nucleus, whereas the s electrons produce very little magnetization but, because of their maximum overlap with the nuclei, they provide most of the nuclear field. The s electron (core) polarization is indeed small and is induced by the d-polarization. The calculation [18] shows that, as a rule-of-thumb, s electrons spending most of their time outside the d orbit contribute positive terms to the nuclear field, while those inside contribute negative terms.

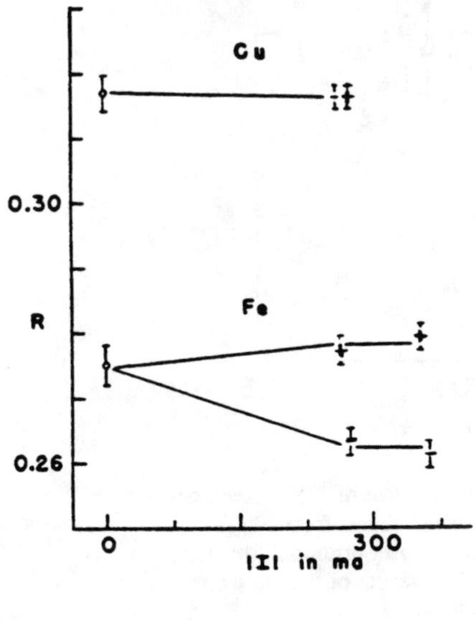


Fig. 14. Normalized two-quantum yield from $\vec{e}^{\,+}\vec{e}^{\,-}$ annihilation in the wings of the angular correlation vs. magnetizing field. For + points the field was parallel to the positron spin; for – points the field was reversed. A large effect was seen for ferromagnetic Fe, none for Cu. See Ref. 21.

## 6. THE NUCLEAR HYPERFINE FIELD IN FREE IONS

In addition to the nuclear Zeeman effect, another very striking phenomenon produced by the nuclear hyperfine field is the Larmor precession of the nuclear magnetic moment (and hence of the nucleus itself) in this field. A measurement of the precession rate yields a value of the nuclear moment if the field is known, and vice versa. This technique was first used for one-electron ions [23] for which the nuclear field can be calculated with high precision. The method consists of exciting and orienting the nuclear state of interest by means of a nuclear reaction which also ejects the atoms from a thin target into vacuum. The velocities imparted to the atoms are selected so that when they emerge from the target most of them have been stripped down to the one electron configuration. By monitoring the time behavior of the radiation from the excited nuclear state, the precession rate of the nuclear moment can be determined and its value measured, since the one-electron field is known.

The result of a recent experiment [24] is given in Fig. 15, which shows the precession of the $3^-$ level of $^{16}N$ in the hyperfine field produced at the nucleus by the one electron in a $N^{6+}$ ion. Details of the method and the experiment can be found in Ref. 24. In this case the nuclear field is calculated to be 5.749 kT.

This method can be extended to atoms or ions which have electronic configurations involving more than one electron but are still simple and well enough established to produce a known magnetic field, consisting of one or more components, at the nucleus – an example would be a configuration of the alkali atom type with one s-electron outside a "closed" core.

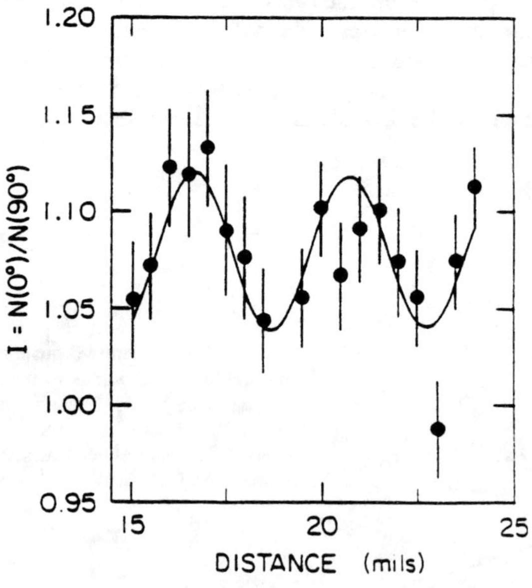


Fig. 15. Precession curve of the 297 keV, $3^-$ state of $^{16}N$ in the one-electron ion $N^{6+}$ [24]. The ratio of gamma-ray counts from $^{16}N(3^-)$ at 0° to counts at 90° is plotted as a function of stopper distance. This ratio determines the orientation of the nucleus. The length of time in the precessing field is determined by the stopper distance.

In this case the precession of the nucleus could be identified predominantly with the field due to the "valence" electron. If the configuration produces more than one dominant field then the precession would consist of several superposed rotations which would need to be separated in the analysis. Experiments of this nature have not yet been pushed to a successful conclusion [25].

A more successful approach has been to apply the method of magnetic decoupling, which has been done for both $^{16}O$ [25,26] and $^{40}Ca$ [25,27] ions. As above, the experiment starts by exciting and orienting the nuclear states of interest by means of a nuclear reaction which also ejects the atoms from a thin target into vacuum. The velocities imparted to the atoms are now selected so that when they emerge from the target they are sufficiently ionized to produce hyperfine fields large enough to cause observable rotations of the nuclear spin and corresponding perturbations of the γ-ray angular distribution of the excited state. An external magnetic field is applied along the orientation axis (i.e., the beam direction) and increased until the nuclear moments are decoupled from the hyperfine fields (i.e., the Larmor precession of each atomic moment becomes so rapid that the nuclear moment can no longer follow it). The decoupling process can be observed through the perturbation of the angular distribution, and its dependence of the external field (the "decoupling curve") is a sensitive measure of the product of the nuclear moment and the atomic hyperfine fields. Decoupling curves for $^{16}O$ ions and $^{40}Ca$ ions are shown in Figs. 16 and 17, respectively, At the present time, the precision of the method for measuring moments rests primarily on the accuracy with which the atomic fields can be determined. Fortunately, it is usually possible to produce ionization states by appropriate velocity selection in which a single atomic configuration dominates the hyperfine interaction. These configurations are usually unpaired s states for which fairly reliable calculations can be made. Moreover, as the decoupling field is increased, the weak fields decouple first and the strong fields later on so that, to some extent, the various hyperfine fields are treated separately. This partial separations of the decoupling fields improves the accuracy of the analysis. It is to experiments of this kind that the work of Tara Das makes such important contributions, as shown in his beautiful paper presented at the *Bangalore Hyperfine Interaction Conference* [28] (see Fig. 18).

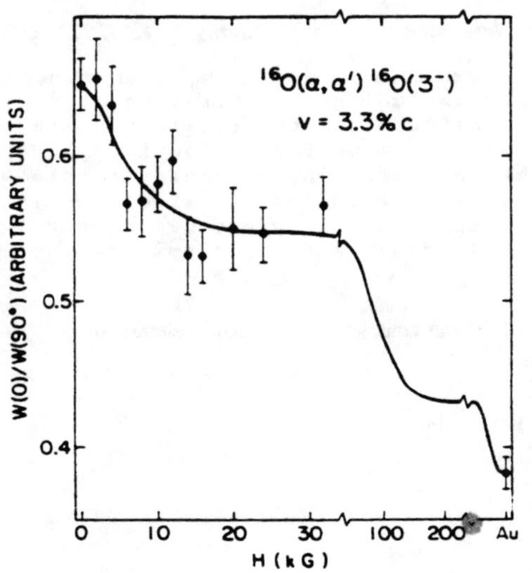


Fig. 16. Decoupling curve for $^{16}O$ ions at velocity 3.3% c [25,26]. The curve is derived from the statistical model of electrons in the L- and K-shells. The point at the far right-hand side labeled "Au" is for implantation into solid Au, where the decoupling process is complete. The first drop is due to the decoupling of 2s electrons, the second to 1s electrons.

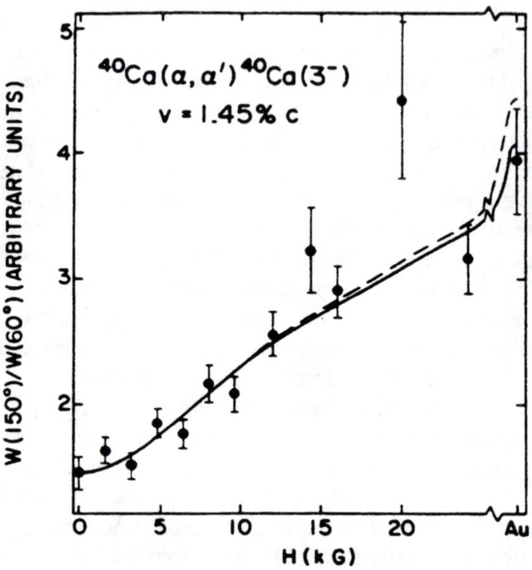


Fig. 17. Decoupling curve for $^{40}Ca$ ions at velocity 1.45% c [25,27]. The curves are obtained with the statistical model and 3s, 3p, and 4s fields. For the solid curve the fraction of perturbed ions is 82%; for the dashed curve the fraction is 89%. The point at the far-right side labeled "Au" is for implantation into solid Au, where the decoupling process is complete.

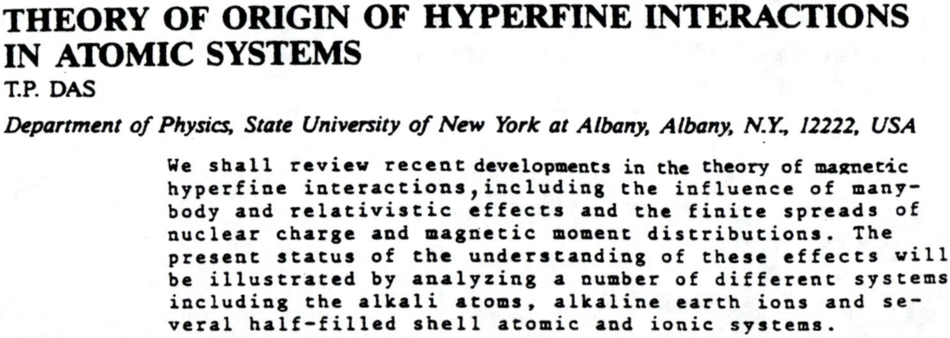

# THEORY OF ORIGIN OF HYPERFINE INTERACTIONS IN ATOMIC SYSTEMS

T.P. DAS

*Department of Physics, State University of New York at Albany, Albany, N.Y., 12222, USA*

We shall review recent developments in the theory of magnetic hyperfine interactions, including the influence of many-body and relativistic effects and the finite spreads of nuclear charge and magnetic moment distributions. The present status of the understanding of these effects will be illustrated by analyzing a number of different systems including the alkali atoms, alkaline earth ions and several half-filled shell atomic and ionic systems.

Fig. 18. Heading of paper [28] in which T. P. Das calculates the hyperfine interaction for various important atomic configurations.

## 7. THE TRANSIENT FIELD

The magnetic dipole HFI can manifest itself in indirect ways to produce interesting phenomena which provide important tools for measuring nuclear magnetic moments. One such phenomenon is the transient field [29] induced at the nucleus of an ion traversing a magnetized medium. This field is large enough to produce a measurable precession of the nuclear moment even though it acts for only a very short time.

Unfortunately, until now the mechanism that produces the transient fields has been imperfectly understood, and consequently there has been no reliable basis for calculating these fields. Nevertheless, many moments of excited nuclear states have been measured by empirically calibrating the fields. One technique that was used on several isotopic sequences was to produce all the available isotopes of a series simultaneously, subject them all to the same transient field, and then measure simultaneously the precessions produced in each isotope. The method depends on already knowing the magnetic moment of one of the isotopes, which then determines the moments of all the others. The technique is attractive, however, since it is an important question how the magnetic moment of (the first-excited state of) a rotational nucleus varies across an isotopic sequence, i.e., as neutrons are added to the nucleus. Classically, one would expect that adding two neutrons with paired-off spins would produce only a *small* effect on a rotating nucleus due to the dilution of the chief effect produced by the rotating protons (see the dashed line in Fig. 19).

Figure 19 shows the result of such an experiment [30] carried out on the isotope sequence $^{120,122,124,126,128}$Te. We will not pursue the interpretation of this experiment further here, but it is clear from the scatter in the points that the empirical knowledge of transient fields is poor. Hence, I would like to express my delight at learning that Professor Das [31]

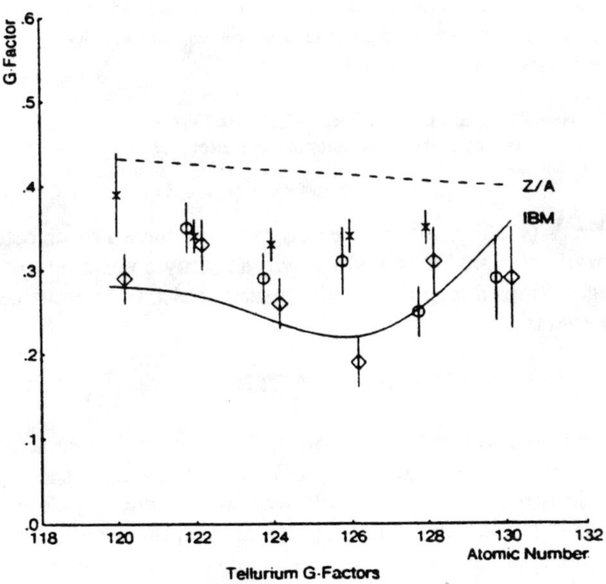


Fig. 19. The g-factors for the tellurium isotopes. The ×'s were measured simultaneously as explained in the text. The other measurements are from various authors. The dashed line represents the classical result; the solid line an Interacting Boson Model calculation. See Ref. 30.

**Spin Exchange between Ion Probes and Localized Moments in Ferromagnets as the Origin of Transient Fields.**

**F.Hagelberg, T.P.Das**
**Department of Physics, State University of New York at Albany, Albany, NY, 12222.**

**K.-H.Speidel**
**Institut fur Strahlen- und Kernphysik, Universität Bonn, D - 5300 Bonn 1, Germany.**

**Abstract.**

The Transient Field phenomenon has been ascribed to a polarization transfer between the electrons of the ionic projectiles and the surplus of majority spin electrons of the ferromagnetic host over the minority spin electrons. Earlier attempts to explain this crucial process failed to account for the order of magnitude of the experimentally observed Transient Field strengths. A recent model which proposes spin exchange scattering between bound projectile electrons and quasi-free host electrons as mechanism of polarization transfer arrives at the correct orders of magnitude but is in conflict with the weak velocity dependence of the experimental polarization, exhibiting a strongly decreasing behavior with increasing velocity. The new model presented here proposes spin exchange between the ionic shell and localized electrons of the ferromagnet as a more adequate approach to the problem. It is shown that calculations involving hydrogen-like ions explain the size of the experimentally observed polarization effects as well as their velocity dependence for various ion probes traversing thin iron foils.

Fig. 20. Paper by F. Hagelberg, K.-H. Speidel, and T. P. Das [31] on transient fields. See also the report by the first author at this conference.

has turned his attention to these fields (see Fig. 20) and we may expect to have a much better understanding of the mechanism of their production which will not only contribute to our basic understanding of the passage of ions through magnetic media but also to improvement in the measurement of nuclear moments.

## 8. THE QUADRUPOLE INTERACTION

As in the case of the monopole and dipole interactions, the quadrupole interaction, Eq. (4), which brings into play the electric field gradient (EFG) at the nucleus, has led to a wealth of atomic and solid-state information. For the early work on the quadrupole resonance the landmark treatise by T. P. Das and E. L. Hahn [32] has been the standard work (see Fig. 21). Research on the quadrupole interaction again became very active when it was observed [4] in the Mössbauer effect, either in pure quadrupole spectra or as a perturbation in dipole spectra. The work was greatly stimulated by a new NMR method [33] that extended quadrupole (and dipole) measurements to unstable nuclei. In this method the unstable nuclei are polarized (by any one of several techniques) and the polarization is held in a magnetic field. NMR in this applied field is achieved when the polarization is destroyed by inducing resonant transitions between the nuclear sublevels by means of a resonant rf field. The destruction of the polarization is detected by observing the change in the radiation pattern of the decaying nuclei, and the resonant frequency gives both the dipole and quadrupole moments if the magnetic field and EFG at the nucleus are known.

For some simple and special configurations the EFGs have been calculated with sufficient precision and reliability to make measurements of nuclear quadrupole moments possible; but, the usual procedure is to calibrate the EFG with a known quadrupole moment and then to measure unknown moments in the same environment. A timely application of this procedure is to the measurement of the quadrupole moments Q of the radioactive $^{8}$Li and $^{8}$B nuclei by use of the polarized nucleus-NMR technique. In the first experiment [34] polarized $^{8}$Li nuclei ($\tau_{1/2} = 0.84$ s) were produced by the reaction $^{7}\mathrm{Li}(\vec{d},p)\,^{8}\vec{\mathrm{Li}}$. An NMR spectrum of

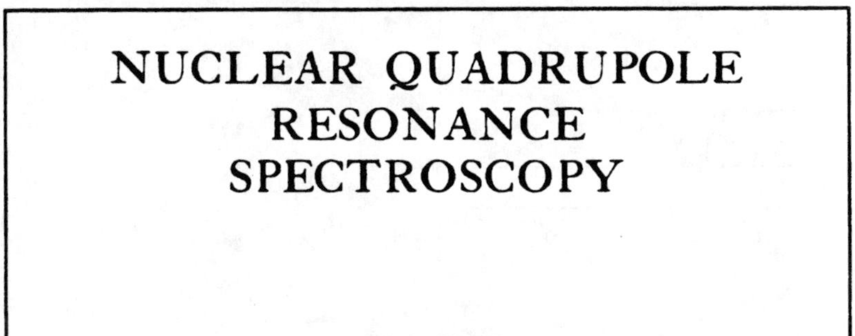

NUCLEAR QUADRUPOLE
RESONANCE
SPECTROSCOPY

T. P. DAS
*Saha Institute of Nuclear Physics, Calcutta, India*

E. L. HAHN
*Department of Physics, University of California, Berkeley, California*

ACADEMIC PRESS INC., PUBLISHERS
NEW YORK • LONDON • 1958

Fig. 21. The landmark paper which covered the early work on nuclear quadrupole spectroscopy and provided great impetus to the field [32].

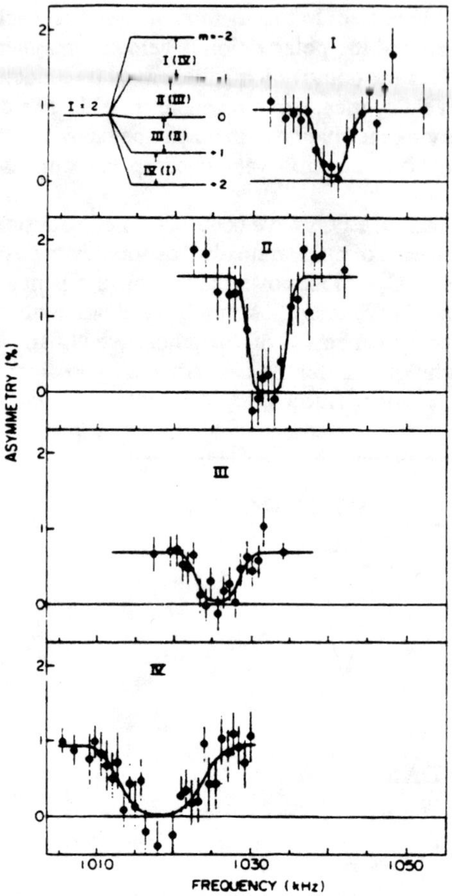


Fig. 22. The four NMR lines observed for $^{8}\vec{Li}$ implanted into a $LiIO_3$ crystal in a field of 1.63 kG [34].

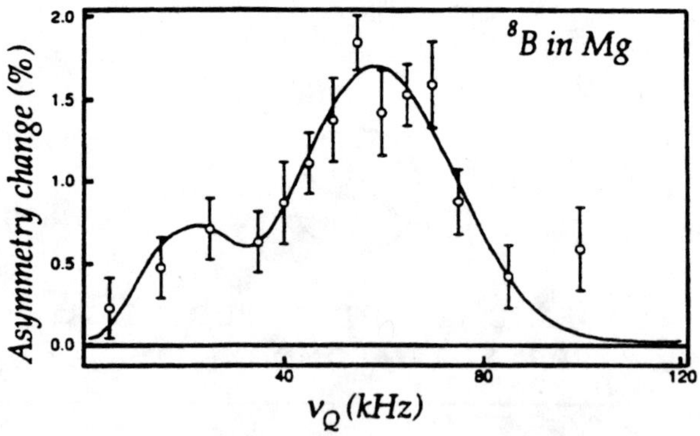


Fig. 23. The NMR of $^{8}\vec{B}$ implanted into Mg obtained with the new method [37] which measures the quadrupole perturbation directly.

$^{8}\vec{Li}$ implanted substitutionally into the compound $LiIO_3$ is shown in Fig. 22. The quadrupole interaction manifests itself in an unequal spacing of the magnetically split level. The four NMR resonances observed as changes in the β-decay radiation pattern are shown. The EFG at the $^{8}\vec{Li}$ nuclei in this compound was obtained from a knowledge of the quadrupole interaction of $^{7}Li$ in the same compound $LiIO_3$ [35] and the known $Q(^{7}Li)$.

The value obtained for $|Q(^{8}Li)|$ in this way was in poor agreement with the result of Ackermann et al. [36] who produced the polarized $^{8}\vec{Li}$ nuclei by the reaction $^{7}Li(\vec{n},\gamma)\,^{8}\vec{Li}$. In a recent measurement of the $^{8}Li$ moment Minamisono [37] has traced the discrepancy to the value [35] used in Ref. 34 for the quadrupole interaction of $^{7}Li$ in $LiIO_3$. All measurements now agree on the value $|Q(^{8}Li)| = 32.7 \pm 0.6$ mb obtained in Ref. 37.

This remeasurement of $|Q(^{8}Li)|$ used a new very sensitive NMR technique [37] which was incorporated into the polarized nucleus-NMR method. The new technique is designed specifically to measure the differences in level spacings caused by the quadrupole perturbation (see Fig. 22). This new technique was also used to measure the Q of $^{8}B$ ($\tau_{1/2} = 0.77$ s) with the result shown in Fig. 23. In this case the EFG in $^{8}B$ implanted in Mg was obtained by measuring the quadrupole interaction of the boron isotopes $^{8}B$ and $^{12}B$ in the same Mg environment and also $^{11}B$ and $^{12}B$ in the same environment in BN [37]. By this series of measurements (all done with the same technique) the Q's of $^{8}B$ and $^{12}B$ were measured relative to the Q of $^{11}B$ which is known. The measurements give [37]

$$|Q(^{8}B)| = 68.3 \pm 2.1 \text{ mb},$$
$$|Q(^{12}B)| = 13.21 \pm 0.26 \text{ mg}.$$

The values of Q for $^{8}Li$ and $^{8}B$ have contributed to the recent excitement over the discovery of neutron haloes in the light nuclei, especially in $^{11}Li$ [38]. If we compare the Q values for $^{8}B$ and $^{8}Li$ with that for $^{12}B$ we see that the former are much larger. Shell model calculations [37] show that the exceptionally large value for $^{8}B$ is the result of the five protons in $^{8}B$ occupying a much larger volume than do the neutrons, as shown in Fig. 24. This is the first that such a *proton skin* has been found in nuclei. Similarly the large Q of $^{8}Li$ can be attributed to a *neutron skin*.

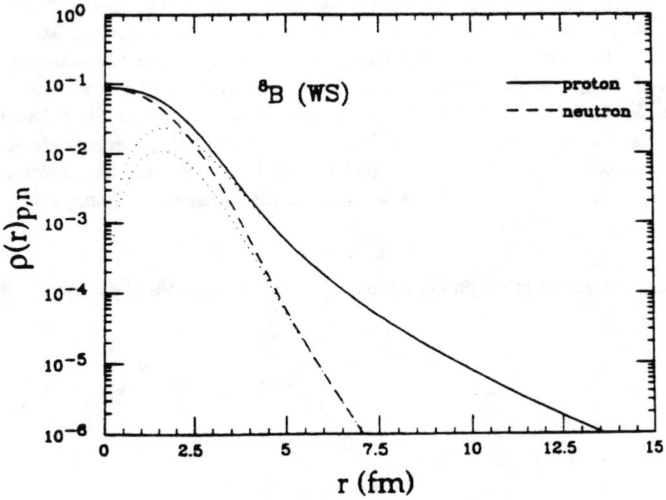


Fig. 24. The proton and neutron density distributions in $^{8}B$ that produce the measured quadrupole moment of $^{8}B$ [37]. The striking difference between the proton and neutron distributions is the first evidence for a *proton skin* [37].

In recent times the physics of surfaces has become a fertile area of investigation. Characteristically, Tara Das has made significant contributions in this area [39], by relating the properties of atoms adsorbed onto silicon surfaces to the microscopic environment of the atoms (see Fig. 25). By the nature of the problem the quadrupole interaction plays an important role in these studies. With his usual insight Tara Das points the way for important experimental research.

In our trip through the region where atomic, nuclear, and solid-state physics is brought together by the hyperfine interaction, we have tried to show that Tara Das is always to be found on the "cutting edge" of the physics. Therefore it is safe to predict that he will continue to shed light on the problems that will arise, stimulating fellow theorists and providing invaluable help and guidance to the toiling experimentalists.

**ELECTRONIC STRUCTURE AND ASSOCIATED PROPERTIES OF ADSORBED ATOMS ON SILICON SURFACE INCLUDING HYPERFINE INTERACTIONS**

S. M. MOHAPATRA,* B. N. DEV,** L. LUO, T. THUNDAT,***
W. M. GIBSON, K. C. MISHRA,**** N. SAHOO, and T. P. DAS
*Department of Physics, State University of New York at Albany, Albany, NY 12222, USA*

Received 6 April 1990

**Using a Hartree-Fock cluster procedure, we have investigated the electronic structures and energetics of adsorbed atoms at the surface of silicon. The systems we have studied involve halogen, chalcogen, the third group metal atoms and alkali metal atoms. Using the results of our investigation, we have determined the locations of the adsorbed atoms, local densities of states, and the vibrational frequencies and amplitudes of the adsorbed atoms. Good agreement is found with experimental data wherever they are avaliable. The adsorption sites obtained for the various classes of atoms are found to be in keeping with expectations from chemical bonding considerations. Predictions are made for the nuclear quadrupole interaction constants associated with the nuclei of the adsorbed atoms which are expected to be measurable by a variety of experimental techniques.**

Fig. 25. Paper on the properties of atoms adsorbed on the surface of silicon. Predictions of quadrupole interactions are included [39].

We have seen the evidence for the creativity, knowledge, skill, and power shown by Tara Das throughout his career. As further testament to his extraordinary productivity we give in Fig. 26 the announcement of the revision of his book [40] dealing with the central theme of his work – the electron.

*Forthcoming...*

**RELATIVISTIC QUANTUM THEORY OF ELECTRONS**

**Second Edition**

**Tara P. Das,** ***State University of New York, Albany***

**June 1993384 pp.; 50 illus.**

**72. 506551-4 $45.00/$36.00**

Fig. 26. Announcement of the forthcoming revision of the book by Tara Das [40].

## REFERENCES


1. H. B. G. Casimir, *On the Interaction between Atomic Nuclei and Electrons*, Prize essay published by Teyler's Tweede Genootschap (De Erven F. Bohn N.V., Haarlem, 1936).
2. M. H. Cohen and F. Reif, *Nuclear Quadrupole Effects in Nuclear Magnetic Resonance*, Solid State Physics, 5 (1957) pp. 326-329.
3. G. Herzberg, *Atomic Spectra and Atomic Structure* (Dover, New York, 1944) pp. 182-196.
4. O. C. Kistner and A. W. Sunyar, *Phys. Rev. Lett.* 4 (1960) pp. 412-415.
5. G. K. Wertheim, *Mössbauer Effect: Principles and Applications* (Academic, New York), see, e.g., pp. 49-52, 73, 78.
6. P. B. Russell, G. L. Latshaw, S. S. Hanna, and G. Kaindl, *Nucl. Phys.* A210 (1973) pp. 133-156.
7. J. Meyer and J. Speth, *Nucl. Phys.* A203 (1973) pp. 17-41
8. G. D. Sprouse, in *Variations on Nuclear Themes Based on a Symposium in Honor of Stanley S. Hanna*, Stanford University, 16 March, 1991, Eds. C. M. Class and L. Cohen (World Scientific, Singapore, 1994).
9. L. R. Walker, G. K. Wertheim, and V. Jaccarino, *Phys. Rev. Lett.* 6 (1961) pp. 98-101.
10. G. M. Kalvius, J. Gal, L. Asch, and W. Potzel, in *Nuclear Zeeman Effect and Recent Advances in Mössbauer Spectroscopy, Proceedings of the International Workshop in Honor of Stanley S. Hanna*, Stanford University, 15 March, 1991, Ed. P. Boolchand (Baltzer, Basel, 1992) pp. 77-95.
11. S. S. Hanna, J. Heberle, C. Littlejohn, G. J. Perlow, R. S. Preston, and D. H. Vincent, *Phys. Rev. Lett.* 4 (1960) pp. 177-180.
12. G. W. Ludwig and H. H. Woodbury, *Phys. Rev.* 117 (1960) pp. 1286-1287.

13. S. S. Hanna, L. Meyer-Schützmeister, R. S. Preston, and D. H. Vincent, *Phys. Rev.* 120 (1960) pp. 2211-2213.
14. L. Meyer-Schützmeister, R. S. Preston, and S. S. Hanna, *Phys. Rev.* 122 (1961) pp. 1717-1720.
15. S. S. Hanna, J. Heberle, G. J. Perlow, R. S. Preston, and D. H. Vincent, *Phys. Rev. Lett.* 4 (1960) pp. 513-515.
16. W. Marshall, *Phys. Rev.* 110 (1958) pp. 1280-1285; private communication.
17. T. P. Das, in *Variations on Nuclear Themes Based on a Symposium in Honor of Stanley S. Hanna*, Stanford University, 16 March, 1991, Eds. C. M. Class and L. Cohen (World Scientific, Singapore, 1994) and references therein.
18. T. P. Das, *Hyperfine Interactions* 6 (1979) pp. 53-62 and references therein.
19. P. Boolchand, M. Tenhover, S. Jha, G. Langouche, B. B. Triplett, S. S. Hanna, and P. Jena, *Phys. Lett.* 54A (1975) pp.293-294.
20. P. Jena and D. J. W. Geldart, *Solid State Comm.* 15 (1974) pp. 139-142.
21. S. S. Hanna and R. S. Preston, *Phys. Rev.* 106 (1957) pp. 1363-1364.
22. S. S. Hanna and R. S. Preston, *Phys. Rev.* 108 (1957) pp. 160-161; ibid 109 (1958) pp. 716-720; ibid 110 (1958) pp. 1406-1408.
23. W. L. Randolf, N. Ayres DeCampos, J. R. Beene, J. Brude, M. A. Grace, D. F. H. Start, and R. E. Warner, *Phys. Lett.* 44B (1973) pp. 36-38.
24. B. T. Neyer, *Lifetimes and Magnetic Moments of States of* $^{16}N$ *and* $^{22}Na$*: The Recoil-into-Vacuum Method*, Ph.D. thesis, Stanford University, 1984.
25. S. S. Hanna, in *Atomic Experiments in Nuclear Physics*, Proceedings of the Haifa International Workshop, Haifa, 1976, Eds. B. Rosner and R. Kalish, (American Institute of Physics, New York, 1976) pp. 577-610.
26. A. Little, H. C. Jain, S. M. Lazarus, T. K. Saylor, B. B. Triplett, and S. S. Hanna, *Hyperfine Interactions* 8 (1980) pp. 3-18.
27. H. C. Jain, A. Little, S. M. Lazarus, T. K. Saylor, B. B. Triplett, and S. S. Hanna, *Phys. Rev. C* 14 (1976) pp. 2013-2015; ibid, *Hyperfine Interactions* 2 (1976) pp. 385-387.
28. T. P. Das, *Hyperfine Interactions* 34 (1987) pp. 149-165.
29. P. Borchers, J. D. Bronson, D. E. Murnick, and L. Grodzins, *Phys. Rev. Lett.*, 17 (1966) pp. 1099-1102.
30. J. Thornton, *Simultaneous Measurements of Nuclear Magnetic Moments by the Transient Field Method*, Ph.D. thesis, Stanford University, 1985.
31. F. Hagelberg, K.-H. Speidel, and T. P. Das, to be published; see the paper by F. Hagelberg in this volume.
32. T. P. Das and E. L. Hahn, *Nuclear Quadrupole Resonance Spectroscopy*, Solid State Phys. Supplement 1 (1958).
33. See review by K. Nakai, in *Variations on Nuclear Themes Based on a Symposium in Honor of Stanley S. Hanna*, Stanford University, 16 March, 1991, Eds. C. M. Class and L. Cohen (World Scientific, Singapore, 1994).
34. T. Minamisono, J. W. Hugg, D. G. Mavis, T. K. Saylor, S. M. Lazarus, H. F. Glavish, and S. S. Hanna, *Phys. Rev. Lett.* 34 (1975) pp. 1465-1468.
35. V. M. Sarnatskii, V. A. Shutilov, T. D. Levitskaya, B. I. Kidyarov, and P. L. Mitnitskii, *Fiz. Tverd. Tela* 13 (1971) pp. 2413-2416 [*Sov. Phys. Solid State* 13, (1972) pp. 2021-2023].

36. H. Ackermann, D. Dubbers, M. Grupp, P. Heitjans, and H.-J. Stockmann, *Phys. Lett.* 52B (1974) pp. 54-56.
37. T. Minamisono, in *Variations on Nuclear Themes Based on a Symposium in Honor of Stanley S. Hanna*, Stanford University, 16 March, 1991, Eds. C. M. Class and L. Cohen (World Scientific, Singapore, 1994).
38. K. Sugimoto and I. Tanihata, in *Variations on Nuclear Themes Based on a Symposium in Honor of Stanley S. Hanna*, Stanford University, 16 March, 1991, Eds. C. M. Class and L. Cohen (World Scientific, Singapore, 1994).
39. S. M. Mohapatra, B. N. Dev, L. Luo, T. Thundat, W. M. Gibson, K. C. Mishra, N. Sahoo, and T. P. Das, *Revs. Solid State Sci* 4 (1990) pp. 873-900.
40. Tara P. Das, *Relativistic Quantum Theory of Electrons* (Harper and Row, New York, 1973).

Local Order
in Condensed
Matter Physics


ISBN 1-56072-220-7

# NOVEL MATERIALS: THE ROLE OF LOCAL ORDER

P. Jena, B. K. Rao, S. N. Khanna

*Physics Department, Virginia Commonwealth University, Richmond, VA 23284-2000*
*and*

S. D. Mahanti

*Physics Department, Michigan State University, East Lansing, Michigan 48824*

**Abstract**

Atomic clusters constitute a new class of matter with unique properties that depend strongly on their size, composition, symmetry, and geometry. The local atomic environment of a probe site in a cluster can be significantly different from that in the bulk because clusters are characterized by a large fraction of surface atoms. This effect can alone change the thermal, electronic, and magnetic properties of clusters and their interactions with each other, as well as with gas molecules. Current understanding of the electronic structure of metal clusters is briefly reviewed. The role of local order is illustrated by concentrating on three different properties of clusters: (1) thermal stability of Be clusters, (2) relationship between magnetism and topological order in alkali metal tetramers, and (3) interaction of hydrogen molecules with transition metal atoms and ions.

## I. Introduction

In metallic systems, any perturbation created by defects can be efficiently screened by conduction electrons within a short distance. Thus, calculations of electronic properties due to point defects can be accomplished by modelling[1] the immediate neighborhood of the probe site as a cluster and embedding the cluster in an environment that simulates the bulk. A cluster that includes the defect site and a couple of shells of host atoms may often be adequate to provide a semiquantitative interpretation of defect related properties. This practice of using clusters as models for bulk systems has provided much insight into many electronic, optical, and magnetic properties of defects in metals and semiconductors over the past two decades.

Recently it has been possible to produce atomic clusters[2] in gas phase by laser vaporizing the target material and allowing the resulting plume to cool by collisions with rare gas atoms, as well as by supersonic jet expansion. The clusters thus formed can be ionized and mass selected using the conventional time of flight or quadrupole mass spectrometry. The properties of these individual clusters can be studied either in beams or by embedding them in various non-interacting matrices. Unlike the clusters used as models to study the properties of bulk materials, the study of gas phase clusters is simpler because one does not have to worry about their interaction with the surrounding host lattice. Thus, free clusters provide a unique ground where the effect of local atomic order on the structural and electronic properties can be probed. The clusters of multi-component species (atoms) can be produced and the effect of impurity can also be studied.

There are two main aspects that distinguishes free clusters from clusters as models of the bulk. (i) The clusters in the gas phase have the freedom to change their atomic structure while clusters used as models of the bulk are confined to the bulk geometry (ii) The wave functions of free clusters are molecular orbitals that decay outside the cluster while for clusters modelling the bulk, these should obey periodic boundary conditions. In spite of these differences, the short range nature of electronic interaction in metals has made it possible to use clusters as models to understand many properties associated with the local atomic order around the probe site.

In this paper we concentrate on the role of local atomic order in understanding the stability and electronic properties of free clusters. Small metal clusters have unique properties for the following reasons. (i) The number of surface atoms are large compared to the bulk atoms. (ii) The atomic arrangements in small clusters are very different from that in the corresponding bulk. (iii) The electronic energy levels in small clusters are discrete while in bulk they form bands. To illustrate the special role these features play, we concentrate here on three separate properties: thermal stability, magnetic properties, and interaction with hydrogen. In section II we provide a brief description of our computational procedure. The results are discussed in section III. A brief summary of our conclusions is given in section IV.

## II. Computational Procedure

An understanding of the properties of clusters requires a knowledge of their atomic structure. This can be achieved by minimizing the total energy of clusters with respect to their various degrees of freedom. Both ab initio methods[3] and dynamic simulations[4] are used to obtain cluster geometry. The former method allows the determination of atomic structure at 0 K while the latter method can yield atomic structures at elevated temperatures. We have studied the thermal stability of Be clusters[5] using the Monte Carlo method, the magnetic properties of alkali clusters using the density functional method, and cluster interaction with hydrogen using ab initio quantum chemical method[3,6].

The interatomic potential used in the Monte Carlo method was expressed in an analytical form whose parameters were determined by a simultaneous fit to the total energy hypersurfaces of $Be_n$ $(2<n<5)$ clusters[7]. Total energies at nearly 500 grid points on the energy hypersurface were obtained from self-consistent density functional calculations within the framework of the linear combination of atomic orbitals-molecular orbital theory (LCAO-MO). The accuracy of the potential was tested[7] by calculating the bulk melting temperature, cohesive energy, and stability of various crystalline phases. Having established the accuracy of the ab initio many-body potential, the finite temperature properties of clusters were studied[5] using over 100,000 Monte Carlo steps per atom. Both heating and cooling runs were performed to ensure thermal equilibrium.

For alkali tetramers, density functional theory with local spin density approximation was used to calculate the total energies for spin singlet and triplet states. For each spin, the clusters were assumed to be two dimensional. The bond lengths and bond angles of the rhombus structure were optimized to study the relationship between geometry and spin multiplicity. The atomic orbitals were constructed from uncontracted gaussian basis functions and norm-conserving pseudopotentials[8] were used to represent the atomic core.

The interaction of hydrogen with neutral and charged transition metal atoms was studied[9] using the quantum chemical technique. The atomic orbitals were again approximated using gaussian basis functions. The exchange interaction among electrons was treated exactly within the unrestricted Hartree-Fock approximation. The correlation correction was included by using fourth order Möller-Plesset perturbation theory (MP4).[3]

The relative merits of density functional theory within the local density approximation (LDA) and the quantum chemical approach have been a longstanding debate. While the LDA is computationally less intensive, it is not as well grounded in the variational theory as is the quantum chemical method. However, the computer intensive nature of the quantum chemical method makes it of limited use in treating large clusters. Comparison between the accuracies

**Table 1**

Comparison of results on $H_2O$ obtained by using quantum chemical (Hartree-Fock + correlation) and density functional (local density approximation) methods.

| Quantity | Quantum Chemical | Density Functional |
|---|---|---|
| Total energy (a.u.) | -76.247 | -76.065 |
| O-H distance (Å) | 0.96 | 0.97 |
| H-O-H angle | 106.34° | 106.26° |
| Charge on O | -0.86 | -0.82 |
| Charge on H | +0.43 | +0.41 |
| Dipole moment | 2.37 | 2.32 |
| CPU time (Silicon Graphics Workstation) | 313.6 s | 94.5 s |

of both the methods is often hampered since the use of differing numerical procedure and atomic basis functions can easily give rise to the same kind of errors as does the use of different ways of treating the many-body problem. To get an accurate handle on the limitations of the two methods, we have carried out a careful study of the total energy, charge state, dipole moment, bond length, and bond angle of $H_2O$ using the LDA and Hartree-Fock-MP4 technique. The results are given in Table 1. We note that the results compare very well with each other and that the small differences are within the error limit in either of the calculations. Thus, our results and conclusions to follow are not affected by the choice of the computational procedure.

## III. Results

### A. Thermal Stability

The mechanism for melting of solids is reasonably well understood. However, the melting of clusters is still somewhat of an enigma. The difficulty arises starting from the definition of melting of clusters. Clearly for a very small cluster there is no melting. This can pass directly from a vibrational state to an evaporated state upon heating. The concept of melting is only meaningful for very large clusters although it is difficult to prescribe criteria for determining the critical size where this comes to play. In clusters, the surface to volume ratio of atoms is rather large. It is also known that the surfaces melt before the bulk due to the low coordination number of surface atoms. Since the number of surface atoms vary as $R^2$ and bulk atoms vary as $R^3$, it is easy to show that the melting temperature of large clusters should scale as

$$T_m = T_b - \frac{C(T_b - T_s)}{R} \tag{1}$$

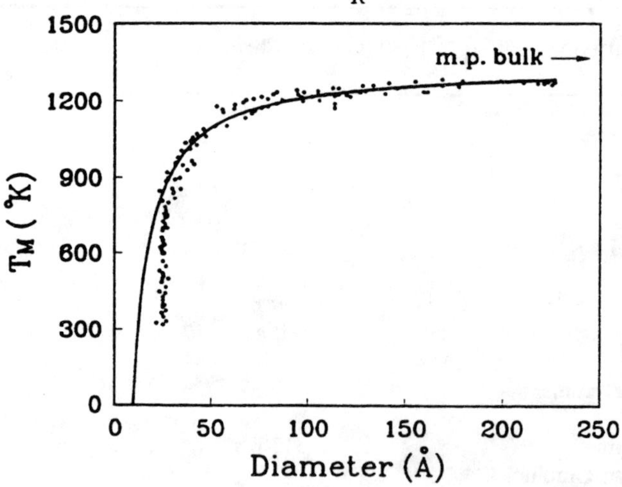


Fig. 1. The dependence of melting point of Au particles on particle diameter. The experimental points are from Ref. 10 and the solid curve is the prediction of Eq. (1).

Here, $T_b$ and $T_s$ are the melting temperature for bulk and surface atoms and R is the radius of the cluster. Note that $T_b > T_s$. We see from Fig. 1 that Eq. (1) describes the melting characteristics[10] of Au particles rather well with R > 15 nm. For small clusters, this simple picture does not work as electronic structure effects become very important.

The melting of small clusters can be monitored[11] by studying the root mean square fluctuation of the inter-atomic distance, δ where

$$\delta = \frac{2}{N(N-1)} \sum_{ij} [\langle r_{ij}^2 \rangle - \langle r_{ij} \rangle^2]^{1/2} \qquad (2)$$

The temperature dependence of δ was calculated for $Be_n$ clusters for $5 < n < 14$ using $10^5$-$10^6$ Monte Carlo steps/atom. Following Berry and co-workers,[11] we define $T^*$ as the temperature at which the microclusters start to soften drastically. In most of the clusters under study we found that δ increases linearly with temperature and at a characteristic temperature, $T^*$, it begins to rise rapidly. In Fig. 2 we plot $T^*$ as a function of cluster size. Note that it is a non-monotonic function of cluster size. The thermal stability of these clusters fall into three main categories. $Be_6$ and $Be_{13}$ are unusually stable and do not soften until the temperature rises above 1300 K. The second group consisting of $Be_7$, $Be_8$, and $Be_{10}$ are characterized as the ones with intermediate stability. The third group consisting of $Be_9$ and $Be_{12}$ are unusually soft with $T^* \sim 200$ K.

The unusually low "melting point" of $Be_9$ can be understood by noting that there are two isomers of $Be_9$ that are energetically degenerate. The oscillation of $Be_9$ between these two isomeric structures gives the appearance of thermal softening. Thus, there are different ways the local atomic order of a cluster controls its thermal stability: (1) The energetic stability of a cluster with specific size may render it unusual stability against thermal fluctuation. (2) The existence of different isomers separated by low energy barriers could make clusters change their structure at elevated temperatures without really "melting."

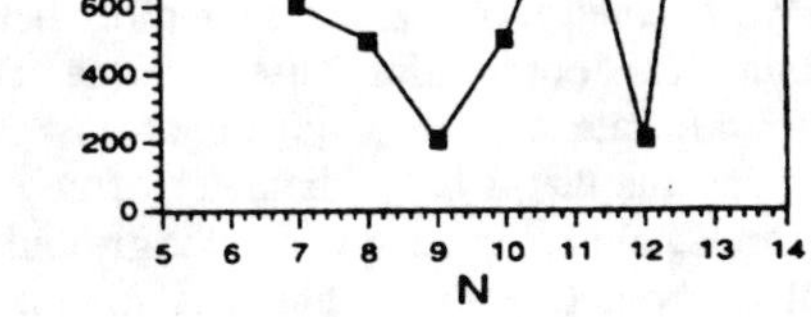


Fig. 2. The "melting" temperature, $T^*$ as a function of size of $Be_n$ clusters.

## B. Effect of Geometry on the Spin Structure of Cluster

One of the most striking effects the geometry of clusters has on their electronic structure can be illustrated by studying their ground state spin configuration. In atoms the ground state spin multiplicity is determined by the Hund's rule. The spherical symmetry of the atomic potential gives rise to degeneracy in the one-electron energy levels. By maximizing the spin multiplicity, the atoms can lower their energies through the exchange interaction. A cluster, however, has two different ways to lower its energy. A cluster with high atomic symmetry can lower its energy by assuming a high spin multiplicity in keeping with the Hund's rule. However, it can also lower energy by the Jahn-Teller mechanism where lowering of the atomic symmetry can split the degenerate energy levels of an otherwise symmetric cluster. Thus, clusters provide a unique ground where the competition between Hund's rule and Jahn-Teller effect can be tested by studying the effect of local atomic order on spin multiplicity.

This interplay between structure and magnetism can be well demonstrated by considering the energetics of a $K_4$ cluster. In Fig. 3 we plot the energy levels of a $K_4$ cluster confined to a singlet spin state and square geometry. The interatomic distances have been optimized in each case. In this configuration, the highest occupied molecular orbital (HOMO) levels are degenerate. This degeneracy can be broken in one of two ways. The cluster can assume a higher spin-multiplicity (namely a triplet) and break the degeneracy of the spin up and down levels. This exchange coupling of the parallel spins lowers the energy of the spin triplet state by 0.31 eV from the spin singlet state having the square geometry. The degeneracy of the HOMO levels of the $K_4$ cluster can also be lowered by distorting the square geometry to a less symmetric rhombus form. The geometry optimized $K_4$ energy levels are also shown for the spin singlet configuration in Fig. 3. The energy of this state is 0.59 eV lower than spin singlet configuration with square geometry. Comparison of these three configurations reveals that the Jahn-Teller distortion is favored over the Hund's rule coupling in making the ground state of the $K_4$ cluster a spin singlet rhombus.

These results indicate that as the $K_4$ cluster is distorted from the square geometry, there must exist certain critical geometries for which the energies of the spin singlet and triplet state must be equal. This is shown in Fig. 4 where we plot the energies of the $K_4$ cluster as a function of bond angle, $\theta$ for both spin singlet and triplet states. For each of these bond angles the bond distances were optimized. We note that for $\theta \approx 72^\circ$, the singlet and triplet states have the same energy. Figs. 3 and 4 also indicate that if the geometry of $K_4$ can be constrained, it would be possible to change its magnetic structure. This appears to be the case in a recent experiment where Nozue et al[12] observed ferromagnetic coupling between $K_4$ clusters isolated in the cages of a K-LTA zeolite. This observation raises an interesting question -how do $K_4$ clusters with nonmagnetic ground state appear magnetic at low temperatures ? A close examination of the zeolite cage reveals that it has a diameter of about 11 Å. This cage is not large enough to accommodate the $K_4$ cluster in its spin singlet ground state as the size of the $K_4$ cluster along the long diagonal is about 12 Å (including 5 Å for the diameter of the K atom). However, if the $K_4$ cluster can be squeezed to form a square, its size would shrink to about 11 Å. This will make it possible to fit into the zeolite cage. As shown above, the $K_4$ cluster with square geometry has a magnetic ground state. Thus, it is likely that the magnetic behavior of $K_4$ clusters in zeolite is due to the effect of geometrical constraints imposed by the dimension

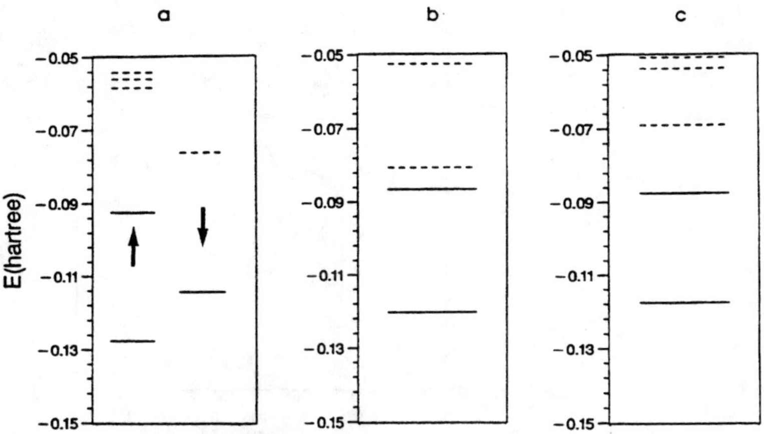


Fig. 3. The one-electron energy levels of $K_4$ clusters with (a) spin singlet square geometry, (b) spin triplet square geometry, and (c) spin singlet rhombus configuration.

of the cages. It is also possible that the $K_4$ cluster may interact chemically with the zeolite cage. It is known that there are charge transfers between clusters and the zeolite. Thus, if $K_4$ remains in $K_4{}^+$ or $K_4{}^{3+}$ configuration, the unpaired electron(s) can give a magnetic fingerprint. For this, it is necessary to study the interaction of $K_4$ clusters with the atoms in the zeolite self-consistently. Such a calculation is difficult, and to our knowledge has not yet been attempted.

### C. Interaction of Hydrogen with Clusters

The interaction of hydrogen with metals has been studied for a long time. Most metals do not form stable metal hydrides. On the other hand, hydrogen chemisorbs dissociatively on all metal surfaces. Since metal clusters are characterized by large surface to volume ratio, it is expected that metal clusters may bind with hydrogen even though the corresponding bulk hydrides may not exist. This has been observed experimentally.[13] Whether the nature of the interaction of hydrogen with clusters is characteristically different from that with the metal surfaces is a less understood problem. The surfaces of clusters are different from bulk surfaces both structurally and electronically. Thus, the effect of local environment on the interaction of hydrogen can be well demonstrated by studying cluster-hydrogen complexes.

One of the first experiments that demonstrated this aspect clearly was carried out by Cox et al.[13] They observed that small clusters of Pt, Rh, and Ni can absorb as much as eight hydrogen atoms per metal atom and that this ratio decreases to unity as clusters become larger in size ($\approx$ 30 atoms). This unique property of small clusters must be strongly influenced by their special electronic structure.

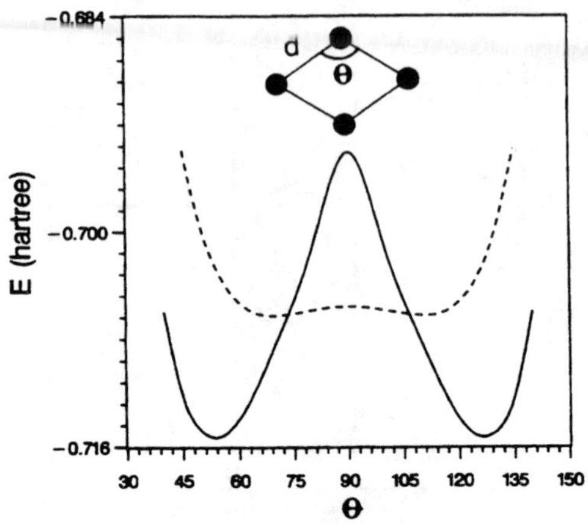


Fig. 4. The binding energies of $K_4$ clusters as a function of bond angle $\theta$. The solid and dashed curve are for the spin singlet and triplet configurations respectively.

We had earlier studied the interaction of hydrogen with Li clusters[14] and observed that clusters did not show any enhanced binding with hydrogen in comparison to their atomic counterparts. In addition, hydrogen atoms bind dissociatively on metal clusters as they do on metal surfaces. We also studied the interaction of $H_2$ with a variety of metal atoms to see if the electronic structure of metal atoms made any difference. Some of these results are presented in Fig. 5. We note that in each case, the equilibrium geometry of the $MH_2$ complex (M = Cu, Al, Ni, Co) occurred with the bond of the hydrogen molecule broken. It was also energetically not favorable to decorate these metal atoms with more hydrogen than what could be permitted from the point of view of the valence of these elements.

To see if the charge state of the atoms makes any difference in the interaction with hydrogen, we have studied the binding of $H_2$ with $Ni^+$, $V^+$, and $Sc^+$ ions. Here, we only discuss[9] the interaction of $Ni^+$ with $H_2$ as the results for other ions are very similar. In Fig. 6 we give the geometries of $Ni^+ (H_2)_n$ clusters obtained by using the quantum chemical technique. The important point to note from these figures is that hydrogen remains in molecular form. The binding energy of a single $H_2$ molecule to a $Ni^+$ ion is 0.55 eV. This indicates that hydrogen is molecularly chemisorbed on a $Ni^+$ ion. It was discussed before that hydrogen interacting with a neutral atom or cluster dissociates and binds in an atomic form. This sharp departure in the nature of interaction of hydrogen with neutral atoms and cations stems from the differences in the origin of hydrogen bonding in each case. In the case of neutral atoms, the bonding arises due to a partial transfer of electron from the metal atom to the antibonding state of the $H_2$ molecule whereas in the case of a cation, the electric field associated with the ion polarizes the $H_2$ molecule causing it to bind in molecular form. The polarization force is long range in nature. Therefore, more than one $H_2$ molecule can be bound to a $Ni^+$ ion. As a

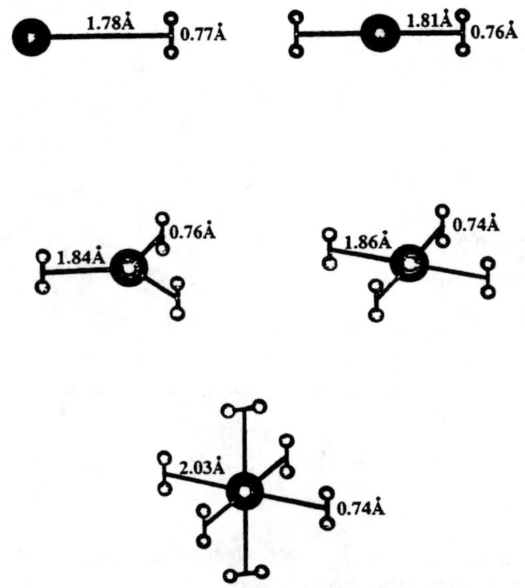


Fig. 6. Equilibrium geometries of $Ni^+ (H_2)_n$ clusters for $n \leq 6$.

## IV. Conclusions

The role of local order on the stability and properties of materials was illustrated by concentrating on small atomic clusters of metallic elements. It was shown that large surface to volume ratio and geometry of clusters are responsible for unique properties of clusters. For example, while the melting point of large clusters increases monotonically with cluster size, their thermal stability can have anomalous behavior in the case of very small clusters. It is possible that certain small clusters can have "melting points" higher than that in the bulk phase while some other clusters may be unusually soft. Similarly, the magnetic properties of small clusters can be very sensitive to cluster size and geometry. A cluster can lower its energy by either increasing its electronic degeneracy or by lowering its atomic symmetry. The former favors a high spin state while the later favors the low spin state. The interaction of clusters with gas atoms is also size specific. For example, hydrogen molecules dissociate and bind in atomic form with neutral metallic clusters, while in small cationic clusters, hydrogen binds molecularly. This distinction between neutral and charged clusters disappears as they grow in size. The above three examples are given as illustrations of the unique role of local order on properties of clusters. In general, study of clusters opens a new area of materials science where materials with unique properties can be custom designed by exploiting the role of the local atomic order.

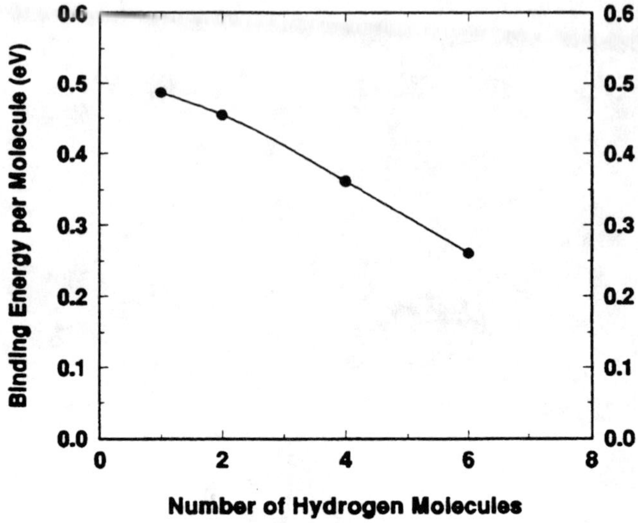


Fig. 7. Binding energy of $H_2$ molecules as a function of the number of hydrogen molecules decorating the $Ni^+$ ion.

**Acknowledgement**

This work was supported in part by a grant from the Army Research Office (DAAL03-91-G-0136). This paper is dedicated to Professor Tara Prasad Das on the occasion of his sixtieth birthday.

**References**

1. B.K. Rao and P. Jena, J. Phys. F16, 461 (1986).

2. P. Jena, B.K. Rao, and S. N. Khanna, eds. Physics and Chemistry of Small Clusters (Plenum, New York), 1987; B.K. Rao, S.N. Khanna, and P. Jena, Phase Transitions 24-26, 35 (1990).

3. W.J. Hehre, L. Radom, P.v.R. Schleyer, and J. Pople, Ab-initio Molecular Orbital Theory (John Wiley, New York), 1986.

4. R. Car and M. Parrinello, Phys. Rev. Lett. 45, 566 (1985).

5. Z. Cai, S.D. Mahanti, A. Antonelli, S.N. Khanna, and P. Jena, Phys. Rev. B46, 7841 (1992).

6. F. Reuse, S.N. Khanna, V. de Coulon, and J. Buttet, Phys. Rev. B41, 11743 (1990).

7. E. Blaisten-Barojas and S.N. Khanna, Phys. Rev. Lett. 61, 1477 (1988).

8. G.B. Bachelet, D.R. Hamann, and M. Schlüter, Phys. Rev. B26, 4199 (1982).

9. J. Niu, B.K. Rao, and P. Jena, Phys. Rev. Lett. 68, 2277 (1992).

10. Ph. Buffat and J.P. Borel, Phys. Rev. A13, 2287 (1976).

11. T.L. Beck, J. Jellinek, and R.S. Berry, J. Chem. Phys. 87, 545 (1987).

12. Y. Nozue, T. Kodaira, and T. Goto, Phys. Rev. Lett. 68, 3789 (1992).

13. D. Cox, P. Fayet, R. Brickman, M.Y. Hahn, and A. Kaldor, Catal. Lett. 4, 271 (1990).

14. B.K. Rao, S.N. Khanna, and P. Jena, Phys. Rev. B43, 1416 (1991); ibid 45, 13631 (1992).

15. P.R. Kemper, J. Bushnell, G. von Helden, and M. Bowers, J. Phys. Chem. 97, 52-58 (1993).

Local Order
in Condensed
Matter Physics


ISBN 1-56072-220-7

# ELECTRONIC STRUCTURE, ELECTRON-PHONON COUPLING, AND SUPERCONDUCTIVITY IN ALKALI-DOPED $C_{60}$

Sashi Satpathy

*Department of Physics and Astronomy*
*University of Missouri, Columbia, Missouri 65211, USA*

## 1. INTRODUCTION

With the discovery of $C_{60}$, first as a molecule [1], and subsequently in condensed solid phases [2], carbon is now known to exist in three distinct crystalline forms: the well-known diamond and the graphite structures, and the new form of carbon, broadly known as the fullerenes. In the diamond form, the carbon atoms are four-fold coordinated, while graphite consists of planar sheets of three-fold coordinated carbon atoms, with the sheets vertically stacked. In contrast to these, fullerene solids are made up of closed-cage carbon molecules, or clusters, where the atoms are arranged in spherical or similar shapes. A large number of stable closed-cage carbon clusters, notably $C_{28}$, $C_{32}$, $C_{50}$, $C_{60}$, $C_{70}$, $C_{240}$, etc., have been found. Much of the work in the recent past has been focused on solids of $C_{60}$ or $C_{70}$.

The recent discovery that solid $C_{60}$ doped with alkali metals are superconductors at moderate temperatures [3] has added enormous enthusiasm to the study of the fullerenes. In view of the several exotic mechanisms that have been proposed for the superconductivity in the high-$T_c$ copper-oxide superconductors, one of the key questions is whether electron-phonon mechanism can adequately describe superconductivity in the fullerenes.

Apart from superconductivity, the fullerenes exhibit scores of novel properties, such as the order-disorder orientational phase transition. Undoped $C_{60}$ forms an fcc crystal above $249K$ with the individual molecules rotating while being pinned to the lattice sites. Below the temperature of about $100K$, a simple cubic lattice forms with 4 molecules per unit cell, where the molecules, now with their rotations stopped, are oriented in different directions with respect to one another [4,5].

In this paper we focus on the $C_{60}$ fullerene solids, both undoped and alkali-dc and examine how various properties are tied to the underlying electronic stru of the single $C_{60}$ molecule. In particular, we focus on the electronic structure, entational ordering, and the electron-phonon mechanism of superconductivity ir alkali-doped fullerenes.

## 2. ELECTRONIC STRUCTURE OF THE $C_{60}$ MOLECULE

The electronic structure of the single $C_{60}$ molecule has been studied by r authors both from simple Hückel molecular orbital theory [6,7] and from ab i density-functional calculations [8]. In the $C_{60}$ molecule the sixty carbon atom arranged in the shape of a soccerball (soccerballene, Buckminster-fullerene, or, sir the buckyball). There are, in total, 30 edges shared by two hexagons (hexagon e and 60 edges shared by a hexagon and a pentagon (pentagon edge). The cl is held together by $sp^2\sigma$ bonds along all edges on the sphere surface, and the additional bonding provided by the $pp\pi$ bonds.

The interaction between the $sp^2$-hybrid orbitals pointing towards each other a the edges is large enough that the bonding and antibonding $sp^2\sigma$ states are rem away from the Fermi energy $E_f$. This leaves the sixty "$p_z$" orbitals, pointed alon radial direction, to form molecular orbitals around $E_f$. The $pp\pi$ interaction bet the sixty $p_z$ atomic orbitals may be described by a simple tight-binding (Hü model,

$$H = \sum_i \alpha|i><i| + \sum_{\langle i,j \rangle} \beta(|i><j| + c.c.) ,$$

where $|i>$ denotes the "$p_z$" orbital on the $i$-th carbon atom, $\alpha$ is the on-site m element, $\beta$ is the nearest-neighbor hopping integral describing the $pp\pi$ interaction $\langle i,j \rangle$ denotes nearest-neighbor pairs. The matrix element $\alpha$ is the same for all a and it defines the zero of energy leaving the single parameter $\beta$ in the Hückel m whose numerical value is about $-2.7$ $eV$, as obtained from the density-funct results discussed later.

That this basic picture of chemical bonding is correct is verified by the resu a density-functional calculation using the linear muffin-tin orbitals method (LM [8]. Fig. 1 shows the excellent agreement between the LMTO results for a $C_{60}$ molecule and the results obtained from the simple Hückel model. The hi occupied molecular orbital (HOMO) is the five-fold $h_u$ and the lowest unocci molecular orbitals (LUMO) are the three-fold $t_{1u}$ and the $t_{1g}$ orbitals.

A group-theoretical analysis of the LMTO results [8] indicates that the molecule is particularly stable because all "$p_z$"-derived occupied orbitals are ı ly bonding on the hexagon edge, while the unoccupied ones are mostly antibon This leads to a simple picture of "double" hexagon bonds and "single" pent bonds. It must be kept in mind, however, that there is not really so much diffe between the hexagon and the pentagon bonds. This may be seen, for instanc calculating for the two bonds the Pauling bond order $n$, which provides a qualit measure of the strength of the bond and is defined as half the difference betwee

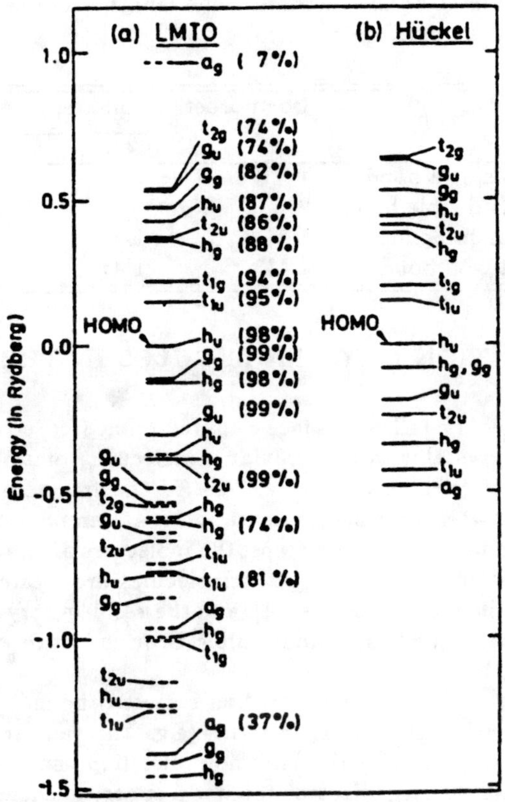


Figure 1: Calculated molecular-orbital scheme along with their orbital symmetries for a single $C_{60}$ molecule: (a) Results of a density-functional LMTO calculation and (b) Results of a nearest-neighbor Hückel model, Eq. (1), with one radially-directed "$p_z$" orbital per carbon atom. In brackets are the carbon $p_z$ orbital characters of the individual molecular orbitals as obtained from the density-functional results [8].

number of electrons in the bonding orbitals minus those in the antibonding orbitals [9].

We summarize in Table 1, the calculated bond orders $n$ and the corresponding bond lengths $D(n)$ predicted from Pauling's empirical formula, $D(n) = D_1 - 0.71 \times \log n$, where all lengths are in Å and $D_1 = 1.545$ Å. By definition, the Pauling bond order is one for the single bond, two for the double bond, etc. The bond-orders, obtained from the LMTO calculation, are 1.54 (hexagon bond) and 1.41 (pentagon bond). This indicates that the hexagon and the pentagon bonds are not very different from each other, a fact also reflected from the lengths of the two bonds being nearly equal in the molecule. The fact that the hexagon and the pentagon bonds are not very different is also verified from the molecular dynamics simulations [10], where there is no noticeable difference between the two bonds at a temperature as low as $300K$.

Table 1: LMTO-calculated bond orders and bond-lengths obtained from the empirical Pauling formula for the single $C_{60}$ cluster.

| Bond | Bond order | Bond length (Å) | |
|---|---|---|---|
| | | Calc. | Expt. |
| C-C single bond | 1 | — | 1.54 |
| C=C double bond | 2 | — | 1.33 |
| hexagon bond | 1.54 | 1.41 | 1.39 |
| pentagon bond | 1.41 | 1.44 | 1.45 |

## 3. ORIENTATION OF $C_{60}$ MOLECULES IN THE CRYSTAL

The $C_{60}$ molecules crystallize in a face-centered-cubic (fcc) structure above 249 $K$. It is a rather unique crystal with the individual molecules rotating freely, while being pinned down to the lattice sites. Below 249 $K$, the free rotation stops, but the molecules flip-flop between equivalent orientations. As temperature is reduced below about 100 $K$, all orientational motion stops, the molecules are now static and become fixed on the lattice sites. Even though the molecules are located on an fcc lattice, they are oriented in different directions [11], and the resulting crystal structure (space group $Pa\bar{3}$) may be described as a simple cubic Bravais lattice with 4 molecules per unit cell [4,5].

There are several pieces to the interaction between the individual $C_{60}$ molecules in the solid: (i) the exchange-correlation terms (e.g., the van der Waals interaction), (ii) the Coulomb forces, and (iii) the intermolecular hopping (kinetic energy) term. Since the $C_{60}$ molecule is charge-neutral and has a spherical shape, it is reasonable to assume that the first two terms have weak directional dependence, and we argue below that the orientation of the molecules in the lattice is, to a large extent, determined by the intermolecular hopping term.

The intermolecular hopping reduces the energy of the occupied MOs of the molecule thereby contributing to the cohesion in the solid. The separation between two atoms on different molecules being large ($\sim$ 3Å) as compared to the C-C distance on the same molecule ($\sim$ 1.4Å), the energy reduction $\Delta E$ per $C_{60}$ molecule may be estimated from second-order perturbation theory [12]:

$$\Delta E = 2\sum_{i}^{\text{occ.}} \sum_{j}^{\text{unocc.}} \frac{|V_{ij}|^2}{E_i - E_j} \quad . \tag{2}$$

Here $E_i$ is the energy of the $i$-th MO and the factor of 2 comes from spin. The intermolecular interaction $V_{ij}$ between the $i$-th MO on one molecule and the $j$-th MO on the second can be calculated from the relevant MO wave functions, once the form of the hopping integral $V_{pp}$ between the $p$ orbitals is known. Gunnarsson et al. [12] have evaluated the role of the intermolecular hopping by adopting the simple TB Hamiltonian, Eq. (1), and the following form for the interatomic hopping integral: $V_{pp\sigma} = v_\sigma d \exp(-\lambda(d - 1.43))$, and $V_{pp\pi} = -V_{pp\sigma}/4$, where $v_\sigma = 6.7eV/$Å, $\lambda = 1.98$Å$^{-1}$, and $d$ is the interatomic distance in Å. The interaction energy per $C_{60}$

molecule for different crystal structures, both for the undoped and the doped cases, were calculated. The key results are that: (a) The experimentally observed crystal structures have the lower hopping interaction energies and (b) The calculated binding energy ($\sim 1.1eV$), taking into account the hopping interaction term only, has the right order of magnitude to explain the measured heat of sublimation ($1.7eV$) [13]. This suggests that the intermolecular hopping term is an important piece for both cluster orientation and cohesion in the solid.

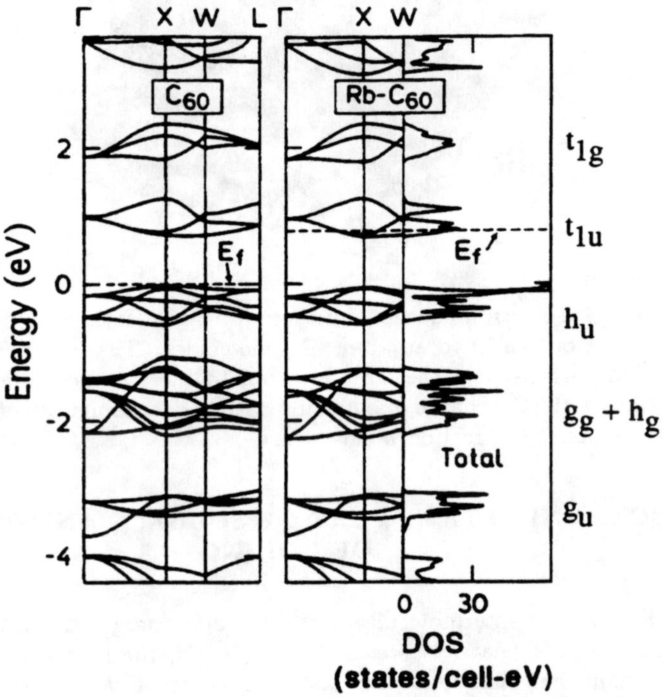


Figure 2: Density-functional band structures of $C_{60}$ and $RbC_{60}$ in the fcc structure with the lattice constant, $a = 14.1$Å. The calculaton used equal C-C bond-lengths on the $C_{60}$ molecule. The figure illustrates the rigid-band fashion in which the outermost $s$ electron of the alkali atom is transferred to the $t_{1u}$ band. The rightmost panel shows the one-electron density-of-states (DOS) for $RbC_{60}$, which is nearly the same as the DOS for undoped $C_{60}$ in the range of energy shown [14].

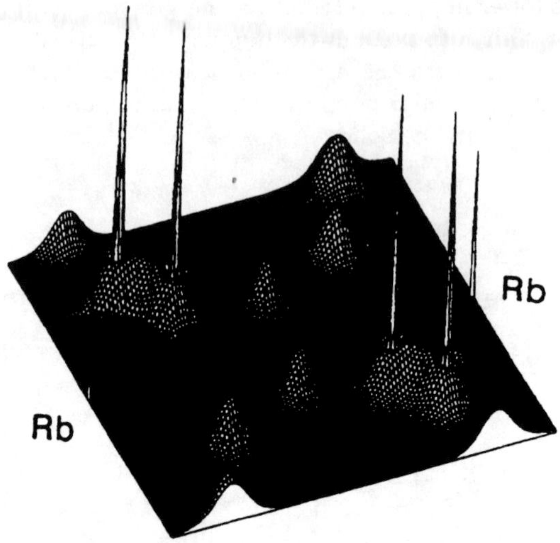


Figure 3: Valence-electron charge-density contour plot for $RbC_{60}$ on a plane passing through a $Rb$ atom and two adjacent $C_{60}$ molecules. The figure indicates the near complete charge-transfer of the $Rb$ $s$ electron to the $C_{60}$ molecule. Spikes indicate atom positions while the charge maxima indicate the accumulation of valence electrons at the mid-points of the carbon bonds [14].

## 4. ELECTRONIC STRUCTURE OF SOLID $C_{60}$ AND THE EFFECT OF DOPING

In solid $C_{60}$ the intermolecular distances being large, the interaction between the $C_{60}$ molecules is relatively weak. Consequently, the identities of the individual MOs are retained in the electronic band structure of the solid. The intercluster interaction merely serves to give a finite dispersion to the MOs of the single cluster, as is characteristic of molecular solids.

Fig. 2 shows the electronic band structure of fcc $C_{60}$ and Rb-doped $C_{60}$ calculated with the density-functional LMTO method [14]. The individual molecular orbitals of $C_{60}$ are broadened by lattice dispersion forming a ladder of bands, each about 0.5 $eV$ wide. The valence bands are formed from the five-fold $h_u$ orbitals and the lowest conduction bands are formed from the three-fold $t_{1u}$ orbitals. The band dispersion is not large enough to obliterate the HOMO-LUMO gap (Fig. 1) in the molecule, making undoped $C_{60}$ a semiconductor. The calculated value of the gap is about $0.8eV$ at the $X$ point as seen from Fig. 2. However, this calculation used equal C-C bond lengths (1.42 Å) in the molecule. With proper bond-lengths, we find that the gap increases to about $1.6eV$ [14,15], a value still too small compared to the experimental value of $3.5eV$. This discrepancy is consistent with the well-known tendency of local-density theory to underestimate band gaps in semiconductors.

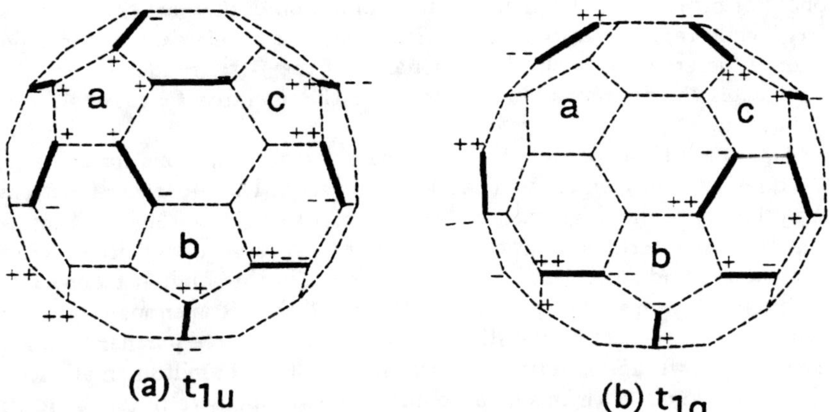


Figure 4: (a) One of the three partner functions for the $t_{1u}$ representation of the icosahedral group, $I_h$, generated from the "$p_z$" antibonding orbitals on the hexagon edge. This partner is symmetric about the pentagon "a". The two other partners are constructed from identical functions, but now located around pentagons "b" and "c", respectively. The three partner functions are subsequently orthogonalized to one another. These three "partners" have predominant ($\sim$ 95%) contribution to the $t_{1u}$ LUMO both in the molecule and in the solid. (b) Same for the $t_{1g}$ representation, relevant for the LUMO +1, i.e., the next higher bands. The "+" denotes an amplitude of +1, while "++" denotes an amplitude of $(\sqrt{5}+1)/2$ , etc.

We have also performed LMTO-LDA calculations for the alkali-doped $M_xC_{60}(M = K, Cs, Rb)$ compounds. The band structure of $RbC_{60}$ is also shown in Fig. 2. We obtain a more-or-less rigid band picture, where the outermost alkali $s$ electron is transferred to the $t_{1u}$ bands of undoped $C_{60}$. The charge-density plots shown in Fig. 3 illustrates the nearly complete charge-transfer of the outermost $s$ electron to the carbon cage. The alkali atom thus becomes positively charged, while the carbon molecule becomes negatively charged, resulting in an ionic compound. Since the $t_{1u}$ band has the capacity to hold six electrons in total, it is now partially filled resulting in a metal, unless exactly six electrons are transferred to the carbon cage. This happens if we have six alkali atom dopants per carbon cage, in which case we have a semiconductor again.

Wave functions for the $t_{1u}$ orbitals

By examining the character of the wave function for the $t_{1u}$ bands we find that antibonding (AB) "$p_z$" orbitals on the hexagon edge contribute about 95% of the charge-density. Therefore if we construct partner functions of $t_{1u}$ symmetry out of such antibonding orbitals then they will span to a very good approximation the 3-dimensional Hilbert subspace of the $t_{1u}$ orbitals at each $k$ point in the Brillouin zone. These partner functions are constructed by projecting the $t_{1u}$ symmetry component of the hexagon edge AB orbitals using the standard projection operator,

$$\Phi^{(j)} = P^{(j)}\Phi \equiv \frac{l_j}{h}\sum_R \chi^{(j)}(R)P_R\Phi. \tag{3}$$

The operator projects out from an arbitrary function $\Phi$ the part that belongs to the $j$-th irreducible representation. Here, $R$ is a symmetry element of the group, $h$ is the order of the group, $l_j$ is the dimensionality of the $j$-th irreducible representation, $\chi^{(j)}(R)$ denotes the appropriate character, and the operator $P_R$ rotates the contours of the function.

In Fig. 4 we show one of the three such partner functions for the $t_{1u}$ LUMO and the other two may be constructed by symmetry. These partner functions are of course a property of the single molecule that carries through to the solid because of the weak intercluster interaction. The partner functions remain valid irrespective of the crystal structure and the orientation of the clusters in the solid. It follows that if one wishes to reproduce just the $t_{1u}$ bands in the solid, then it is enough, to a very good approximation, to restrict to the Hilbert space of these three partner functions. One can then write down an analytical expression for a $3 \times 3$ hamiltonian at each $k$ point of the Brillouin zone. With this hamiltonian, one can calculate the $t_{1u}$ band structure and the Fermi energy density-of-states $N(0)$ for a variety of $C_{60}$ structures. The typical calculated value of the density-of-states is $N(0) \sim 10$states/$eV$/spin/molecule. The analytical form of the $3 \times 3$ hamiltonian and the detail results of band calculations may be found in Ref. [14].

## 5. ELECTRON-PHONON MECHANISM OF SUPERCONDUCTIVITY

Solid $C_{60}$ is a semiconductor, but when doped with alkali atoms, it becomes a metal and exhibits superconductivity. The superconducting transition temperature $T_c$ depends on the type and concentration of the dopant (see Table 2). Currently the compound $RbCs_2C_{60}$ has the highest $T_c$ of 33 K among the fullerenes. The key question we address here is whether the electron-phonon (e-ph) coupling mechanism is a plausible mechanism for superconductivity in the fullerenes.

The presence of a phonon changes the distance between carbon atoms and therefore the electronic hopping matrix elements between the neighbouring orbitals, which in turn affects the electronic energies. The basic quantity describing this electron-phonon interaction is the electron-phonon matrix element [16,17]

$$g_{v,nk,m(k+q)} = \langle nk| \sum_i e_{i\nu q} \sqrt{\frac{\hbar}{2M\omega_{\nu q}}} \frac{\partial H}{\partial R_i} |m, k+q\rangle \tag{4}$$

where $|m, k\rangle$ is a band electron state, $\omega_{\nu q}$ is the phonon frequency, and the $i$ summation is taken over all atoms and xyz coordinates. The quantity $e_{i\nu q}$ is the phonon

Table 2: Superconducting transition temperatures $T_c$ of some alkali-doped fullerenes.

| Compound | $T_c$ |
|---|---|
| $K_3C_{60}$ | 19.0 |
| $Rb_3C_{60}$ | 28.6 |
| $RbCs_2C_{60}$ | 33.0 |

eigenvector, and it is the variation of the electronic hamiltonian- caused by the phonon induced lattice distortion- of which the matrix element is being taken. The dimensionless electron-phonon coupling constant $\lambda$ is then expressed as a double integration over the Fermi surface and a summation over the phonon modes:

$$\lambda = \sum_{\nu q} \frac{2}{\hbar\omega_{\nu q} N(0)} \sum_{nmk} |g_{\nu,nk,m(k+q)}|^2 \delta(\epsilon_{nk}) \delta(\epsilon_{m(k+q)} - \epsilon_{nk} - \hbar\omega_{\nu q}). \tag{5}$$

Here the electronic energies are measured with respect to the Fermi energy $E_f$ and $N(0)$ is the density-of-states per spin at $E_f$. If we neglect the weak intercluster coupling as compared to the strong intracluster coupling, then the double integration in (5) may be performed analytically to yield the equation

$$\lambda = N(0) \sum_{\nu} V_\nu, \tag{6}$$

where

$$V_\nu = \frac{2}{9\hbar\omega_\nu} \sum_{nm} |g_{\nu,nm}|^2 \quad . \tag{7}$$

The quantity $V_\nu$ corresponds to the coupling of the $\nu$-th phonon and the subscripts $n, m$ of $g_{\nu,nm}$ denote the electronic states of the single molecule. Now, since the relevant electron states for superconductivity in the alkali-doped fullerenes is the partially-filled $t_{1u}$ band and, furthermore, since the higher bands are separated from $E_f$ by much larger energies compared to the phonon energies, the summation over the band indices $n, m$ in Eq. (7) is to be taken over the $t_{1u}$ bands only.

Not all phonons can couple to the $t_{1u}$ electrons. The electron-phonon coupling, which depends on the symmetries of both the electron and the phonon, is non-zero only if $(\Gamma_{el} \times \Gamma_{el})_{sym} = \Gamma_{phonon}$. The allowed electron-phonon coupling for the icosahedral group $I_h$ are summarized in Table 3. First of all, we notice from the Table that only $A_g$, $G_g$, and $H_g$ phonons can ever couple to the electron states of the $C_{60}$ molecule. The $t_{1u}$ electrons are affected only by the phonons of $A_g$ or $H_g$ symmetries.

To compute the e-ph coupling parameter quantitatively, we need models for (i) the phonon frequencies and eigenvectors (ii) the electronic states of the cluster and (iii) scaling of the electronic hamiltonian matrix elements with distance. With these, the e-ph coupling strength can be calculated by displacing the carbon atoms according to

Table 3: Symmetry-allowed electron-phonon coupling for the icosahedral group, $I_h$.

| phonon | electron symmetry | | | | | | | | | |
|---|---|---|---|---|---|---|---|---|---|---|
| symmetry | $a_g$ | $t_{1g}$ | $t_{2g}$ | $g_g$ | $h_g$ | $a_u$ | $t_{1u}$ | $t_{2u}$ | $g_u$ | $h_u$ |
| $A_g$ | 1 | 1 | 1 | 1 | 1 | 1 | 1 | 1 | 1 | 1 |
| $G_g$ | | | | 1 | 1 | | | | 1 | 1 |
| $H_g$ | | 1 | 1 | 1 | 1 | | 1 | 1 | 1 | 1 |
| $A_u$, $T_{1u}$, $T_{2u}$, $G_u$, $H_u$, $T_{1g}$, and $T_{2g}$: | | | | | | no coupling | | | | |

the various normal modes and then by evaluating how the electron states are affected by the displacement according to Eq. (4).

The vibrational modes of the $C_{60}$ molecule have been calculated by several authors using a variety of methods, viz., (a) the ball-and spring model [18], (b) the bond-charge model [19], (c) quantum-chemical methods [20], and (d) density-functional methods [21]. We calculate the e-ph coupling strength with the first two models. The simplest of the above is the ball-and-spring model, first applied by Weeks and Harter to the $C_{60}$ molecule. In this model, the carbon atoms are replaced by point masses coupled by two spring constants: a bond-stretching spring constant and a bond-bending spring constant, whose values were taken in the original work of Weeks and Harter [18] to be $7.6 \times 10^5 dyne/cm$ and $0.7 \times 10^5 dyne/cm$, respectively. However, since this produces vibrational frequencies, too high in magnitude as compared to the observed values, we use somewhat softer spring constants, of magnitude 5.3 and $0.50 \times 10^5 dyne/cm$, respectively. The vibrational modes are then obtained by a single diagonalization of the $180 \times 180$ dynamical matrix. This simple model is actually not too bad in describing the phonon frequencies of the experimentally observed dipole active and Raman modes.

The second model we use is the bond-charge model of Onida and Benedek [19]. Ninety bond charges (BC), each of zero mass, representing the electrons involved in the C-C bonds, are placed between the atoms. In addition to the bond-stretching and bond-bending forces between atoms, a bond-bending force corresponding to the BC-atom-BC angle and the Coulomb forces are included in the model. The calculated vibrational frequencies agree quite well with the experimental values as seen from Table 4.

To describe the coupling of the electron states to the phonons, we use here the simple one-parameter tight-binding model, Eq. (1), involving the "$p_z$" orbitals only. This is reasonable since, as indicated in Fig. 1, the "$p_z$" orbitals contribute about 95% of the wave function character of the $t_{1u}$ bands.

We finally need to determine how the electronic matrix element $\beta$ scales with the separation between two adjacent carbon atoms. A rough estimate of this scaling is given by Harrison's universal scaling formula [22],

$$\beta_{ll'm} = \eta_{ll'm} \frac{\hbar^2}{md^{\eta}} \tag{8}$$

where $\eta_{pp\sigma} = 3.24$ and $\eta_{pp\pi} = -0.81$ and the exponent $\eta = 2$ for $p$ orbitals. However, this $d^{-2}$ scaling of the matrix elements is only approximate. This scaling will be changed in the fullerenes because of admixture of orbitals other than the $p$ orbitals.

We have used the LMTO method to compute the matrix element scaling by performing a series of density-functional band calculations for a number of interatomic distances. In the one-parameter tight-binding theory, the energy separation $\Delta E$ between the average band positions is proportional to the hopping integral $\beta$, and, therefore, the scaling of $\beta$ with distance $d$ may be inferred from the relationship

$$\Delta E = const. \times d^{-\eta} \tag{9}$$

The band positions were calculated by averaging the band structure energies over ten special $k$ points in the fcc Brillouin zone. The calculated values of $\Delta E$, for

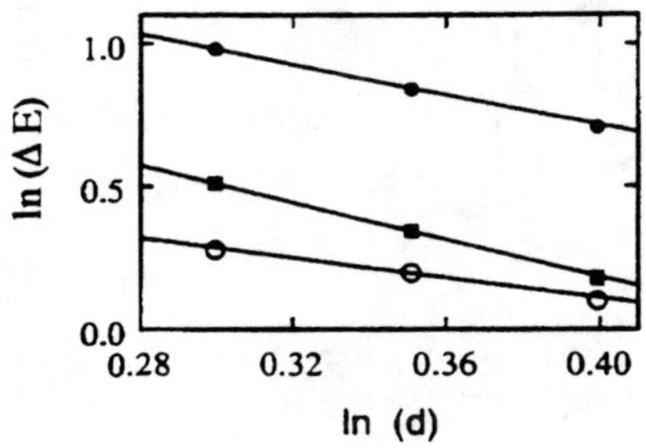


Figure 5: Variation of energy separation $\Delta E$ of various bands from the $h_u$ band in fcc $C_{60}$ as a function of the intracluster C-C distance $d$ as obtained from density-functional LMTO calculations: $t_{1u}$ band (open circle), $t_{1g}$ band (solid circle), and $h_g + g_g$ band (solid square).

the separations of the $t_{1u}$, $t_{1g}$, and the $g_g + h_g$ bands with respect to the valence $h_u$ bands, are plotted on a log-log scale in Fig. 5. The value of the scaling parameter $\eta$, calculated from the slope in the figure, is about $2.5 \pm 0.5$.

The electron-phonon coupling parameter for various phonon modes are then calculated according to the expression (4), by examining the change of energy of the $t_{1u}$ molecular orbital when atoms are displaced according to the individual phonon eigenvectors. The results are summarized in Table 4. The calculated phonon frequencies, especially those obtained from the bond-charge model, agree well with the experimental values.

We observe that in general tangential phonons, i.e., phonons with large atomic displacements on the sphere surface, couple most strongly to the electron states. This is expected, since the tangential vibrations cause the largest deviation in the carbon-carbon distances on the molecule thereby maximally affecting the electronic states. The importance of the tangential phonons is also confirmed from the first-principles calculation of the Hopfield factor reported by Mazin et al. [23] From Table 4, we obtain the total e-ph coupling strength V = $\sum_\nu V_\nu =\sim$ 40 - 80 meV, which is similar to the values obtained independently by other authors [20,24]. Taking a value of $N(0) \approx 10$ electrons / eV / spin / molecule as obtained from the band calculations, we then obtain the magnitude of the electron-phonon coupling parameter to be $\lambda = N(0) \times V \approx 0.4$ - 0.8. The superconducting transition temperature may be estimated from McMillan's formula,

$$T_c = \frac{\langle \omega_{log} \rangle}{1.20} exp[\frac{-1.04(1+\lambda)}{\lambda - \mu^*(1+0.62\lambda)}], \tag{10}$$

which holds if the lattice is not too strongly coupled to the electron ( $\lambda \lesssim 1.5$). Here, $\langle \omega_{log} \rangle$ is the logarithmic average of the phonon frequencies and $\mu^*$ is the effective Coulomb pseudopotential parameter, reduced down from the bare value $\mu = UN(0)$ because of retardation effects, where U is the screened repulsion between two elec-

Table 4: Strength of electron-phonon coupling $V_\nu$ of the vibrational modes (phonons) to the $t_{1u}$ electron state of the $C_{60}$ molecule. BC referes to the bond-charge model and BS refers to the ball-and-spring model for the vibrational modes. The phonon frequencies are in units of $cm^{-1}$. The tangential (T) component of the individual phonon eigenvectors, i.e., the component corresponding to the atomic motion on the surface of the carbon sphere, are listed, with the balance being the radial component. Thus, for instance, the $A_g(2)$ mode is a pure tangential mode, $A_g(1)$ is a pure radial mode, while $H_g(1)$ is a mixed mode. The parameter $\eta$ refers to the scaling of the electron hopping integral, $\beta = const. \times d^{-\eta}$.

| phonon | $H_g(1)$ | $H_g(2)$ | $H_g(3)$ | $H_g(4)$ | $H_g(5)$ | $H_g(6)$ | $H_g(7)$ | $H_g(8)$ | $A_g(1)$ | $A_g(2)$ | $\langle\omega_{\log}\rangle$ ($cm^{-1}$) | $V$ ($meV$) |
|---|---|---|---|---|---|---|---|---|---|---|---|---|
| $\omega$(expt) | 273 | 437 | 710 | 744 | 1099 | 1250 | 1428 | 1575 | 497 | 1469 | 817 | |
| $\omega$(calc. BC) | 271 | 409 | 716 | 791 | 1154 | 1215 | 1449 | 1687 | 497 | 1451 | 826 | |
| $\omega$(calc. BS) | 230 | 347 | 442 | 692 | 1081 | 1315 | 1596 | 1745 | 425 | 1527 | 749 | |
| $T(BC)$(%) | 24 | 4 | 4 | 75 | 95 | 99 | 100 | 100 | 0 | 100 | | |
| $T(BS)$(%) | 28 | 6 | 3 | 72 | 94 | 97 | 100 | 100 | 0 | 100 | | |
| Electron-phonon coupling strength, $V_\nu(meV)$ | | | | | | | | | | | | |
| Bond-charge model (BC) | | | | | | | | | | | | |
| $\eta = 2$ | 7.4 | 0.0 | 0.0 | 2.7 | 0.1 | 1.8 | 8.3 | 8.9 | 0.0 | 5.5 | | 34.7 |
| $\eta = 2.5$ | 11.7 | 0.0 | 0.0 | 4.2 | 0.1 | 2.8 | 13.0 | 13.9 | 0.0 | 8.5 | | 54.2 |
| $\eta = 3$ | 16.7 | 0.0 | 0.0 | 6.0 | 0.2 | 4.1 | 18.7 | 19.9 | 0.0 | 12.3 | | 78.1 |
| Ball-and-Spring model (BS) | | | | | | | | | | | | |
| $\eta = 2$ | 4.4 | 0.0 | 0.0 | 1.7 | 0.1 | 0.2 | 1.1 | 14.3 | 0.1 | 4.8 | | 26.9 |
| $\eta = 2.5$ | 6.8 | 0.0 | 0.0 | 2.6 | 0.1 | 0.4 | 1.7 | 22.4 | 0.1 | 7.5 | | 41.9 |
| $\eta = 3$ | 9.8 | 0.0 | 0.0 | 3.8 | 0.2 | 0.5 | 2.4 | 32.3 | 0.2 | 10.8 | | 60.4 |

tronson the Fermi surface. For a wide-band metal, the reduction has been shown to be

$$\mu^* = \frac{\mu}{(1 + \mu\, ln(W/\omega_{ph}))}, \tag{11}$$

where W is a typical electronic energy. The prevalent opinion seems to be that W is not the typical energy of the $t_{1u}$ bands, which is only about 0.5 eV wide, but rather, W should be taken to be $\sim$ 15 eV, considering that band calculation gives a number of subbands spread over an energy range of $\sim$ 30 eV. This together with the value $\omega_{ph} \simeq 0.1 eV$ results in a value of $\mu^* \sim 0.2$. With this estimated value, McMillan's formula yields $T_c$ values in the range of 5 K and 35 K, in agreement with the observed superconducting transition temperatures.

Now, the separation between the $C_{60}$ molecules can be reduced by either applying external pressure or by choosing smaller alkali atoms. This will increase the width of the $t_{1u}$ bands, reducing thereby the density-of-states $N(0)$. The relevant phonon quantities, $\langle\omega_{log}\rangle$ and V, on the other band, are essentially unaffected. The transition temperature will therefore decrease in both cases as indicated from the expressions for $T_c$. This is more-or-less observed supporting further the electron-phonon mechanism of superconductivity in the fullerenes.

## 6. SUMMARY

In summary, we examined the nature of electron states in undoped and alkali-doped solid $C_{60}$ starting from the electronic structure of a free $C_{60}$ molecule. The basic picture of chemical bonding was extracted from a series of density-functional calculations of the electronic structure for the free molecule as well as for the solid fullerenes. Undoped solid $C_{60}$ is an insulator with the conduction bands of $t_{1u}$ symmetry being completely empty. In the alkali-doped $C_{60}$, the dopant atoms are fully ionized with the alkali electrons occupying the $t_{1u}$ bands in a "rigid band" fashion. The analytical wave functions for the $t_{1u}$ LUMO and the next higher $t_{1g}$ MO of the molecule were constructed from group theory. This allows one to construct a $3 \times 3$ analytical Hamiltonian for the conduction bands in the solid.

The strength of electron-phonon coupling in the solid was evaluated by adopting two diffrent models to describe the phonons, viz., the Weeks-Harter "ball-and-spring" model and the Onida-Benedek "bond-charge" model, and by using a scaling of the hamiltonian matrix elements with distance estimated from density-functional calculations. The sizeable strength of the calculated electron-phonon coupling parameter, $\lambda \simeq 0.4 - 0.8$, indicates that the superconductivity in the doped fullerenes can be explained within the framework of the conventional superconductivity theory.

## REFERENCES


1. H. W. Kroto et al., Nature **318** 162 (1985).
2. W. Krätschmer, L. D. Lam, K. Fostiropoulos, and D. R. Huffman, Nature **347** 354 (1990).
3. A. F. Hebard et al., Nature **350** 600 (1991) ; K. Holczer et al., Science **252** 1154 (1991); M. J. Rosseinsky et al., Phys. Rev. Lett. **66**, 2830 (1991).
4. R. Sachidanandam and A. B. Harris, Phys. Rev. Lett. **67**, 1467 (1991).
5. W.I.F. David et al., Nature **353**, 147 (1991).
6. A. D. J. Haymet, Chem. Phys. Lett. **122**, 421 (1985).
7. R. C. Haddon, L. E. Brus, and K. Raghavachari, Chem. Phys. Lett. **125**, 459 (1986).
8. S. Satpathy, Chem. Phys. Letts. **130**, 545 (1986).
9. L. Pauling, The Nature of the Chemical Bond (Cornell Univ. Press, Ithaca, New York, 1980).
10. Q. Zhang, J. Y. Yi, and J. Bernholc, Phys. Rev. Lett. **66**, 2633 (1991).
11. P. A. Heiney et al., Phys. Rev. Lett. **66**, 2911 (1991).
12. O. Gunnarsson, S. Satpathy, O. Jepsen, and O.K. Andersen, Phys. Rev. Lett. **67**, 3002 (1991).
13. C. Pan et al., J. Phys. Chem. **96**, 2944 (1991).
14. S. Satpathy, V. P. Antropov, O. K. Andersen, O. Jepsen, O. Gunnarsson, and A. I. Liechtenstein, Phys. Rev. B **46**, 1773 (1992).
15. S. Saito and A. Oshiyama, Phys. Rev. Lett. **66**, 2637 (1991).
16. C. O. Rodriguez, A. I. Liechtenstein, I. I. Mazin, O. Jepsen, and O. K. Andersen, Phys. Rev. B **42**, 2692 (1990).
17. D. Rainer, Prog. Low Temp. Phys. **10**, 371 (1986).
18. D. E. Weeks and W. G. Harter, Chem. Phys. Lett. **144**, 366 (1988).
19. G. Onida and G. Benedek, Europhys. Letts. **18**, 403 (1992).
20. C. M. Varma, J. Zaanen, and K. Raghavachari, Science **254**, 989 (1991).
21. G. B. Adams, J. B. Page, O. F. Sankey, K. Sinha, J. Menendez, and D. R. Huffman, Phys. Rev. B **44**, 4052 (1991).
22. W. A. Harrison, *Electronic Structure and the Properties of Solids: The Physics of the Chemical bond* (Freeman, San Francisco, 1979).
23. I. I. Mazin, S. N. Rashkeev, V. P. Antropov, O. Jepsen, A. I. Liechtenstein, and O. K. Andersen, Phys. Rev. B **45**, 5114 (1992).
24. M. Schluter, M. Lannoo, M. Needels, G. A. Baraff, and D. Tomanek, Phys. Rev. Lett. **68**, 526 (1992).

Local Order
in Condensed
Matter Physics


ISBN 1-56072-220-7

# GENERALIZED TIGHT-BINDING MOLECULAR DYNAMICS FOR CLUSTERS

Madhu Menon and K. R. Subbaswamy

*Department of Physics and Astronomy, University of Kentucky*
*Lexington, KY 40506-0055, U.S.A.*

## A. INTRODUCTION

The importance of local order in determining certain global properties in solids has been reemphasized in recent years with the discovery of cluster-based solids such as the fullerene solids. The electronic band structure of solid $C_{60}$ bears a very close resemblance to the electronic levels of the $C_{60}$ molecule. The recently proposed "crystals of clusters" idea[1] might lead to the synthesis of more solids in this class. Thus, it is important to have accurate knowledge of the geometric and electronic structure of clusters of atoms that are precursors to such solids.

Unfortunately, *ab initio* methods cannot yet handle the full geometric optimization of even moderate size clusters. Thus, a semi-empirical interface between the more accurate *ab initio* methods on the one hand and classical n-body potential treatments that work well in the crystalline environment is needed. With this in mind we have developed a generalized tight-binding molecular dynamics (GTBMD) method[2, 3] based on van Schilfgaarde and Harrison's nonorthogonal generalization of tight-binding electronic structure theory.[4, 5] We will give a brief overview of the method emphasizing application to fullerenes.

## B. NON-ORTHOGONAL TIGHT-BINDING MOLECULAR DYNAMICS

The details of this new non-orthogonal technique have been published elsewhere.[2, 3] Here we give only an outline. As indicated earlier, this formulation is based on van Schilfgaarde and Harrison's non-orthogonal orbital tight-binding electronic structure calculation.[4, 5] It is necessary to take into account the non-orthogonality explicitly, because for systems with higher coordinations, the non-orthogonality cannot be separated out as a simple two-body repulsion as in the tetrahedral case. Here, the simplicity of the original theory is maintained and, the only additional parameter entering into calculations is the non-orthogonality constant $K$ (see below), which has been fit to give the correct equilibrium spacing for the dimer.

The eigenvalue equation for a system is given by

$$(\underline{H} - E_n \underline{I})\psi_n = 0. \tag{1}$$

Let $\{\phi_a\}$ be a complete set of non-orthogonal atomic orbitals. Then one can write the eigenstates $\psi_n$ of this system as a linear combination of $\phi_a$,

$$\psi_n = \sum_j c_j^n \phi_j \tag{2}$$

The eigenvalue problem (1), then becomes

$$\sum_j (H_{ij} - E_n S_{ij}) c_j^n = 0. \tag{3}$$

where

$$H_{ij} = \int \phi_i^* H \phi_j d\mathbf{r}, \tag{4}$$

and,

$$S_{ij} = \int \phi_i^* \phi_j d\mathbf{r}. \tag{5}$$

In matrix form eq.(3) becomes

$$(\mathbf{H} - E_n \mathbf{S})\mathbf{C}^n = 0, \tag{6}$$

where $\mathbf{C}^n$ is a column vector with coefficients $c_i^n$ 's. Differentiating this with respect to $x$, and making use of the fact that both **H** and **S** are hermitian, one can obtain the Hellmann-Feynman theorem for the electronic part of the force:[2, 3]

$$\frac{\partial E_n}{\partial x} = \frac{\mathbf{C}^{n\dagger}(\frac{\partial \mathbf{H}}{\partial x} - E_n \frac{\partial \mathbf{S}}{\partial x})\mathbf{C}^n}{\mathbf{C}^{n\dagger}\mathbf{S}\mathbf{C}^n}. \tag{7}$$

The remaining contribution to the force is given by a repulsive pair potential which includes the ion-ion interactions and the electron-electron interaction, which is doubly counted in the electronic part of the force. The Hamiltonian matrix elements as well as the repulsive pair potential are taken to scale exponentially with the distance.[3] There is no cutoff in the interactions between atoms. In this form, the molecular dynamics theory is applicable to systems with arbitrary coordination. In this scheme the two disposable parameters in the force calculation are determined by fitting to the experimental bond length and vibrational frequency of the dimer. We have tested our scheme by determining the structures of small Si[3] and C[6] clusters for which *ab initio* results are available. The agreement is excellent.

# C. APPLICATION TO FULLERENE

## C.1 Structure and vibrational modes of $C_{60}$

We first obtained the equilibrium structure by starting from a nearly correct configuration and allowing all atoms to relax by removing 1% of each velocity at each time step. The optimization is stopped when the largest velocity component of any atom falls below 54 m/s (corresponding to a temperature of about 10 K). In the final equilibrium configuration for $C_{60}$ two distinct C-C bond lengths were obtained; these were 1.38 Å for the bonds that are part only of hexagonal rings ("double bonds") and 1.41 Å for the pentagonal rings ("single bonds").

Setting our parameter values to reproduce results for the carbon dimer, we obtained agreement for the vibrational modes of $C_{60}$ comparable to those obtained by LDA methods as shown in the table below.

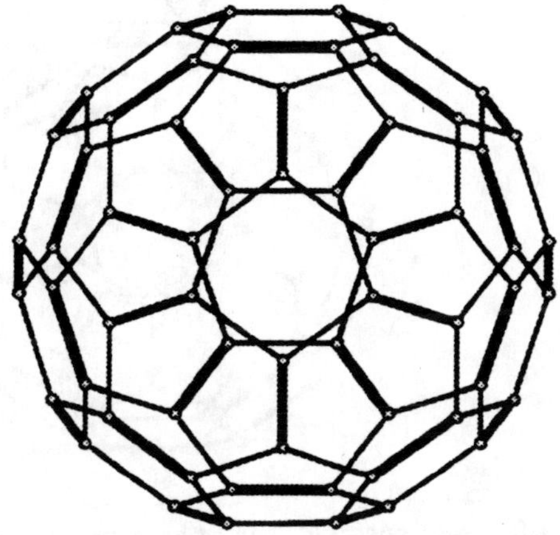

Fig.1 $C_{60}$ molecule

Table 1: Measured and calculated frequencies (in $cm^{-1}$)for the normal modes of $C_{60}$

| Mode | Experimental[1] frequency | Present work | Car-Parrinello Ref.[8] | Real-space LDA Ref.[9] |
|---|---|---|---|---|
| $A_g(1)$ | 491 | 529 | 454 | 537 |
| $A_g(2)$ | 1469 | 1717 | 1368 | 1680 |

[1]Ref. [7],

## C.2 Interaction with Oxygen

We next describe the results for the bonding of a single oxygen atom to $C_{60}$. In the cases reported below, the oxygen atom is introduced approximately 1.6 Å from the surface of a fully relaxed $C_{60}$molecule, and a molecular dynamics optimization is carried out, annealing as described earlier. The lowest energy geometry obtained in this manner is shown in Fig. 2. In this configuration, the oxygen is bonded to two carbon atoms over the short bond in $C_{60}$ that is shared by two adjoining hexagons (as shown in Fig. 2); the underlying double bond is weakened into a single bond in the process. This is termed the "epoxide" structure, and its formation has now been established experimentally.[10, 11]

The Mulliken population analysis shows that there is almost the complete transfer of an electron to the oxygen atom, while for other bonding geometries there is almost no electron transfer to the oxygen. This is consistent with the known difficulty in oxidizing $C_{60}$. All C-C bonds in the vicinity of the chemisorbed oxygen are stretched; in particular, the distinction between single and double bonds disappears for these bonds. The binding energy for epoxide structure is almost 1 eV higher than the case where the oxygen is bonded directly over a C atom.

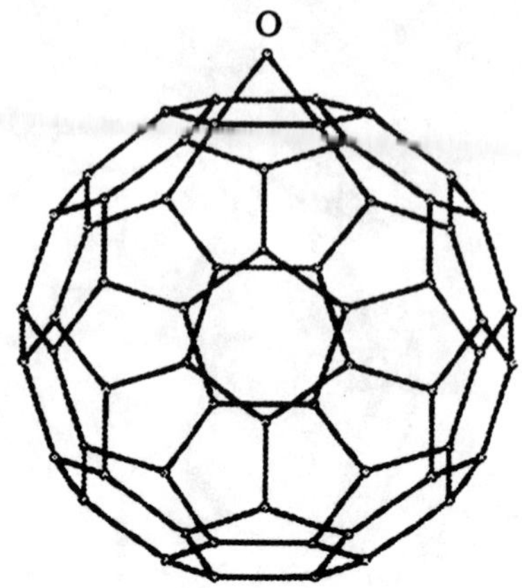


Fig.2 Lowest energy geometry for a single oxygen bonded to $C_{60}$

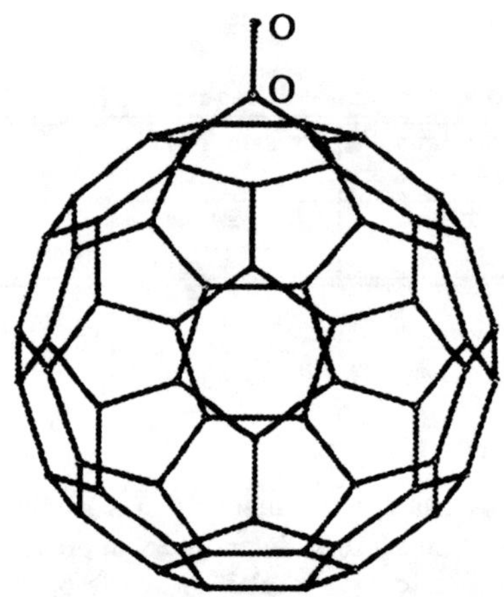


Fig.3 $O_2$ on $C_{60}$

We now turn to the consideration of $C_{60}O_2$. The optimizations are begun by introducing two oxygen atoms in close proximity to each other, in different positions and orientations relative to $C_{60}$, and annealing as described earlier. When the pair is released over a double bond, it breaks the C-C double bond. The resulting broken six-fold ring is transformed into a seven-fold ring with one of the oxygen atoms at one vertex, with the second oxygen atom once again giving rise to the "hairy" structure (Fig. 3). This structure was obtained irrespective of whether the $O_2$ was released in a transverse or radial orientation over a double bond. This is our lowest energy configuration for two oxygen atoms bonded to $C_{60}$. Releasing $O_2$ over a single bond also results in the breaking of a C-C bond, with the oxygen atoms incorporated such that the

O-O bond is pointing radially outward from the $C_{60}$ cage (forming a "hairy" structure). When $O_2$ was released symmetrically straddling a single bond it migrates to and breaks an adjacent single bond. The resulting broken five-fold ring now becomes a six-fold ring with the oxygen atom at one vertex. The second oxygen atom is bonded to the first oxygen atom in such a way that the O-O bond is pointing radially outward.

Based on these findings we predict that the optical spectrum of the stable $C_{60}O_2$ complex should be substantially different from that of pristine $C_{60}$. We are not aware of any reports of experimental results for isolated $C_{60}O_2$ complexes.

## C.3 The "Bucky Dumb-Bell"

In a recent paper by Rao *et al.*[7] it was reported that exposure to visible or ultraviolet (UV) light causes solid $C_{60}$ to polymerize, where the molecules cross-link via covalent bonds. After phototransformation the solid $C_{60}$ film was found to be no longer soluble in toluene. Raman scattering studies[12] have found Ar-ion laser radiation to initiate an irreversible transformation of fcc $C_{60}$ to a different solid phase with a richer Raman spectrum. Further, Raman Scattering studies of the polymeric phase revealed a new Raman-active mode at 116 $cm^{-1}$, which was identified with an inter-molecular vibrational mode.[7, 12] While the pristine spectra for the solid $C_{60}$ exhibit primarily intramolecular modes (ten Raman active modes of $A_g$ and $H_g$ symmetry and four IR active modes of $F_{1u}$ symmetry), consistent with the weak inter-molecular interaction, the phototransformed phase was found to exhibit many more lines indicating a lowering of the icosahedral symmetry. Furthermore, the laser desorption mass spectrum (LDMS) of a $C_{60}$ film phototransformed with UV-visible radiation revealed a succession of 20 regularly spaced clear peaks in the vicinity of $C_{120}$, $C_{180}$ and so on with diminishing intensities.[7] Based on these observations the authors of Ref.[7] proposed that a 2+2 cyclo addition photochemical reaction[13] had taken place where parallel double bonds from adjacent $C_{60}$ molecules were broken and reformed as a four-sided ring.

Motivated by this work we proceeded to investigate the various possible covalent $C_{60}$ dimer bonding configurations to determine the most stable bonding geometry using our MD scheme.

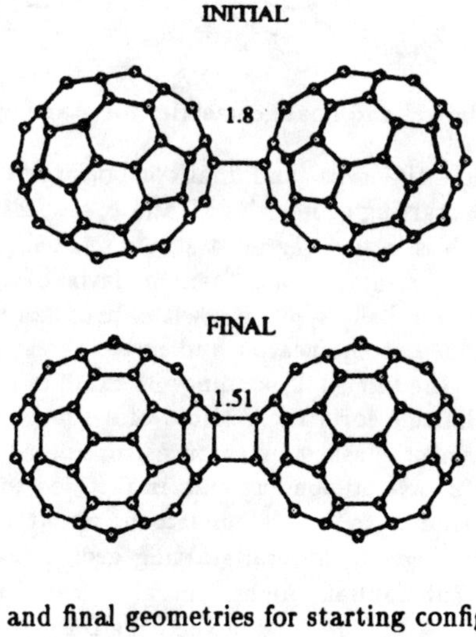


Fig.4 Initial and final geometries for starting configuration 7

We do a detailed computational analysis of covalent bonding between $C_{60}$ dimers by considering the following possible relative orientations as the initial configuration in our molecular dynamics relaxations: $C_{60}$ units facing each other with the hexagonal edges closest to each other in parallel (**1**) or perpendicular (**2**) orientations of the double bonds; the pentagonal edges parallel to each other (**3**); the pentagonal faces of the dimers facing each other in an eclipsed (**4**) or staggered (**5**) geometry; hexagonal faces parallel (**6**); the corners of the units as the closest approach (**7**); the corner of one $C_{60}$ near a bridge site over the double bond of the other molecule (**8**); the corner of one $C_{60}$ over the center of a hexagonal face of the other (**9**), and the $C_{58}$-$C_2$-$C_{60}$ geometry that was discussed as a possible alternative linkage in Ref. [7] and later rejected by the same authors (**10**). In all our simulations the initial configurations were chosen in such a way that the $C_{60}$ dimers were within covalent bonding range of each other (≈2Å). Damped molecular dynamics relaxation was then performed by removing 1% of each velocity at each time step. No symmetry restriction is imposed while the optimization is being carried out. The process is stopped when the largest velocity component of any atom falls below 54 m/s (corresponding to a temperature of about 10 K).[3]

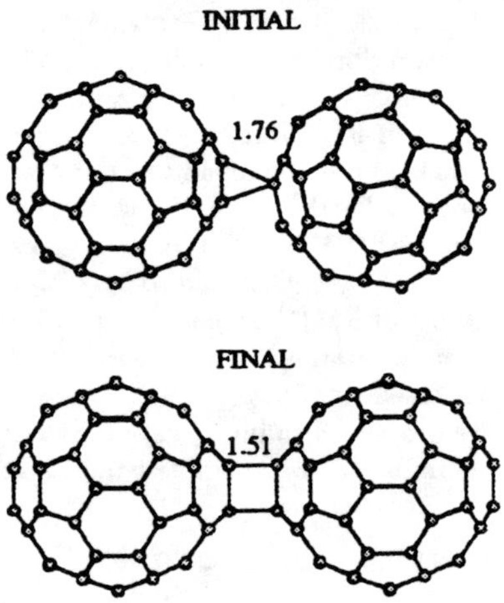


Fig.5 Initial and final geometries for starting configuration **8**.

In Figs. 4 and 5 we show the initial and final configurations from our molecular dynamics relaxation for two different starting geometries (**7** and **8**, respectively). In both cases, the final geometry is the same, which is in fact the most stable relaxed geometry obtained. The initial configurations were taken so as not to bias them in favor of symmetric bonding. The final geometry shows a "bucky dumb-bell" structure, with almost all the deviations from the perfect icosahedral symmetry confined to the nearest and next-nearest carbon atoms surrounding the four atoms participating in the inter-ball bonding. It exhibits $D_{2h}$ symmetry. The four atoms involved in the inter-ball linkage form the vertices of a square of side 1.51 Å. In Table I the total energies are listed for all the relaxed structures considered here.

The starting geometry **2** gave no bonding and, in fact, led to a repulsion between the balls. The geometry **3** which is similar to **1** was found to be about 1 eV higher in energy than the relaxed structure shown in Fig. 1. Several starting geometries (e.g., (**4**, **5**, **6**, and **9** )) led to relaxed structures with substantially higher energies (See Table I). The relaxed $C_{58}$-$C_2$-$C_{60}$ geometry (**10**), which involves the breaking of the cage framework, is found to be significantly higher in energy than the bucky dumb-bell (by 4.8 eV). Subsequent experimental analysis by

**the authors of Ref. [7] has ruled out this structure. Considering the extensive sampling of initial orientations, it is reasonable to conclude that the 4-fold cyclic linkage shown in Figs. 1 and 2 is the minimum energy configuration.**

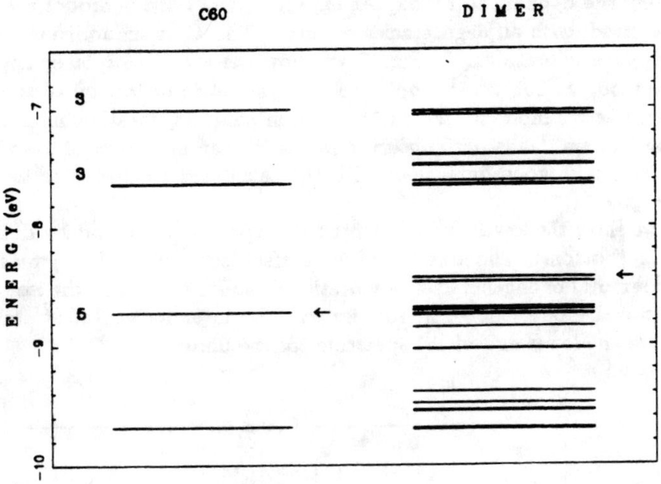


**Fig.6 Energy levels in the vicinity of the Fermi level for an isolated $C_{60}$ and the bucky dumb-bell.**

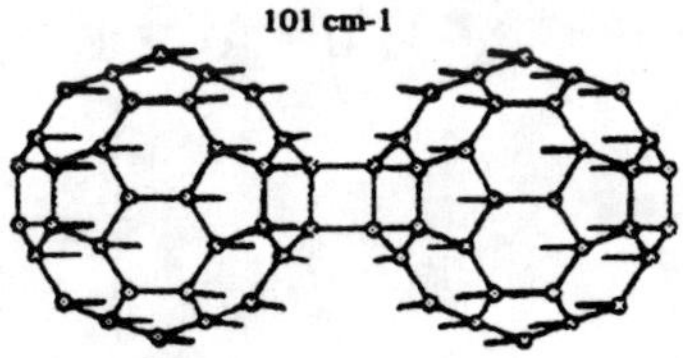


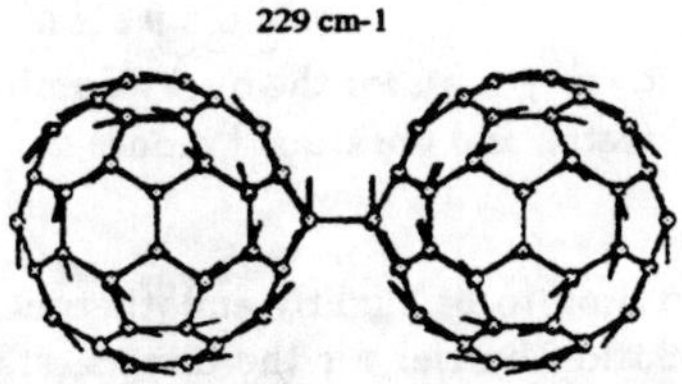


**Fig.7 Vibrational displacement patterns for the bucky dumb-bell for a symmetric distortion mode at 101 $cm^{-1}$ (upper panel) and for a libration mode at 227 $cm^{-1}$.**

In Figs. 6a and 6b we show the energy levels in the vicinity of the Fermi level for an isolated $C_{60}$ and the bucky dumb-bell (our lowest energy configuration for the dimer). An isolated $C_{60}$ has a five-fold degenerate highest occupied molecular orbital (HOMO) and a three-fold degenerate lowest unoccupied molecular orbital (LUMO) with a gap of 1.10 eV. As seen in Fig. 5b, the formation of the dimer has removed the degeneracy of energy levels near the Fermi level as a consequence of the lowered symmetry. The narrowing of the optical gap is consistent with the observed darkening of the photo-induced film.[7]

We have also determined the vibrational mode frequencies for the bucky dumb-bell by explicitly constructing the dynamical matrix. As expected, most of the modes are easily recognized as monomer-derived, with all degeneracies removed. There are six additional modes associated with inter-molecular vibrations. In Fig. 7 we show the atomic displacement patterns for one Raman active mode at 101 $cm^{-1}$ (top panel: a symmetric distortion of the dumb-bell), and for one infra-red active mode at 228 $cm^{-1}$ (bottom panel: a libration mode). Experimentally a new mode at 116 $cm^{-1}$ has been observed in the Raman spectrum of the photo-polymerized $C_{60}$ film,[7] which is in good agreement with the calculated frequency of the distortion mode displayed in Fig. 7.

In Fig. 8 we show the low frequency vibrational spectra for the bucky dumb-bell (top) and the isolated $C_{60}$ (bottom). The lowering of the perfect icosahedral ($I_h$) symmetry for the dimer gives rise to a removal of degeneracies of vibrational modes present in the isolated $C_{60}$ resulting in many additional modes shown in the figure. The inter molecular stretch and librational modes dominate the lower end of the spectrum for the dimer.

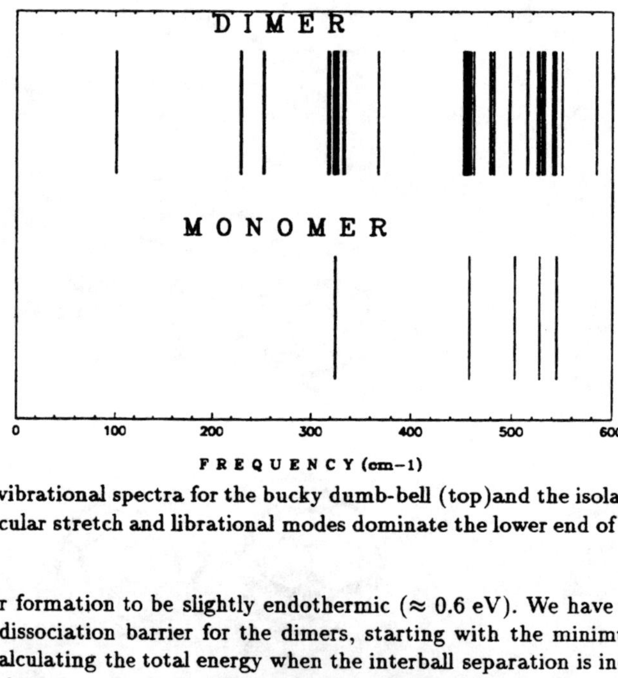


Fig.8 Low frequency vibrational spectra for the bucky dumb-bell (top)and the isolated $C_{60}$ (bottom). The inter molecular stretch and librational modes dominate the lower end of the spectrum for the dimer.

We find the dimer formation to be slightly endothermic ($\approx 0.6$ eV). We have estimated an upper bound to the dissociation barrier for the dimers, starting with the minimum geometry shown in Fig. 4, by calculating the total energy when the interball separation is increased while keeping all other bonds and angles frozen. From this and other supplementary calculations we estimate the barrier to be in the range of 2-4 eV.

## C.4 Comparisons of $C_{60}$ and $Si_{60}$

Determination of the ground state geometry of the $Si_{60}$ cage has been the subject of some recent works.[15, 16] Piqueras *et. al*[15] have made use of the AM1 method[17] to investigate the ground state structure of the icosahedral $Si_{60}$ cage. Based on their symmetry restricted optimization they found this geometry ($I_h$ symmetry) to be an energy minimum with two

**distinct bond lengths, namely the same qualitative feature that characterizes the $C_{60}$ cage. Using a conventional orthogonal tight-binding molecular dynamics optimization scheme the authors of Ref. [16] have performed a relaxation of an $Si_{60}$ icosahedral cage structure. Their results showed this icosahedral geometry to be unstable, relaxing into a structure resembling a puckered ball.**

**For a $Si_{60}$ cage we first obtained the equilibrium structure by starting from a perfect icosahedral configuration (Fig. 9, initial) and allowing all atoms to relax by removing 1% of each velocity at each time step. No symmetry restriction is imposed while the optimization is being carried out. We find the initial configuration with $I_h$ symmetry to be unstable, distorting to a lower symmetry geometry without any change in the coordination. The silicon atoms that formed planar hexagons and pentagons in the perfect icosahedral structure now move radially inward or outward on relaxation, with atomic arrangement tending toward tetrahedral geometry. The relaxed structure has $C_{2h}$ symmetry (Fig. 9, final). Of the 60 vertices, 10 have the ideal tetrahedral angle. Another 36 vertices have at least one angle within 2 degrees of the ideal tetrahedral angle. This suggests that the Jahn-Teller distortions favor displacements of atoms in the direction of increasing coordination. Several distinct bond lengths are obtained for the optimized structure, ranging from 2.34 Å to 2.53 Å with an average bond length of 2.40 Å.**

**We have also performed a detailed comparisons of electronic level structures near the Fermi level for the undistorted and distorted $Si_{60}$ cages. The perfect icosahedron (our starting configuration) has metallic character with a partially filled five-fold degenerate HOMO level. In the relaxed structure the degeneracies are completely removed by a symmetry lowering Jahn-Teller distortion and a HOMO-LUMO gap of 0.42 eV opens up. Although this value is less than half the gap found in bulk crystalline silicon, the tendency toward semiconducting character is clear.**

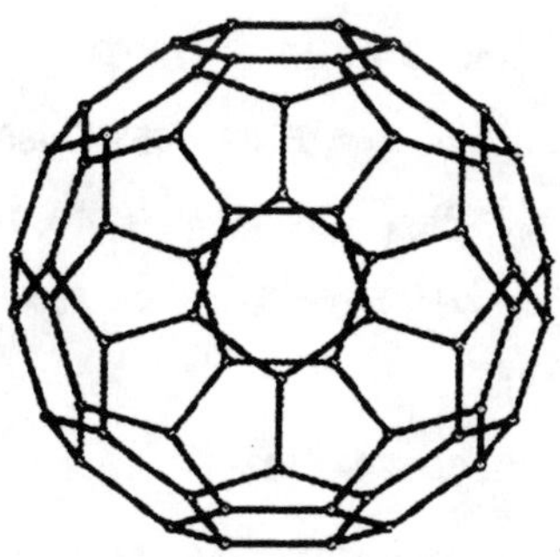

INITIAL

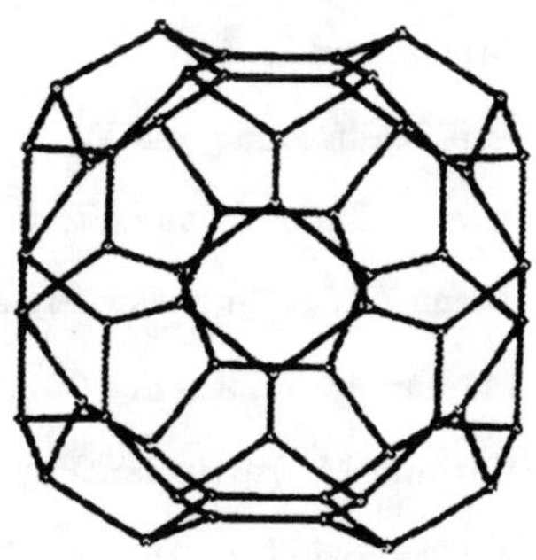

FINAL

Fig.9 Initial and final configurations of molecular dynamics optimization. The final relaxed structure shown has a $C_{2h}$ symmetry.

This result for $Si_{60}$ is in striking contrast to that for $C_{60}$ which has a perfect icosahedral symmetry. This difference must be understood on general grounds by a consideration of the special chemistry of the C-C bond. Although Si and C are both isovalent, their behavior in forming chemical bonds is quite different. The smaller covalent radius of the carbon atom can generate sufficient strain energy to offset any energy gain by forming more closed packed structures for carbon clusters. This explains the persistence of linear and ring structures with low coordination for small carbon clusters with $N \leq 10$. As is well known, the three-fold coordinated graphitic structure is nearly isoenergetic with the four-fold coordinated diamond. Silicon on the other hand, readily forms multicoordinated (coordination in excess of 4) structures for $N \leq 10$.[3, 18, 19] It should perhaps be not surprising that a three fold coordinated silicon cage, when allowed to relax, distorts to more tetrahedral type geometry.

# D. SUMMARY AND CONCLUSIONS

In conclusion, we have discussed non-orthogonal molecular dynamics methods based on the tight-binding theory and shown its application to clusters. The non-orthogonal scheme has been used successfully to treat isolated clusters of covalent atoms.

# E. ACKNOWLEDGEMENTS

This research was supported in part by the US National Science Foundation through grant EHR-91-08764 and by the University of Kentucky Center for Computational Sciences.

## References

[1] S.N. Khanna and P. Jena, Phys. Rev. Lett. **69**, 1664 (1992)

[2] M. Menon and K.R. Subbaswamy, Phys. Rev. Lett. **67**, 3487 (1991).

[3] M. Menon and K.R. Subbaswamy, Phys. Rev. **B 47** 12754 (1993).

[4] M. van Schilfgaarde and W. A. Harrison, Phys. Rev. B **33**, 2653 (1986).

[5] M. van Schilfgaarde and W. A. Harrison, J. Phys. Chem. Solids, **46**, 1093 (1985).

[6] M. Menon, K.R. Subbaswamy and M. Sawtarie, Phys. Rev. B**48** 8398 (1993).

[7] A. M. Rao, et al., Science **259**, 955 (1993).

[8] R. Car and M. Parrinello, Phys. Rev. Lett., **60**, 204 (1988).

[9] D. A. Drabold, J. D. Dow, P. A. Fedders, A. E. Carlsson, and O. F. Sankey, Phys. Rev. B **42**, 5345 (1990).

[10] K. M. Creegan, J. L. Robbins, W. K. Robbins, J. M. Millar, R. D. Sherwood, P. J. Tindall, D. M. Cox, A. B. Smith, III, J. P. McCauley, Jr., D. R. Jones, and R. T. Gallagher J. Am. Chem. Soc. **114**, 1103 (1992).

[11] Y. Elemes, S. K. Silverman, C. Sheu, M. Kao, C.S. Foote, M. M. Alvarez, and R. L. Whetten, Angew. Chem. Int. Ed. Engl. 31 (1992) 351.

[12] P. Zhou et al., Appl. Phys. Lett. **60**, 2871 (1992).

[13] J. D. Coyle, *Introduction to Organic Photochemistry* (Wiley, New York, 1986), p. 61.

[14] See the special issue Acc. Chem. Res. **25**, 98 (1992) for a review of fullerene-based solids.

[15] M. C. Piqueras, R. Crespo, E. Orti and F. Tomas, Chem. Phys. Lett. , **213**, (1993) 509.

[16] F.S. Khan and J.Q. Broughton, Phys. Rev. B**43**, (1991) 11754.

[17] M. J. S. Dewar, E. G. Zoebisch, E. F. Healy and J. J. P. Stewart J. Am. Chem. Soc. **107**, (1987) 3902.

[18] K. Raghavachari, J. Chem. Phys. **84**, (1986) 5672.

[19] K. Raghavachari, and C. M. Rohlfing, J. Chem. Phys. **89**, (1988) 2219.

Local Order
in Condensed
Matter Physics


ISBN 1-56072-220-7

# ELECTRONIC EXCITATIONS IN NANOSCALE GRAPHITIC TUBULES WITH HELICAL SYMMETRY

P. J. Lin-Chung and A. K. Rajagopal
*Naval Research Laboratory*
*Washington, D.C. 20375*

Abstract

A complete set of wave functions having helical symmetry is constructed to describe the electronic states of helical arrangement of atoms on a cylinder. The effects of a constant magnetic field along the tubule axis is also examined. In contrast to the often used cylindrical wave states, it is suggested that the use of these helical states are useful for studies of electronic properties of graphitic tubules in much the same way as the plane waves appropriate to the empty lattice incorporates the salient crystal symmetry features of the system.

The discovery of fullerine tubules [1,2] with diameters ranging between 20 and 300 A has led to interesting research into the electronic structures of periodic arrangement of atoms on a two-dimensional sheet wrapped on a cylinder [3,4,5]. These tubules may occur either as a single cylinder or as multiple concentric cylinders. As shown by White and his colleagues in Refs. 3, 4, and 5, rolling up of a two-dimensional honey-comb lattice sheet into a cylinder leads to many possible types of graphitic tubule characterised by the helical and rotational symmetry; they found all tubules have helical structure. The concentric cylinders are then obtained by rolling a parallel set of two-dimensional sheets. They carried out electronic band structure calculations for a single tubule based on tight-binding model which showed that these tubules will display properties ranging from semiconducting to metallic depending on the radii and the helical structure.

A simple model of interacting electrons moving on the surface of a cylinder has been studied [6,7] by way of investigating the elementary excitations in the system with a view that it may be applicable for discovering some of the electronic properties of electrons associated with graphitic tubules. By its very nature, this model has no helical symmetry built in so that the interesting features found by White et. al. [4] are not incorporated.

The purpose of the present work is to offer an alternate model of interacting electrons on a cylinder but now described in terms of helical symmetry states characterised by two parameters of the helix, the pitch and the pitch angle, instead of the cylindrical states as was done in Refs. 6 and 7. In spirit, this is similar to using the empty lattice to glean the salient symmetry features of a crystal system instead of a plane wave description-free electron model. We will show that the energy level structure of the electrons on a cylindrical surface appear to be different from that in Ref. 7 since we employ the appropriate quantum numbers which incorporate the characteristic helical symmetry. It should however be pointed out that without putting in the actual atomic structure, this model will give the same results as for the cylindrical symmetry model when bulk properties such as the Fermi energy, magnetization, etc., are computed, as expected. On the other hand properties which involve the electronic structure directly, as for example, Kohn anomalies, plasma mode dispersions, etc., will be different in appearance from the corresponding properties of the simpler model. We can relate the two by an analysis of the results for the cylindrical model using an appropriate mathematical transformation. This does not mean that we do not gain new and important insight into the properties of the helical system. In fact, when the actual band structures are incorporated, the features due to helical symmetry will be found in our results and those due to the details of atomic structure will be additional to these.

We begin by first giving a brief discussion of the helical waves and contrast it with the cylindrical waves. We also incorporate the effect of a constant magnetic field along the tube axis. We present here typical results illustrating our viewpoints described above, and more detailed results and discussion may be found in our work to be published elsewhere.

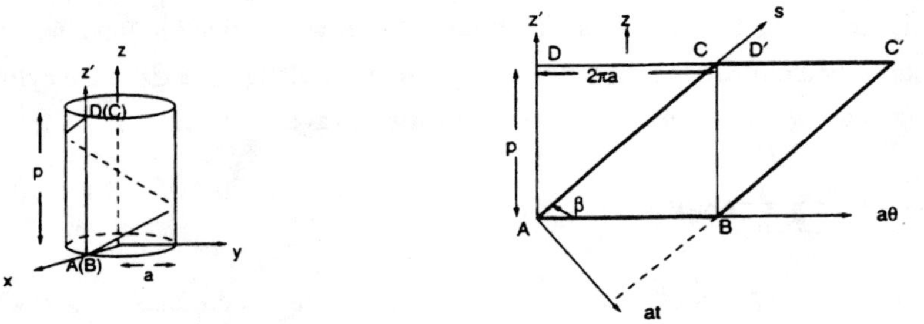


Fig.1 Unrolling of the cylindrical surface containing a spiral into a rectangle ABCD and a parallelogram ABC'D'

The basic property of a helix with pitch, $p$, wound on a right circular cylinder of radius $a$, is that any translation along the axis of the cylinder by an amount $h$ is accompanied by a rotation, $\alpha(h) = (2\pi h)/p$, about the axis. This is in contrast to the symmetry of the cylinder where the rotation about the axis and translation along the axis are independent operations. Consider now a cylinder containing one full turn of the spiral. Let the radius of this cylinder be $a$, so that its height, $p$, is the pitch of the spiral. When this cylinder is cut along a vertical line on the cylinder surface and spread into a plane we obtain a rectangle with edges $2\pi a$ and $p$. On the other hand, if we cut along the helix and spread it into a plane we obtain a parallelogram with the common base edge of length $2\pi a$, but with the diagonal of the rectangle obtained in the first construction forming the other side of the parallelogram (see Fig. 1). We choose to employ a rectangular helical coordinate system, ($s$, $at$) with $s$-axis along the diagonal which corresponds to the spiral line and the$at$-axis perpendicular to it. The pitch angle, $\beta$, is given by

$$\tan\beta = \frac{h}{a\alpha(h)} = \frac{p}{2\pi a} \qquad (\beta \neq 0) \tag{1}$$

The transformation relating the two systems of coordinates is

$$s = z\sin\beta + a\theta\cos\beta$$

$$at = -z\cos\beta + a\theta\sin\beta \tag{2}$$

The rectangular region corresponding to $z=-\infty$ to $+\infty$ and $\theta=0$ to $2\pi$, becomes $s=-\infty$ to $+\infty$ and $at=0$ to $2\pi a\sin\beta$. The proper boundary conditions on the wave function of a free particle moving on the surface of an infinite cylinder as described in Refs. 6 and 7 are then the periodicity around the cylinder and free wave along the axis of the cylinder, leading to the well-known cylindrical (normalised) wave functions :

$$\psi_{kl}(z,\theta) = \frac{\delta(\rho - a)}{2\pi\sqrt{a}} \exp i(kz + l\theta) \tag{3}$$

with k=(-∞, +∞) and $l$=0, ±1, ±2,........ The δ-function in ρ represents the surface of the cylinder of fixed radius, $a$.

The corresponding spiral waves constructed here obey the following boundary conditions:

$$\psi(s + 2\pi a\cos\beta, t = 2\pi\sin\beta) = \psi(s, t = 0) \tag{4}$$

$$\psi(s + s_p, t) = \exp(i\phi)\psi(s,t) \tag{5}$$

Here $s_p = \frac{p}{\sin\beta}$ , and φ is a phase, dependent on $s_p$ .

The first relation corresponds to the periodicity in θ-direction of the cylinder, and the second corresponds to the definition of the pitch. The corresponding normalised wave functions in the ($s$, $at$)-space are

$$\psi_{MM'}(s,t) = \frac{\delta(\rho - a)}{2\pi\sqrt{a\sin\beta}} \exp i(Ms + M't) \tag{6}$$

The quantum numbers M, M' correspond to momenta along the $s$ and the$at$ directions respectively, and the boundary conditions lead to M:(-∞, +∞) and the relation

$$M'\sin\beta + aM\cos\beta = m, \quad m = 0, \pm1, \pm2, \cdots. \tag{7}$$

The quantum numbers, $k$, $l$, associated with the cylindrical states are related to M, M' by a direct comparison of the two wave functions:

$$M = k\sin\beta + \frac{l}{a}\cos\beta, \qquad M' = l\sin\beta - ka\cos\beta \tag{8}$$

Thus we obtain the identification of m in Eq. (7) to be just $l$ . Henceforth we shall use (M,m) as the quantum numbers for representing the helical states. These wave functions form a complete orthonormal system and incorporate the helical symmetry of the problem. We therefore suggest their use in describing the quantum mechanics of electrons on graphitic tubules. We must point out that we can construct Bloch-like states which take into account the periodic arrangement of atoms forming the cylindrical motif by employing these states in much the same way as one does with plane waves out of which the Bloch states in three-dimensional lattices are constructed[8].

We now consider the Hamiltonian for a system of electrons moving on coaxial tubules so as to keep the discussion general. Let μ* be the effective mass of the electron on the j-th cylinder of radius, $a_j$, in the presence of a magnetic field, B along the tube axis (z). We use the dimensionless flux unit, $\tilde{\Phi}_j = \Phi_j / \Phi_0$ where $\Phi_j = \pi a_j^2 B$ is the magnetic flux enclosed by the j-th cylinder, $\Phi_0 = 2\pi c / e$ is the flux quantum in units where $\hbar = 1$. Let $\beta_j, p_j$ be the respective pitch angle and pitch of the j-th tubule. The many-electron Hamiltonian is then written in the (M,m)-representation in the second quantized form:

$$H = H_0 + H_{\text{int}} \tag{9}$$

$$H_0 = \sum_{Mm} \sum_{j\sigma} E_{Mmj\sigma} C^{\dagger}_{Mmj\sigma} C_{Mmj\sigma} \tag{10}$$

$$\begin{aligned} H_{\text{int}} = \frac{1}{2} \sum_{M_1 m_1} \cdots \sum_{M_4 m_4} \sum_{\sigma_1 \sigma_2} \sum_{j_1 j_2} & \tilde{V}'_c(M_1 - M_4, m_1 - m_4; j_1, j_2) \\ & \times \delta\left(k_{j_1}(M_1 m_1) + k_{j_2}(M_2 m_2) - k_{j_2}(M_3 m_3) - k_{j_1}(M_4 m_4)\right) \\ & \times \delta_{m_1 + m_2, m_3 + m_4} C^{\dagger}_{M_1 m_1 j_1 \sigma_1} C^{\dagger}_{M_2 m_2 j_2 \sigma_2} C_{M_3 m_3 j_2 \sigma_2} C_{M_4 m_4 j_1 \sigma_1} \end{aligned} \tag{11}$$

Here
$$E_{Mmj\sigma} = \frac{k_j^2(Mm)}{2\mu *} + \frac{1}{2\mu * a_j^2}\left(m - \tilde{\Phi}_j\right)^2 + \frac{\sigma g \hbar^2}{4\mu * a_j^2}\left(\tilde{\Phi}_j\right) \tag{12}$$

$$\tilde{V}'_c(M, m; j_1, j_2) = \left(\frac{e^2}{\pi \sin\beta_{j_1} \sin\beta_{j_2}}\right) I_m\left(a_< \left|k_{j_1}(Mm)\right|\right) K_m\left(a_> \left|k_{j_1}(Mm)\right|\right) \tag{13}$$

$$k_j(Mm) = \frac{a_j M - m\cos\beta_j}{a_j \sin\beta_j} \tag{14}$$

The last term in Eq.(12) is the spin-paramagnetic term. $I_m$ and $K_m$ are the modified Bessel functions. $a_<$ ($a_>$) is the smaller (larger) of $a_{j_1}$ and $a_{j_2}$.

In Fig. 2, we exhibit the changes in the nature of the one-particle energies for a given spin state that occur when we go from the cylindrical case (β=π/2) in the absence of the field ( $\tilde{\Phi}$ =0) to the spiral case (β=π/3) with nonzero magnetic field ( $\tilde{\Phi} \neq 0$). The energy dispersions in Eq. (12) at zero field are shown by solid curves. For m=0, these

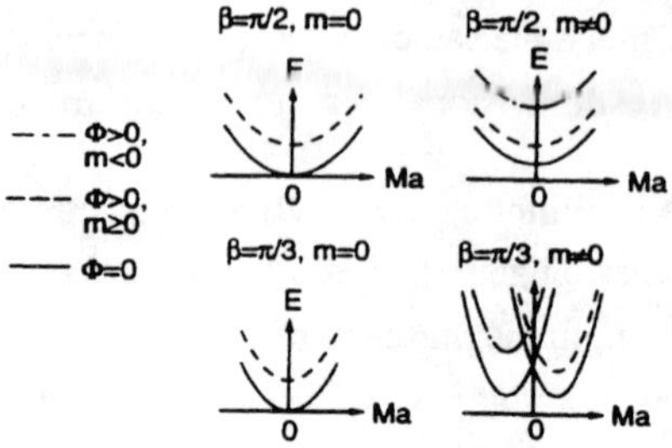


Fig.2 Schematic diagram of the electron energies for a given spin state.

curves which are shown on the left part of Fig. 2, exhibit changes in the curvatures for different pitch angles, β. For m≠0, the degeneracy of the energy states for ±m is lifted in the case of β=π/3 as can be seen on the lower right panel of Fig. 2. The magnetic field shifts each of the zero field dispersions as given by the dashed curves in the left panels of Fig. 2, and it also lifts the degeneracy of the ±m levels for both the cylindrical and the spiral cases as shown on the right panels of Fig. 2 by the dashed (m>0) and dash-dotted (m<0) curves.

In Fig. 3a,b we display the effects of helical structure on the two-particle intra- and inter-Coulomb interaction matrix elements calculated from Eq.(13) for the case $a_<$=11Å, $a_</a_>$ =0.76 and $\beta_{j_1}=\beta_{j_2}$ to contrast with that given in Ref. 7 for the cylindrical case ( β=π/2). The dashed curves in Fig. 3a represent the intra-tubule Coulomb matrix elements for the spiral system and have been compared with the cylindrical system (solid curves).

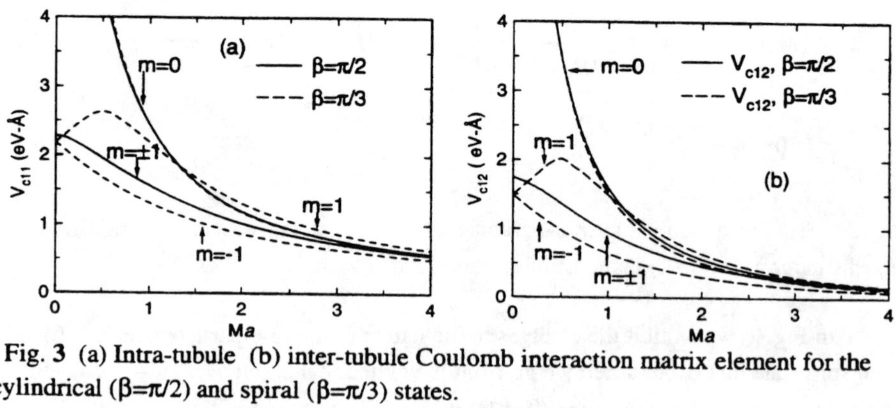


Fig. 3 (a) Intra-tubule (b) inter-tubule Coulomb interaction matrix element for the cylindrical (β=π/2) and spiral (β=π/3) states.

The solid curves (dotted curves) in Fig. 3b are the inter-tubule Coulomb matrix for the cylindrical (spiral) system. The common features of all the curves are the one-dimensional behavior (logarithmic for m=0 and finite for m≠0) for $a$M<<1 and 2-dimensional behavior for $a$M>>1. These behaviors resemble that found for the semiconductor quantum wires [9,10]. The intertubule interactions resemble those of layered structure [11] and approach $M^{-1}\exp[-M(a_{>}-a_{<})]$ for $(a_{>}-a_{<})M>>1$. For both intra- and inter-tubule Coulomb interactions, the degeneracies of $V_C$'s for ±m are lifted for β=π/3. The intertubule interactions are found to be weaker than the intra-tubule interactions as expected. In Figs. 2 and 3 the different features for the cases β=π/2, π/3 confirm the basic difference in the two empty lattices.

We now focus on a single tubule case with the following set of system parameters: electron density per unit area, $2.27\times10^{14}cm^{-2}$, the cylinder radius $a$=11Å, the static dielectric constant $\varepsilon_0$=2.4, and effective mass of the electron $\mu^*=0.25m_0$. In Fig.4., we depict the Fermi-energy as a function of the magnetic field exhibiting three features: In the absence of any explicit potential leading to the spiral structure, the Fermi energy $E_F$ is independent of the spiral property, as is expected for any macroscopic average property, in view of the transformation, Eq.(7). With only diamagnetic contributions, the one-particle properties of this system are periodic in $\tilde{\Phi}$ (dotted curve) while this periodicity is lost when the paramagnetic contribution is included (solid curve). The origin of the peak structures in this figure can be understood by observing that as one goes from $\tilde{\Phi}$ to $\tilde{\Phi}$+1/2 one less m-subband is occupied; when one goes further to $\tilde{\Phi}$+1, one more subband is occupied. Thus for each state crossing the Fermi level, a peak in this figure is obtained.

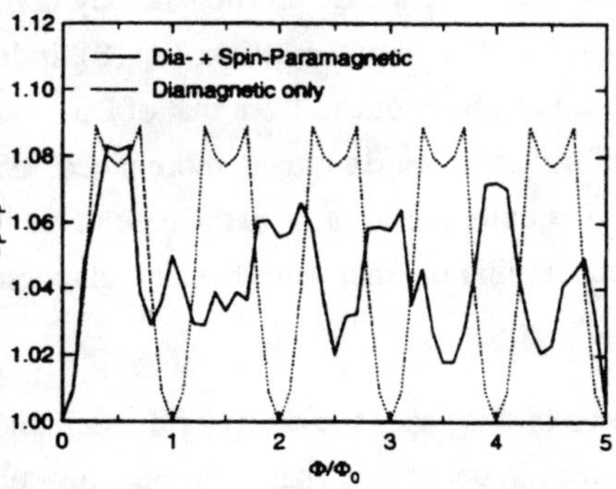


Fig. 4 Fermi energy as a function of $\tilde{\Phi}$ with (solid curve) and without (dotted curve) spin-paramagnetic contribution.

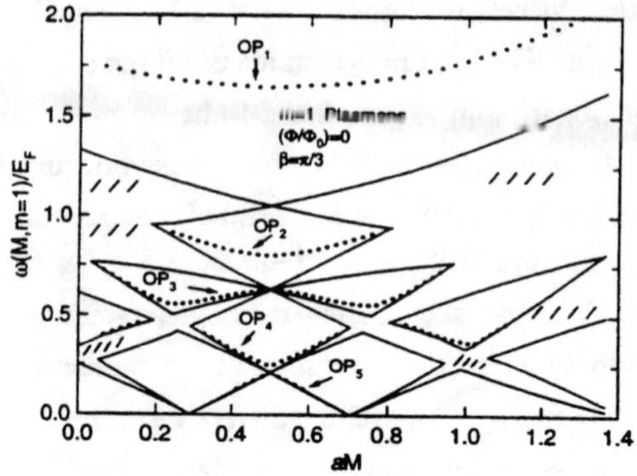


Fig. 5 Optical plasmon dispersion curves (dotted) for a $\beta=\pi/3$ helical system with angular momentum transfer, m=1, and in the absence of the magnetic field. The shaded region bounded by solid curves is the region allowed for *e-h* excitations.

In Fig. 5., we exhibit the plasmon dispersion in zero field for a spiral system with $\beta=\pi/3$. We should first mention that plasmon dispersion is a collective effect corresponding to density fluctuations in this system and is not a macroscopic property. We therefore expect the spiral structure to have an effect on this property. There are two main types of plasmons in this tubule system; those whose frequencies vanish for zero wave vectors and are thus named acoustic plasmons (AP) which occur for m=0, and those which have finite frequency for zero wave vector and hence called optical plasmons (OP), which occur for $m\neq0$. In Fig. 5 we display optical plasmons for a spiral system with $\beta=\pi/3$ when $\tilde{\Phi}=0$. This should be contrasted with those given by Lin and Shung [7] who calculated these for the cylindrical case. These authors found several OP's, which were the same for m=1 and -1. For our present case this degeneracy is lifted, and we have given in Fig. 5 only the m=1 OP's. The transformation, Eq. (8), indicates that in the absence of spiral potential, our result can be deduced from that of Lin and Shung [7]. Because of the nature of the transformation, we find many more pieces of the mode for the spiral case. The result for m=-1 is quite different from the case of m=1. The effect of magnetic field and other results concerning the magneto-plasmon dispersions are given in more detail ina separate paper [12].

In summary, we have shown in this paper the utility of the helical coordinate system and the helical wave functions in describing many physical properties of systems possessing helical symmetry. Symmetry considerations alone give rise to several interesting features, and the inclusion of atomic structure leading to the helical configuration would further modify these results. A discussion of some of these effects may be found in Ref.12. It should be stressed that the mathematical transformation shown here from cylindrical to helical symmetry has given us important insight into the properties of the helical system not explicitly evident in the cylindrical representation. The helical wave functions constructed here are also the proper eigenfunctions to describe systems with helical symmetry but not possessing the higher cylindrical symmetry, e.g., polymer chains and other tubules. When opened into a two dimensional plane these do not form two-dimensional Bravis lattices.

## ACKNOWLEDGEMENT

This work is supported in part by the Office of Naval Research.

## REFERENCES

1. S. Ijima, Nature (London) 354, 56 (1991)
2. T. W. Ebbesen and P. M. Ajayan, Nature (London) 358, 220 (1992)
3. J. W. Mintmire, B. I. Dunlap, and C. T. White, Phys. Rev.Lett. 68, 631 (1992)
4. C. T. White, D. H. Robertson, and J. W. Mintmire, Phys. Rev. B47, 5485 (1993)
5. For an extensive review of theoretical work on Fullerines and Graphitic Tubules, see C. T. White, J. W. Mintmire, R. C. Mowrey, D. W. Brenner, D. H. Robertson, J. A. Harrison, and B. I. Dunlap, in Buckminsterfullerines, edited by W. E. Billups and M. A. Ciufolini (VCH Publishers, Inc.,New York, 1993) p.125
6. L. Wang, P. S. Davids, A. Saxena, and A. R. Bishop, Phys. Rev. B46, 7175 (1992)
7. M. F. Lin and P. W. Shung, Phys. Rev. B47, 6617 (1993) and B48, 5567 (1993); see also P. S. Davids, L. Wang, A. Saxena, and A. R. Bishop, (Los Alamos Preprint) (1993)
8. P. J. Lin-Chung and A. K. Rajagopal, to be published
9. S. Das Sarma and W. Y. Lai, Phys. Rev. B32, 1401 (1985)
10. G. Y. Hu and R. F. O'Connel, Phys. Rev. B42, 1290 (1990)
11. T. Ando, A. B. Flowler and F.Stern, Rev. Mod. Phys. 54, 537 (1982)
12. P. J. Lin-Chung and A. K. Rajagopal, to be published

Local Order
in Condensed
Matter Physics


ISBN 1-56072-220-7

# HYDROGREN IN METALLIC SUPERLATTICES

E. Karlsson, B. Hjörvarsson, S. Olafsson and F. Stillesjö

*Department of Physics*
*Uppsala University, Box 530,*
*S-75121 Uppsala, Sweden*

## 1. INTRODUCTION

Superlattices are artificially produced structures consisting of thin layers of two or more materials which alternate in a regular fashion throughout a sample. Superlattices made up of semiconducting materials, like GaAs/AlGaAs, have been in the focus of interest for a relatively long time because of their technical importance, as well as their interest for basic science; both having to do with their very specific electronic structure.

Metallic superlattices, on the other hand, have only recently attracted attention as physically and technologically interesting materials, but the preparation conditions, being quite different from those for semiconductor superlattices, are now reasonably well mastered for many metal/metal combinations and epitaxial growth has been achieved for several combinations of metals. Most probably, metal-semiconductor and metal-insulator superlattices (even with some of the metals superconducting) will also be of importance in the future. The present report will be concerned with the measurement on hydrogen in metallic superlattices, a topic with both practical and fundamental aspects.

In the absorption from the gas phase the first step is a dissociation of $H_2$ molecules which is followed by a diffusion on the surface and into the bulk material. The kinetics and the final population of the various possible H-sites will then depend on the absorption potentials for protons in the different regions of the sample. A useful starting point for discussions is the diagram by Nörskov [1]) shown in Fig.1. It is based on the dependence of the H-binding energy on the local conduction electron density (from a jellium model) indicated in the right hand side if the Figure, from which hydrogen potential energies (aclthough only the so-called "embedding" energy part and not binding due to hybridization) can be predicted at the surface, at regular interstitial sites and in vacancies if the electron density variation is known.

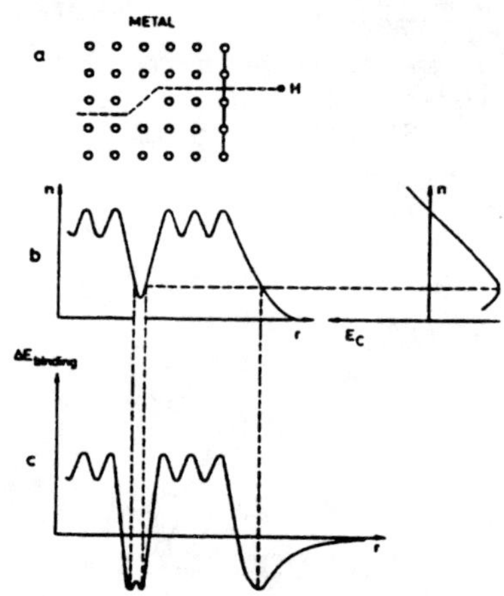


Figure 1. Potentials for hydrogen absorption, as predicted from the "jellium" model. Reproduced from Nörskov [1].

Some typical features are:

(a) for protons in the tapering-off tail of the electron distribution just outside the surface layer of the metal there is a minimum, called the *chemisorption minimum*,

(b) if the potential minima for interstitial sites in the bulk of the metal lie below the vacuum level we have a hydrogen absorbing metal (in a realistic case there are more than one set of potential minima, as for instance for the O- and T-sites in the cubic metals; some of these may be favourable for hydrogen absorption and others not),

(c) the center of a vacancy may not always be the most stable position for protons, because off-center positions can have a more favourable electron density for H absorption.

In the following, it will be shown how the N-15 method for hydrogen profiling can be used to deduce the hydrogen absorption selectively when the samples are built up of layers of different metals. Measured changes in hydrogen concentration reflect indirectly the distribution of local electron density when approaching an interface between two metals.

## 2. HYDROGEN PROFILING WITH THE N-15 METHOD

The N-15 method belongs to a relatively large class of methods where the concentration of certain elements can be determined by measuring the yield of a nuclear reaction when the sample is bombarded with high-energy ions from an accelerator. It is used to determine hydrogen concentrations as function of depth from the surface with a depth resolution on the nm scale. The method is based on the sharp resonance around 6.385 MeV of the nuclear reaction $^{1}H(^{15}N, \alpha\gamma)^{12}C$ . $^{15}N$-ions (usually with +2 charge) are accelerated up to energies in excess of the resonance energy, 6.385 + $E_x$ , and are directed towards the target which consists of a thin foil (of the order of a few μm thickness). At $E_x = 0$, H-atoms on or near the surface respond to the exciting particles since the reonance condition is fulfilled. This can be detected by the characteristic γ-quantum (with energy $E_\gamma = 4.43$ MeV) from the reaction, while H-atoms at deeper lying layers remain uneffected.

The search for hydrogen at greater depths x from the surface is accomplished by succesively raising the bombarding energy. The $^{15}N$ ions loose energy during the penetration of the foil at a rate given by the stopping power dE/dx for the material. This quantity is to a good approximation constant over a large range of x, such that for an initial energy of $E_{res} + E_x$ the resonance occurs at the depth

$$x = (E_{res} - E_x)/ (dE/dx) \qquad (1)$$

The H-concentration profile can therefore be recorded by registering the yield of the reaction for different initial beam energies. The depth resolution can be expressed as

$$\Delta x = \Delta E/(dE/dx) \qquad (2)$$

where the resonance width has contributions from the natural width Γ of the nuclear resonance (1.8 keV for the N-15 resonance), the Doppler broadening $(\Delta E)_D$ arising through the thermal motion of the H-atoms , and the fluctuations in beam energy during the experiments (which usually are in the range 1 - 5 keV). The Doppler broadening is the dominating term and usually brings the total observed width at the surface up to 10 - 15 keV. For a typical medium-weight metal hydride with a stopping power of 4 MeV/ μm for N-ions, such a resonance width corresponds to a depth resolution near the surface of about 15 Å . Beam straggling increases this width almost linearly with depth such that in a typical material it amounts to about 50 Å at a depth of 100 Å and to 200 Å at 1000 Å depth.

The gamma ray yield is proportional to the H-concentration at the depth in question. It should be noted that the number of H-atoms per depth interval, when expressed in energy terms as the scanning passes the depth interval $\Delta x$ is given by the quantity $x/(dE/dx)$. A calibration experiment can be made using a metal hydride with well known and uniformly distributed H-concentration, for example a $TaH_x$ sample [2]. The sensitivity of the detection is such that, given a sufficiently low background in the gamma ray detector, hydrogen concentrations down to 50 ppm can be analysed.

The hydrogen uptake is usually limited by a dissociation barrier at the surface. It is therefore essential to work with clean and well-defined surfaces. Experiments are carried out with the samples placed at the end of a differentially pumped (< $10^{-9}$ torr) beam tube in the accelerator. In the specialised equipment at the Uppsala Tandem Laboratory [3] it is also possible to carry out surface treatments and to perform the hydrogenation directly in a measuring chamber connected to the accelerator. After each hydrogenation sequence, carried out at specified temperatures and hydrogen pressures, the H-atmosphere is disconnected and the sample kept at a low temperature during the H-profiling by cooling with liquid nitrogen. This prevents hydrogen from being released during the measurement.

## 3. HYDROGEN PROFILING IN SUPERLATTICES

### 3.1 Analysis methods; examples

Superlattices of the type A/B can be built up with metal A hydrogen-absorbing and metal B non-absorbing . Examples are the combinations Ti/Fe and V/Mo. A hydrogen profile is shown in Fig. 2 [4] for a Ti/Fe superlattice with a layer thickness of 300 Å for Ti and 100 Å for Fe (the sum of these numbers is the modulation wavelength $\lambda$ of the superlattice). The total thickness of the sample was 5000 Å, which corresponds to an $^{15}N$ beam energy of 8.3 MeV. Except for the outermost Ti-layer that had lost part of its H-content before the measurement started, its shape is characteristic for a sample that has been hydrogenated to a uniform concentration for all Ti layers; the H-rich Ti layers are well resolved from the H-deficient Fe-layers, at least near the surface, and the oscillating amplitude is damped further into the sample because of the energy straggling of the beam with consequent loss of depth resolution.

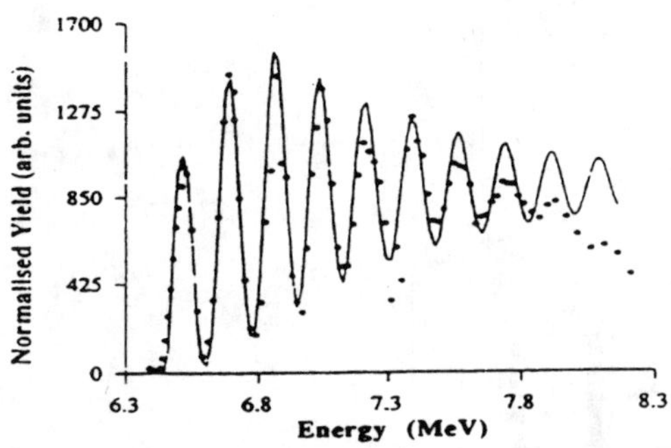


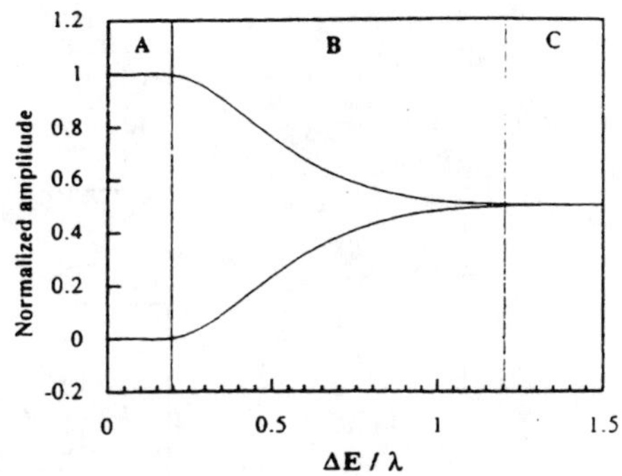


Figure 2 (left). Results of hydrogen profiling in a hydrogenated Ti/Fe superlattice
The solid line is a simulation assuming a rectangular H-profile for each V-layer

Figure 3 (right). Envelope functions for different ratios between depth resolution $\Delta E$ and superlattice wavelength $\lambda$

When the modulation wavelength $\lambda$ is much shorter than the depth resolution $\Delta E$, individual layers cannot be resolved in a profiling experiment, but the *average* N-15 reaction yield can still be used to derive the hydrogen concentration $c_H$ in the absorbing layers as can be seen from the asymptotic behaviour for the theoretical amplitude as function of $\Delta E/\lambda$ in Fig. 3.

When studying layer thickness effects on the hydrogen absorption capacity it is important for the comparison that the different layers have been hydrogenated under the same conditions. Since it is always difficult to obtain exactly the same surface conditions for two different samples Hjörvarsson et al. [5] developed the so-called *multi-layered superlattice* technique for the comparison of superlatttices with different wavelength $\Lambda$.

Such a sample is built, on a common backing material, as a series of superlattices (each with about 10 periods) with succesively increasing λ' s and has, consequently, a common external surface to be exposed to the hydrogen atmosphere. Fig. 4 illustrates the principle of the multi-layered superlattice. It can also be built with the layers in the reverse order, i e with the lowest λ' s closest to the backing. The latter can be used for control measurements, for instance to verify that the diffusion of the hydrogen has been effective to reach equilibrium throughout the whole sample.

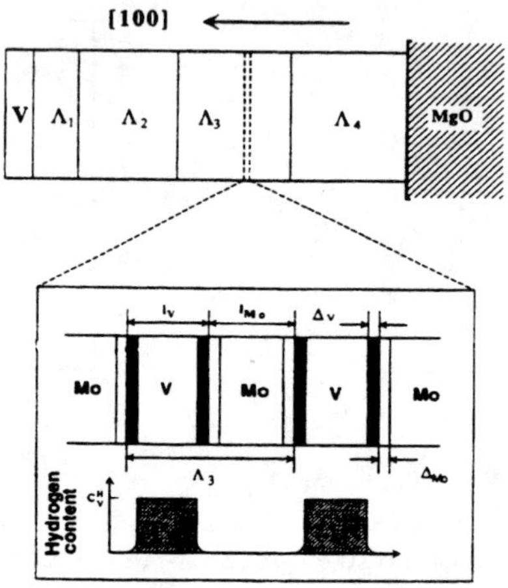


Figure 4. A multi-layered superlattice

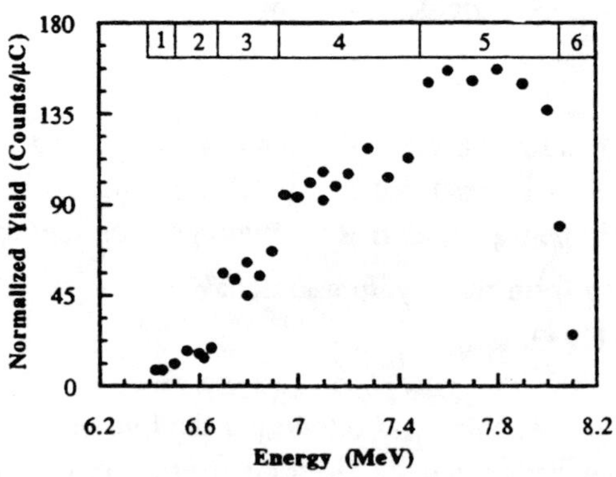


Figure 5. Hydrogen concentration for five different Λ's in a V/Mo superlattice

Fig. 5 shows the average hydrogen concentration at different depths for a multi-layered V/Mo superlattice epitaxially grown on an (100) oriented MgO substrate with a V-layer chosen as external surface to facilitate hydrogen uptake [5]. There is a constant ratio between Mo and V layers ($l_{Mo}/l_V = 0.73$) for all separate superlattices marked 1 to 5, and the wavelength $\Lambda = l_{Mo} + l_V$ increases in the order 3.4 , 10, 23, 33 and 67 Å, respectively. Although the proportion of vanadium in all parts of the sample is the same, the hydrogen content for the thinnest layers is very low. This indicates that thin layers of vanadium surrounded by layers of another metal differ essentially from bulk metalas far as hydrogen absorption is concerned. The quality of the V/ Mo superlattices used in these investigations was known to be very high, with little crystallographic disorder, and interface roughness limited to one monolayer, as seen from X-ray diffraction [5] and TEM.

It is reasonable to assume that the hydrogen absorption anomaly is concentrated to regions close to interfaces in the superlattices. If, as a first approximation, an interface layer of thickness $\Delta_V$ is assumed to be completely hydrogen-free, then the average concentration $\langle c_H \rangle$ of hydrogen in the vanadium layer can be written

$$\langle c_H \rangle = \eta \langle c_H \rangle_{interior} [\, 1 - 2\, \Delta_V / l_V] \qquad (3)$$

where $\langle c_H \rangle_{interior}$ is the H-concentration in the inner, bulk-like region of the V-layer; $\eta$ is zero for $l_V < 2\Delta$ and unity otherwise. The full treatment presented in ref. [5] contains correction terms dependent on $l_V$ and $l_{Mo}$, which however can be neglected for $l_V \gg \Delta_V$. In a hydrogen profiling experiment the $^{15}N$ reaction yield will be proportional to $\langle c_H \rangle$ (although with a proportionality constant depending on the ratio $l_{Mo}/l_V$), and it is therefore of interest to plot the yield as function of $(l_V)^{-1}$ for further analysis.

In a second approximation one distinguishes the bulk-like interior region of the H-absorbing layers with the concentration $\langle c_H \rangle_{interior}$ and a small, remaining concentration $\langle c_H \rangle_{interface}$ in the layers of thickness $\Delta$ closest to the non-absorbing metal. The latter will add an extra term $\langle c_H \rangle_{interface} \Delta / L_V$ to the expression (3) for the total measured average concentration $\langle c_H \rangle$. Such representations of the H-profiling data are discussed in the following example.

### 3.2 Interface effects in V/Mo superlattices

As mentioned already, the V/Mo superlattices can be built up with high perfection and are therefore suitable as test materials for basic physical phenomena. The lattice constants for the respective bulk metals (both b.c.c) do not differ by more than 3 %, and alloy formation at the interfaces is negligible below 980 K. If the individual layers are sufficiently thin the structure will stabilize as a coherent superlattice, which means that the lattice distances perpendicular to the growth direction are adjusted in both types of layers and show a common value, which is the mean of the two bulk values, 3.04 Å and 3.14 Å. The upper limit for coherent lattice growth in the V/Mo superlattices is 8 lattice distances (which is 16 monolayers) for $l_V = l_{Mo}$. Compared to the bulk materials one can expect changes in hydrogen binding energies both from the lattice expansion of V and from the influence of the electron-rich Mo-layers near interfaces.

Hydrogen profiling experiments by Hjörvarsson et al. [5] for multilayered V/Mo superlattices, taken at different hydrogenation conditions are presented in $l_V^{-1}$ diagrams in Fig. 6. All data for (001) grown lattices are well represented by straight lines when decreasing $l_V$ from 50 to 20 Å ($1/l_V$ increasing from 1/16 to 1/5 $(ML)^{-1}$), which shows the validity of an equation of type (3). For $l_V > 50$ Å the superlattice growth is no longer coherent; the V-layers have their regular, bulk lattice parameters and interface effects are also negligible, which is reflected in a kink in the lines in plots of the type shown in Fig. 4 [7].

Extrapolation to $\langle c_H \rangle = 0$ using eq. (3) would mean a hydrogen-free interface region in vanadium of $\Delta_V \approx 5$ Å for the (001) orientation, which is about three atomic monolayers. However, deviations from straight-line behaviour is observed for $1/l_V = 1/4$ ML and taking into account an $\langle c_H \rangle_{interface}$ term shows that the total thickness of layers (at both interfaces to Mo) having non-bulk like behaviour is $\approx$ 7 Å.. Variation of hydrogenation pressure and temperature allows a further distinction of the filling of different V-layers in the interface region. A comparison of data showed that for a high content of hydrogen ($\langle c_H \rangle_{interior} = 0.97$) the amount of hydrogen in the third atomic plane (counted from the Mo/V-interface) was also close to 1.0, as compared to only 0.03 when $\langle c_H \rangle_{interior}$ was 0.08 [8].

Fig. 7 suggests a theoretical interpretation within the framework of the Nörskov model. The potential well distribution for the protons is modulated by an assumed electron density distribution (high for Mo, low for V) with the step at the metallic interface smeared out to conform with the observed H-absorption data. It should be kept in mind that the Nörskov theory also has a hybridization term in the H-binding energy, which however is almost the same for these two metals [6]. In this picture, the three monolayers of the H-sublattice closest to the interface are unoccupied when hydrogen is loaded at low pressures, but one of them starts to be filled with increasing pressure. A thermodynamic equilibrium in a two-level system representing the third and the fourth atomic V-plane results in an difference of 25 ±5 meV for the binding energy for the sites in (or between) these planes. Planes closer to the interface are not populated at these temperatures and pressures.

This qualitative picture of the electron-structural basis for H-absorption in metallic superlattices was later supported by a first principle calculation of the electron density distribution near interfaces of V/Mo superlattices by Papadia et al. [7]. The H-absorption experiments have therefore been able to show that in these nearly two-dimensional

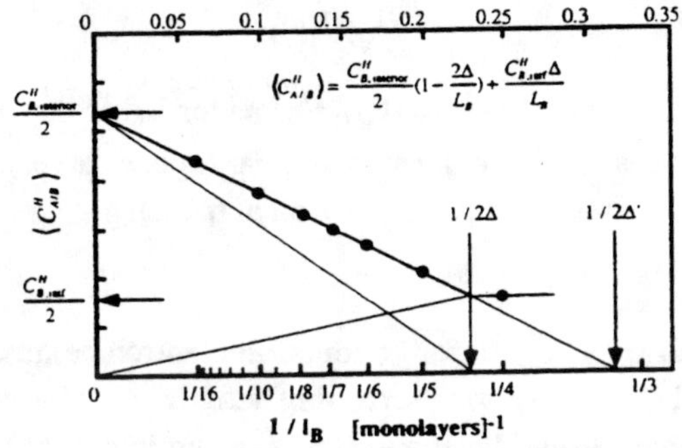


Figure 6. Hydrogen content as function of the inverse layer thickness of V

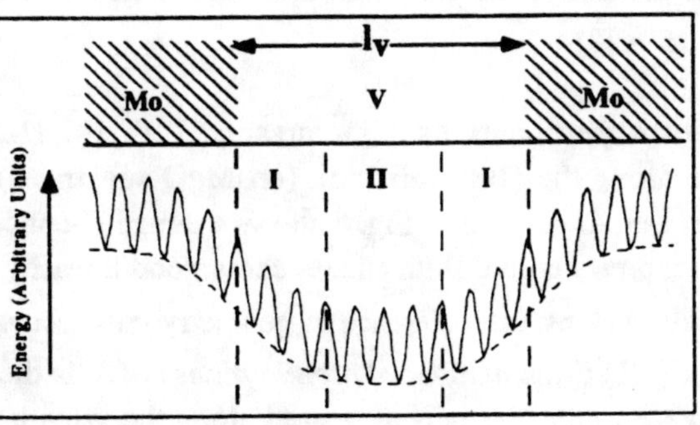


Figure 7. Schematic picture of hydrogen absorption potentials for thin V-layers

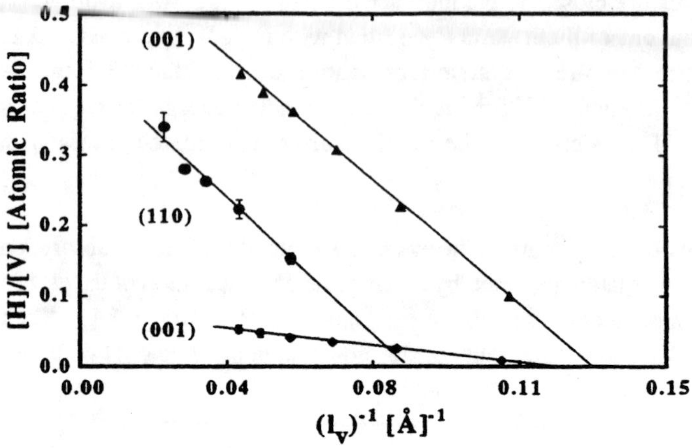


Figure 8. Hydrogen contents shown as $(l_V)^{-1}$ plots for V/Mo superlattices grown in (001) and (110) directions and hydrogenated at 1 bar (upper curves). Shown is also the result of hydrogenation at 0.1 bar.

metallic systems the spill-over region for conduction electrons extends over much larger distances than the usual Thomas-Fermi screening lengths usually assumed for bulk (3-dimensional) metals (which would be about 0.5 Å using interstitial electron densities of 0.035 and 0.037electrons/$a_o$ for V and Mo ). Experiments with the V-layers kept constant but Mo-layer thicknesses varied confirm also the long range of the band bending in metallic interfaces since the H-absorption capacity for vanadium is still sensitive to $l_{Mo}$ when it is increased from 10 to 15 Å [8].

In another series of experiments by Hjörvarsson et al. [9] H-absorption in V/Mo superlattices grown along the (100) direction (on MgO substrates) and (110) direction (grown on sapphire) were investigated. Fig. 8 shows average H-concentrations as function of $1/l_V$ in the usual representation. Both curves show good linearity but are shifted by 40 % on the $(l_V)^{-1}$-scale. Taking the difference in geometry into account this is exactly what would be expected ($\sqrt{2}/1$) if the *number of atomic planes* of V is the important parameter rather than the distance to the V-atoms measured along the growth direction. These data support therefore the picture of available H-sites being filled or unfilled according to their order when counted from the interface. Analoguos effects are observed when (110) and (100) grown V layers are tested for their superconductivity: the $T_C$ vs. thickness curves scale with the factor $\sqrt{2}$. Comparisons of this kind point at interface screening effects as a common origin.

## 4. CONCLUSIONS

It is evident that protons, when inserted into well-defined metallic environments, can be used as probe particles for local electronic structure. Further information obtainable from experiments on multi-layered superlattices includes effects of the growth mode and changes of binary phase diagram boundaries when going from bulk metal to metal layers confined by the superlattice structure. For example, one conclusion drawn from the experiments on hydrided vanadium layers is that the solid solution limit for hydrogen in in Mo/V superlattices at room temperature must be essentially higher (≈10%) than the ≈3 % observed for hydrogen in bulk vanadium. This has to to with the strong, negative internal pressure experienced by the vanadium layer in the superlattice. In this sense, metallic multi-layers can be said to serve as laboratories for testing materials under extreme conditions. A technical motivation for such interface studies is that they can also serve as idealized models for grain boundaries in alloys.

## REFERENCES

1. J.K.Nörskov, *Europhysics News* 19 (1988) 65
2. B.Hjörvarsson, J.Rydén, T.Ericsson and E.Karlsson, *Nucl. Instr .& Meth. B* 42 (1989) 257
3. B.Hjörvarsson, J.Rydén, L.E.Svensson and E.Karlsson, *UUIP (Uppsala University, Institute of Physics) Report* 1181, Febr. 1988
4. J.Rydén, B.Hjörvarsson, T.Ericsson, A.Chamberod and B.Rodmacq, *J. Less Common Metals* 172 -174 (1991) 784
5. B.Hjörvarsson, J.Rydén, E.Karlsson, J.Birch and J.-E.Sundgren, *Phys. Rev. B* 43 (1991) 6440
6. J.K.Nörskov, *Phys.Rev.B* 75 (1989) 2875
7. S.Papadia, K.Karlsson, P.O.Nilsson and T.Jarlborg, *Phys.Rev.B* 41 (1990) 1857
8. B.Hjörvarsson, F.Stillesjö, S.Olafsson, E.Karlsson, J.Birch and J.-E.Sundgren, *J.Magn. and Magn. Mat.* 126 (1993 612
9. B.Hjörvarsson et al., to be published.

Local Order
in Condensed
Matter Physics


ISBN 1-56072-220-7

# HYDROGEN TRAPPING IN VACANCIES IN METALS STUDIED BY PAC

Gary S. Collins and Steven L. Shropshire[1]

*Department of Physics, Washington State University*
*Pullman, WA 99164-2814, USA*

**ABSTRACT**

The interaction between hydrogen atoms and lattice vacancies in metals has been studied with atomic-scale resolution using perturbed angular correlations of gamma rays (PAC). The hydrogen atoms and vacancies have been detected through changes in nuclear quadrupole interactions at adjacent $^{111}$In/Cd probe nuclei. Results are summarized for Pt samples which had been doped with complexes of probe atoms containing 1-4 vacancies. Hydrogen (H) was introduced by cathodic charging. Transformations were observed between H-free and H-decorated complexes. During charging, signals were identified with complexes decorated first with one and later with more than one H atom (possibly two). During desorption of H, decorated sites were observed to transform into undecorated sites. Binding enthalpies of H to the vacancy complexes were determined from changes of site fractions during desorption and were all found to be close to 0.25 eV. Extremely large changes in the quadrupole coupling parameters of the 3V and 4V complexes were observed when H trapped, but not of the 1V and 2V complexes. This is attributed to atomic reconstruction of the multivacancy complexes, that is, to large local lattice relaxations. This indicates that the relative stability of small multivacancy complexes against rearrangement is small. Detailed structural models of the complexes are described.

## 1. Introduction

Hydrogen (H) is a common impurity in metals, affecting many physical properties. H atoms diffuse rapidly at ambient temperature by jumps between interstitial sites and "trap" at lattice defects such as grain boundaries, dislocations and lattice vacancies. Activation enthalpies for interstitial migration of H are in the range 0.1-0.4 eV,[2] while activation enthalpies for binding with lattice defects are larger, of the order of 0.2-1.0 eV.[3,4] Such values often lead to a trapping regime, frequently near room temperature, in which H can diffuse rapidly and saturate the traps, but is able to desorb, reversibly, on heating at a mildly elevated temperature. Defect traps thus increase the solubility of H while reducing its diffusivity and permeability. There is considerable technological interest in understanding the trapping at defects from a fundamental point of view in order to model transport of hydrogen isotopes, for example in the first-walls of fusion reactors.

We have applied a microscopic method to monitor the lattice interactions between individual defects and H atoms. The technique, perturbed angular correlations of gamma rays (PAC), is applied to measure interactions of nuclear moments of radioactive probe nuclei with fields arising from spin and charge densities within 1-2 surrounding atomic shells. Considering nonmagnetic, cubic metals, there will be an electric-field gradient (efg) at probes whose point symmetry is lower than cubic, such as when a defect is trapped next to the probe. The efg produces a quadrupole interaction characterized by a coupling frequency and an efg asymmetry parameter which provides a measure of the symmetry of local charge disturbances due to defects. Using the $^{111}$In probe, PAC has been extensively applied over the past 15 years to study the properties of intrinsic point defects in metals.[5] Vacancies trap next to the oversized $^{111}$In impurity probe due to a strain interaction, producing efg's for which coupling frequencies of order 100 Mrad/s are observed. Resolution of different complexes is excellent and, for example, in our own past work signals from In probes in complexes with 1 to 4 vacancies have been detected in Pt and other metals and identified with structural models.[6]

In this paper, an overview of our studies of H-defect interactions in Pt is presented, including published[6,7,8,9,10] and previously unpublished results.[11,12] PAC has also been applied to study H-vacancy interactions in Ni[13], Mo[14] and W[15]. Our experimental approach is as follows:[13] (1) Probe atoms are diffused in high-purity samples, coming to rest on defect-free, substitutional lattice sites. (2) Samples are damaged by plastic deformation to produce vacancies and annealed at an elevated temperature chosen on the basis of previous measurements so as to maximize the site fraction of selected vacancy-probe complexes. (3) Hydrogen is introduced at this point by cathodic charging of the samples during electrolysis, leading to replacement of signals of undecorated complexes by decorated ones. (4) Defect interactions are analysed by changes in site fractions during charging and desorption and by changes in the magnitude and symmetry of the efg. Using this approach ensures that vacancies are immobile during H trapping and detrapping, so that changes of site fractions and signal parameters can occur only as a consequence of hydrogen motion and interactions.

After a brief description of PAC, we consider the structures of the defect complexes and how they have been identified. Results of studies of H trapping and detrapping are then presented from which information is obtained about the fraction of occupied traps, the numbers of H atoms trapped, and binding enthalpies of H to the different complexes. Finally, unexpected findings about the microscopic structures and dynamics of H-decorated complexes are described: atomic reconstruction of the structures of multivacancy complexes which occurs when H atoms trap, and localized "caged" motion in a decorated complex.

## 2. Perturbed Angular Correlations

Lattice sites of the In/Cd probe atoms in Pt are detected via the nuclear quadrupole interaction. $^{111}$In decays with a mean lifetime of 4.0 d to the second excited state of $^{111}$Cd, followed by a gamma-gamma cascade to the ground state. The two gammas each have an angular correlation with respect to the alignment of the spin (I= 5/2) of the long-lived ($\tau$= 120 ns) intermediate nuclear level, and thus with respect to each other. When the nucleus precesses due to hyperfine perturbations, the coincidence decay curve of the intermediate level is modulated by an envelope function which can be analysed to yield the PAC perturbation function $G_2(t)$, which contains all information about the hyperfine interactions. For quadrupole interaction of a spin I= 5/2 level, it is given by

$$G_2(t) = s_0 + \sum_{n=1,3} s_n \cos(\omega_n t) \exp(-\sigma_n t), \quad (1)$$

in which the three frequency harmonics $\omega_n$ are constrained so that $\omega_3 = \omega_1 + \omega_2$, and the amplitudes $s_n$ plus $s_0$ sum to one. The independent experimental frequencies $\omega_1$ and $\omega_2$ may be expressed equivalently in terms of the quadrupole coupling frequency $\omega_Q = (3\pi/10)\ eQV_{zz}/h$ and efg asymmetry parameter $\eta = |(V_{xx} - V_{yy})/V_{zz}|$. When $\eta = 0$, $\omega_1 = \omega_o$ and $\omega_1{:}\omega_2{:}\omega_3$ are in the proportion 1:2:3. Q is the nuclear quadrupole moment and $V_{ii}$ are components of the diagonalized, traceless efg tensor. The principal component $V_{zz}$ is given by a sum over charges $q(r,\theta)$ external to the probe atom: $V_{zz} = (1 - \gamma_\infty)\ \Sigma\ q(r,\theta)\ (3\cos^2\theta - 1)/r^3$, in which $\gamma_\infty$ is the Sternheimer antishielding factor arising from polarization of closed shells of the probe atom by external charges. The exponentials in Eq. 1 reflect inhomogeneous broadening attributed to distant defects and/or lattice strains; good fits to the data were obtained using the empirical constraints $\sigma_n = (\omega_n/\omega_1)\sigma_1$. Finally, the amplitudes $s_0$ and $s_n$ sum to one and are slowly varying functions of the asymmetry parameter $\eta$ for the random polycrystalline samples studied here.

Several different probe-defect complexes were usually present during a measurement, so that experimental spectra were analysed using superpositions of $G_2$ functions, out of which site fractions f and characteristic hyperfine parameters $\omega_1$, $\sigma_1$ and $\eta$ serving to label the site were determined. The reader is urged to consult Ref. 6 for further details about the measurement methodology.

## 3. Structures of Undecorated and Decorated Vacancy-Probe Complexes in Pt

Undecorated complexes are considered first. Vacancies and small mobile vacancy clusters were produced in Pt by plastic deformation at 77 K. PAC measurements made following isochronal annealing steps at temperatures up to ~500 K revealed clearcut signals of 3 defect-associated sites.[16] Two complexes were observed to form during annealing at about 290 K and to dissolve at about 350-400K; their formation has been attributed to migration of divacancies (2V's) and trapping of one or two at defect-free probe sites to make 2V or 4V complexes.[6] A third complex identified as the 1V complex[5] forms athermally during deformation at 77K and is immobile up to our highest annealing temperature. Finally, a fourth complex was identified whose symmetry is cubic in the undecorated state ($\omega_1 = 0$) but which is detectable when H atoms trap because the point symmetry becomes noncubic. Experiments indicate that the 3V complex forms at 380K by annealing out and trapping of 3V's.[8] The 3V+H complex disappears at 490 K, either by detrapping of H or by detrapping of both H and the vacancies.

Signal parameters and identifications for the various undecorated and decorated complexes which have been observed are listed in Tab. 1. A complex formed by trapping n vacancies at a defect-free substitutional probe atom is here designated nV. Columns 2-5 contain signal parameters and maximum observed site fractions of undecorated complexes. Columns 6-11 contain signal parameters determined for singly decorated (1H) and multiply-decorated (mH) complexes. The

*Table 1. Hyperfine parameters of Cd-vacancy complexes in Pt measured at 77 K*

| *Site* | $\omega_0$ *Mrad/s* | $\eta$ | $\sigma_1$ *Mrad/s* | $f_{max}$ | $\omega_1^{1H}$ *Mrad/s* | $\eta^{1H}$ | $\sigma_1^{1H}$ *Mrad/s* | $\omega_1^{mH}$ *Mrad/s* | $\eta^{mH}$ | $\sigma_1^{mH}$ *Mrad/s* |
|---|---|---|---|---|---|---|---|---|---|---|
| 0V | 0 | - | ~20 | 1.0 | 96(1) | 0.0(1) | ~10 | ------ *not observed* | ------- | |
| 1V | 94(1) | 0.0(1) | ~20 | 0.05 | ~99 | 0.0(1) | - | ~104 | ~0 | ~50 |
| 2V | 226(1) | 0.0(1) | ~20 | 0.20 | 178(1) | 0.46(1) | ~10 | 155(5) | 0.68(5) | ~50 |
| 3V | 0 | - | ~20 | 0.25 | 257(1) | 0.004(4) | 2(2) | 257(1) | 0.004 | 2(2) |
| 4V | 55(1) | 0.0(1) | ~20 | 0.15 | 180(2) | 0.52(2) | ~15 | 171(3) | 0.51(3) | ~50 |

parameters[12] and identifications[6,12] supercede an earlier tabulation.[8] Structural models[6] are shown in Fig. 1. Identifications and structures were determined using a variety of inputs:[6,12] • Observed trapping behavior was checked against recovery behavior and interpretations of annealing stages developed from studies of macroscopic properties such as resistivity. • Vacancy character was confirmed for the 1V and 2V complexes from

observations of the complexes after quenching. • The vacancy character and the number of vacancies in the complexes were also checked via defect annihilation experiments, in which mobile interstitials were used to annihilate vacancies at the complexes, causing the reaction $InV_n + I \rightarrow InV_{n-1}$. • Identification was further guided by comparison of $\omega_1$ and $\eta$ with results of simple point-charge calculations of efg's for a broad range of possible structures.[6] • Finally, the "invisible" cubic-symmetric site was identified as a 3V complex in analogy with a similar site in Ni.[13] For further details, the reader is referred to Ref. 6.

## 4. Hydrogen Trapping

Foil samples 100-200 $\mu$m thick were used as cathodes in electrochemical cells and charged with H atoms during electrolysis at room temperature. The electrolyte

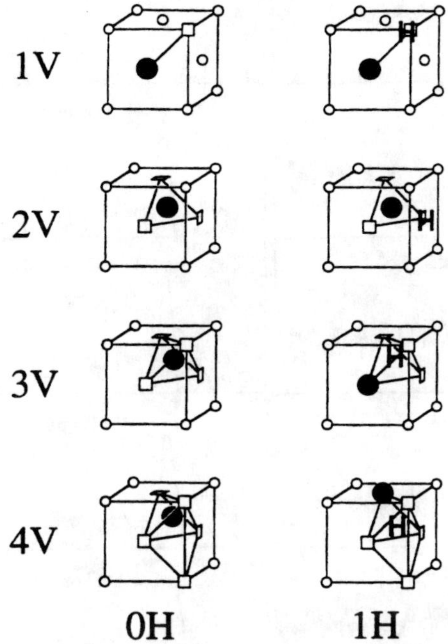


*Fig. 1. Structures of Cd-vacancy complexes in Pt containing 0 and 1 hydrogen atom. Host atoms, Cd probes, vacancies and H atoms are indicated, respectively by small circles, large filled circles, squares, and the letter H. Unlike 1V and 2V complexes, the 3V and 4V complexes undergo structural transformations.*

was 0.05 M $H_2SO_4$ or $D_2SO_4$ to which a small amount of an As salt was added which strongly increased the flux of H atoms into the metal.

Figure 2 shows a series of PAC time-domain spectra measured at 77 K after predoping a Pt sample with 2V and 4V complexes and then charging it with H at room temperature for the cumulative times shown.[11] On the right are frequency spectra obtained by Fourier transformation which exhibit triplets of peaks at frequencies $\omega_1:\omega_2:\omega_3$ for each complex. In (a), triplets for the 2V (226:452:678 Mrad/s) and 4V (55:110:165 Mrad/s) complexes are visible. Following charging for 5 min at 4 mA/cm$^2$ (b), about half of the site fractions of the 2V and 4V signals has disappeared and been replaced by two signals with quite similar $\omega$'s and non-zero $\eta$'s which produce composite peaks visible near 220, 330 and 550 Mrad/s. The two decorated signals have been resolved by careful fits in the time domain.[6] Because inhomogeneous broadening of the decorated signals is no greater than for the undecorated complexes after charging for 5 min, the sites are considered to each have a unique structure; they are therefore attributed to complexes decorated with single H atoms: 2V+1H and 4V+1H. The correspondence between the 2V and 2V+1H and 4V and 4V+1H signals was established in other measurements in

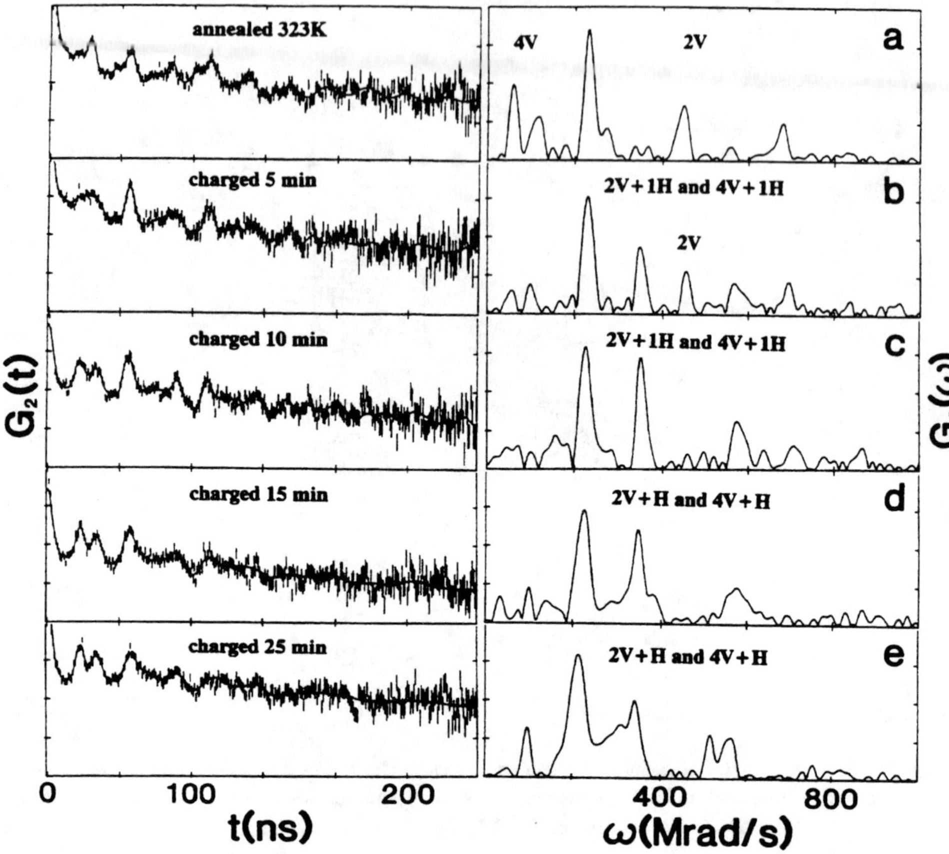


*Fig. 2. PAC spectra of $^{111}Cd$ for Pt measured after deformation at 77 K and annealing at 323 K (a), and after charging cathodically with hydrogen for progressively longer times (b, c, d and e). At left are the time-domain PAC spectra (0-300 ns), with their frequency spectra at right (0-1000 Mrad/s). The frequency spectra exhibit "triplet" signals for 2V and 4V complexes in (a and b) and for H-decorated complexes in (c, d and e).*

which the 2V complex alone was initially present.[9,12] Longer charging times (c, d and e) lead to significant broadening of the frequency distributions widths of the 220 Mrad/s triplet; this is attributed to decoration of the two complexes by more than one H atom, and not to H atoms at interstitial sites in the lattice. The latter alternative is ruled out because the defect-free probe signal remains unbroadened. Multiply-decorated sites are denoted as 2V+mH and 4V+mH. Finally after prolonged charging (e), a new signal ($\omega_1 = 96$ Mrad/s, $\eta=0$) begins to be observed; this site has been identified with H atoms at an interstitial site near defect-free In probes because it is the only site observed after charging well-annealed Pt.[9]

The evolutions of site fractions of the 2V and 4V complexes from the spectra in Fig. 2 are exhibited in Fig. 3 (in which the 2V+mH and 4V+mH site fractions are summed together). The trends are consistent with the identifications described above, with the sites progressively becoming singly and multiply decorated. It is possible that the signals fitted for the mH complexes correspond to 2H complexes, but a firm conclusion is difficult because there is an increased frequency broadening which could either reflect a superposition of similar signals for complexes having H atoms in inequivalent locations or for a single complex with varying occupation numbers. On the other hand, additional experiments showed that samples can be annealed in such a way as to produce decoration of all 2V and 4V complexes with a single H atom, each;[12] such a partitioning of H among the traps suggests that binding of the second H atom is smaller than the first. Similarly, if binding of the third H is even weaker, the mH complexes are likely to correspond to 2H sites.

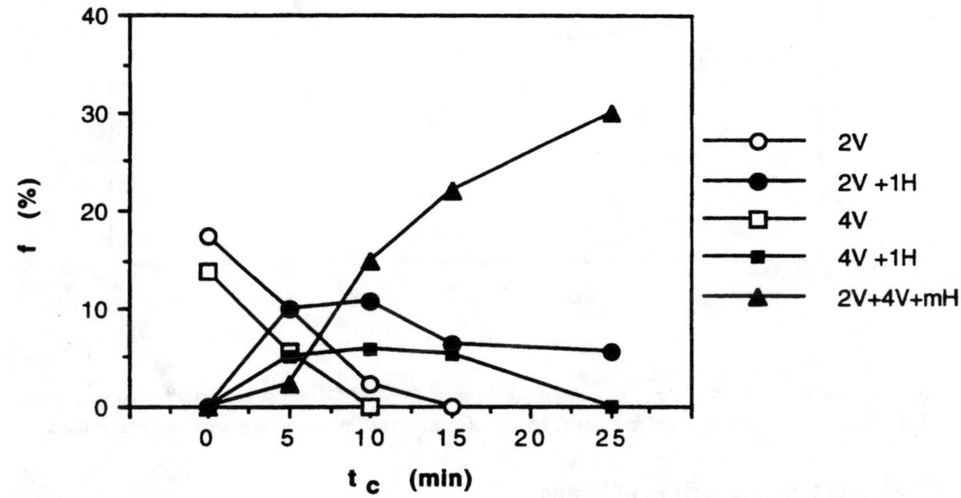


*Fig. 3. Site fractions of 2V and 4V complexes in Pt, from fits of Fig. 2. During charging with hydrogen, the undecorated complexes transform first into singly-decorated complexes, and later into mutiply-decorated complexes.*

**5. Hydrogen Detrapping during Desorption**

Detrapping during annealing and desorption of H is observed via decreasing site fractions of H-decorated complexes and the reappearance of undecorated sites.[13] Samples which had been doped with selected complexes were charged with H and then annealed at 294 K in a series of steps, after each of which PAC measurements were made to determine the site fractions of the complexes. Results for H detrapping from 2V+1H and 4V+1H complexes measured in this way[12] are shown in Fig. 4. The decay of the 2V+1H site fraction is exponential, as in the case of the 3V+H complex,[10] and has a halflife of 110(2) min. Binding enthalpies have been determined using a model for diffusion of H in a medium of traps.[17] Analysis indicates that the binding enthalpies for 1V, 2V, 3V and 4V complexes are 0.23(2), 0.28(1), 0.24(1) and >0.2 eV, respectively.[10] For reasons that are unclear, these values disagree with the value 0.5 eV measured by ion-beam profiling[18] and attributed to 1V+1H complexes.

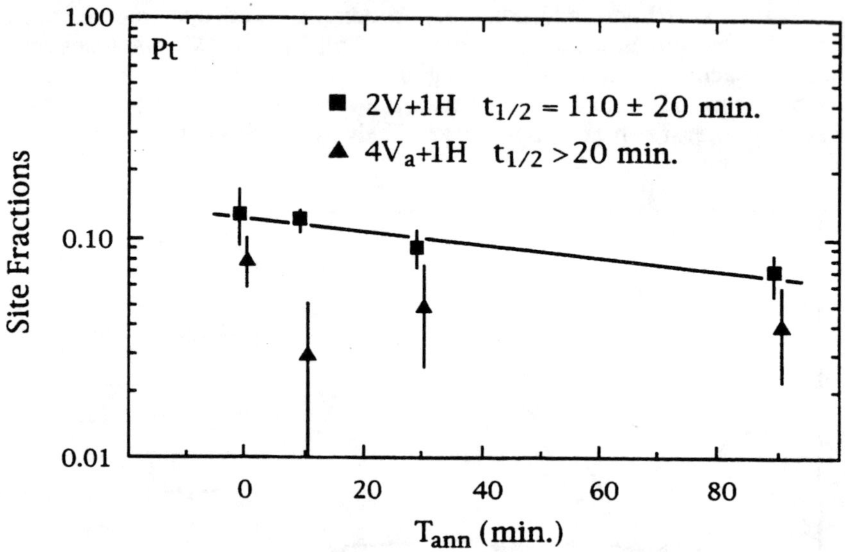


*Fig. 4. Decrease of site fractions of hydrogen-decorated 2V and 4V complexes in Pt during desorption at room temperature. The 2V+1H fraction is observed to decrease exponentially with annealing time.*

## 6. Atomic Reconstruction of Traps

Because the screening radius of H in metals is unusually small, it can produce only a relatively small "direct" contribution to the efg. In agreement with this expectation, the small changes of the hyperfine interaction parameters for the 1V and 2V complexes (and also of the 96 Mrad/s interstitial H site) can be explained using an effective charge of a H atom equal to 1/8 of the effective charge of a vacancy.[6] Similar small disturbances have been observed for 1V complexes with In/Cd probes in other metals[14,15] and for adjacent, interstitial H atoms.[19]

The efg changes observed when H traps at 3V and 4V complexes, on the other hand, are very large and therefore must have a different origin. We attribute them to large lattice relaxations which result in a completely transformed structure.[6] This is most clearly illustrated by in the course of trapping H at the 3V complex, for which we proposed that the probe atom gets pushed back from a tetrahedral interstitial site onto a regular lattice site by the "pressure" exerted by the trapped H atom.[6,7] Such a model explains the precise axial symmetry of the efg ($\eta=0$). Moreover, the remarkably small inhomogeneous broadening of the 3V+H signal measured at 77 K can be understood to result from the relaxation of lattice strains due to distant defects around the probe. With three vacancies all on one side, the probe is effectively sitting on a very small surface (see Fig. 1).

**7. Localized Motion**

While the 3V+H signal is unbroadened when measured at 77 K, there is significant broadening at room temperature. Excellent fits were obtained using a dynamically broadened $G_2(t)$ function derived from Eq. 1 by setting $\sigma_n = 0$ (no static inhomogeneities) and multiplying the entire function by $\exp(-\lambda t)$.[20] This nuclear relaxation is attributed to hopping of the probe atom from corner to corner in a tetrahedron of vacancies (see Fig. 1). The temperature dependence $\lambda(T)$ was measured over the experimental range 275-325 K in which relaxation was detected, and was found to follow an Arrhenius law with an activation energy of 0.34(5) eV.[12] This is interpreted as the activation enthalpy for hopping of the probe atom between the four equivalent sites in the tetrahedron.

**8. Discussion and Summary**

Hydrogen-vacancy interactions in metals can be usefully studied using PAC, a microscopic method giving appreciable insight into trapping phenomena. Because of the excellent resolution of signals from different complexes, it is possible to observe to what extent the structures of complexes affect the interactions. This is an advantage over diffusion measurements, for which assumptions have to be made about the types of defects present in a damaged sample which may be uncertain.

In the present work it was found that the binding enthalpy is largely independent of the detailed structure of the complex. Similar observations have been made for Mo and W by PAC.[14,15] This is perhaps to be expected since all the small complexes considered here consist of one or more vacancy-sized open volumes. The energetics of H appear to depend much more on the electron density distributions near the vacancies, as has been emphasized in effective medium calculations of binding of H in vacancies on metals.[21,22] While the presence of the probe atom is likely to affect the geometry of the complexes, it is less likely to modify the interactions unless there is a strong chemical interaction with H. This is because most of the vacancy surfaces remain surrounded by host atoms.

Lastly, the measurements provide evidence that H can induce atomic reconstructions of multivacancy complexes. This may indicate that the observed structures of multi-vacancy complexes are only slightly more stable than various alternative structures.

## 9. Acknowledgements

Hwa-Jae Jang made a significant contributions to this work in its early stages. This work was supported in part by National Science Foundation under grants DMR 86-19688 and 90-14163 (Metals Program).

## 10. References

1. Present Address, Department of Physics, Idaho State University, Pocatello, ID, USA
2. J. Volkl and G. Alefeld, in *Hydrogen in Metals I*, ed. G. Alefeld and J. Volkl (Springer, Berlin, 1978) p. 321.
3. Ch. A. Wert, in ref. 2, vol. II, p. 305.
4. S. M. Myers, P. M. Richards, W. R. Wampler and F. Besenbacher, J. Nuclear Materials **165** (1989) 9.
5. see, e.g., F. Pleiter and C. Hohenemser, Phys. Rev. **B25** (1982) 106; E. Recknagel, G. Schatz and Th. Wichert, in *Hyperfine Interactions of Radioactive Nuclei*, ed. J. Christiansen (Springer, Berlin, 1983) p. 133.
6. G. S. Collins, S. L. Shropshire and J. Fan, Hyperfine Interactions **62** (1990) 1.
7. G. S. Collins, H.-J. Jang and S. Shropshire, in *Nuclear Physics Applications on Materials Science*, ed. E. Recknagel and J. C. Soares, NATO Series E144 (Kluwer, Dordrecht, 1988) p. 415.
8. G. S. Collins, S. L. Shropshire and H.-J. Jang, Defects and Diffusion Forum **66-69** (1989) 335.
9. G. S. Collins, G. McGhee, S. L. Shropshire, H.-J. Jang, J. Fan and R. B. Schuhmann, Hyperfine Interactions **60** (1990) 663.
10. S. L. Shropshire and G. S. Collins, Hyperfine Interactions (in press).
11. Hwa-Jae Jang, M.S. thesis, Washington State University, 1988 (unpublished).
12. Steven L. Shropshire, Ph.D. dissertation, Washington State University, 1991 (unpublished).
13. G. S. Collins and R. B. Schuhmann, Phys. Rev. **B34** (1986) 502.
14. M. S. Abd El Keriem, D. P. van der Werf and F. Pleiter, J. Phys.: Condensed Matter **5** (1993) 1801.
15. J. R. Fransens, M. S. Abd El Keriem and F. Pleiter, J. Phys.: Condensed Matter **3** (1991) 9871.
16. G. S. Collins and R. B. Schuhmann, Hyperfine Interactions **15/16** (1983) 395.
17 A. McNabb and P. K. Foster, Trans. Metall. Soc. AIME **227** (1963) 618.
18. J. P. Bugeat and E. Ligeon, Phys. Lett. **71A** (1979) 931.
19. e.g., J. Trager, M. Karger, T. Butz and F. E. Wagner, Hyperfine Interactions **15/16** (1983) 795.
20. G. S. Collins, S. L. Shropshire and H.-J. Jang, Hyperfine Interactions **60** (1990) 651.
21. J. K. Norskov and F. Besenbacher, J. Less Common Metals **130** (1987) 475.
22. M. J. Puska, R. M. Nieminen and P. Jena, Phys. Rev. **B35** (1987) 6059.

Local Order
in Condensed
Matter Physics


ISBN 1-56072-220-7

# METASTABLE DEFECTS OF HYDROGEN ION-IMPLANTED INTO SILICON

K. Bonde Nielsen, B. Bech Nielsen, F. Borum, J. Byberg[1] and B. Holm

*Institute of Physics and Astronomy, Aarhus University, Denmark DK-8000 Arhus C, Denmark*

**The properties of hydrogen embedded in silicon have been studied by a number of techniques. In this review, we shall attempt to correlate available experimental data and compare with theoretical calculations. The isolated hydrogen impurity can be observed after ion implantation at low temperature and displays metastability in n-type material. We shall review the defect scenario that evolves after implantation and show that the observed metastability may be the signature of hydrogen, occupying the position midway between two relaxed bonded silicon atoms. According to theory, this bond-center configuration is stable (metastable) depending on Fermi level position.**

## 1. Introduction

Hydrogen in silicon has been studied intensively over the past decade by a large number of experimental techniques [1]. Many steps advancing our understanding of this unique system of impurity and host have been taken thanks to a fruitful interaction between theoretical calculations and experiment. Hydrogen has a profound effect on the electrical properties of semiconductors, and investigations of hydrogen in silicon are obviously well motivated solely from an application point of view. In addition, many such studies have their own right, addressing the basic concepts in the physics of defects in semiconductors; e.g. the crucial role that charge-state controlled lattice relaxations play for stability and metastability of hydrogen sites, for diffusion of hydrogen, and for reactions with other defects. In this review we shall attempt to correlate observations that are presumably directly connected with properties of isolated hydrogen (primary hydrogen centers) in silicon. Throughout we shall take guidance from the impressive theoretical work [2] developed parallel to the experimental progress, but which presently seems to have moved somewhat ahead. An important theoretical achievement is the prediction and explanation of a stable structure, bond-center hydrogen, in which the hydrogen atom is positioned midway between two bonding silicon atoms. A major contribution to the development of this model comes from studies of muonium in silicon [3].

In the following sections we shall first compare the structural information available from muon spin techniques (LF-$\mu$SR, TF-$\mu$SR and $\mu$LCR), electron spin resonance (ESR) and channeling. Then we shall examine the dynamics of a metastable hydrogen defect which has been observed recently [4] by deep level transient spectroscopy (DLTS). Finally we shall correlate the available data and compare with theory in an attempt to show that the observed DLTS-signal may indeed represent the deep donor state of bond-center hydrogen. We shall conclude the presentation by drawing attention to other related observations of hydrogen, and discuss some prospects for further progress, in particular as far as ion-implantation studies are concerned.

---

[1] Institute of Chemistry, Aarhus University

## 2. Structural observations.

Apart from dynamics (which may be influenced by quantum effects) muonium acts as a light hydrogen isotope and should show analogous equilibrium properties. It is well established that muonium in silicon is fully accounted for in three coexisting forms [3]; diamagnetic muonium ($\mu^+$), resembling ($H^+$) ( or possibly ($H^-$)), and the two paramagnetic forms, normal muonium (Mu) and anomalous muonium ($Mu^*$), resembling $H^0$ in two different local environments. Most large-scale total-energy calculations (supercell or cluster) carried out in recent years [2] agree to the fact that the lowest-energy position for hydrogen (muonium) in silicon, in a positive or neutral charge state, is the bond center (BC); i.e. midway between two silicon atoms each of which is relaxed outward by about 0.4Å. This bond center model can be visualized in a simple molecular orbital picture [5], and predicts features in the nuclear hyperfine structure which distinguish it from a possible vacancy associated center [6]. Displaying anisotropic hyperfine interaction, $Mu^*$ is believed to represent the bond-center configuration. Unambiguous proof of this assignment has been provided by $\mu$LCR results [7] in connection with theory. In particular, detailed first principle calculations using density functional theory in a supercell approximation [8] explains the <111> axially symmetric hyperfine interaction of $Mu^*$ remarkably well.

Normal muonium (Mu) differs from $Mu^*$ by having much larger $\mu$SR-frequency and isotropic hyperfine interaction. The tetrahedral interstitial position of low electron density is the natural intuitive choice for the Mu-site. However, theory predicts that site to be a saddle point. Mu is therefore envisaged to move rapidly between sites of lower symmetry in such a way as to average out any anisotropy; e.g. (quantum) diffusion between back bonding positions [3].

Extensive studies of paramagnetic hydrogen centers in silicon have mostly revealed secondary defects such as hydrogen-vacancy type centers. Only recently a primary center that can be compared to the muonium data has been found by Gorelkinskii and Nevinnyi [9] who performed "in situ" ESR-spectroscopy after low-temperature proton implantation. This center denoted (AA9) is paramagnetic under illumination and has hyperfine parameters that agrees well with those of $Mu^*$. The center anneals at ~200K with an activation energy of ~0.48 eV. It is perhaps not surprising that the experimental conditions chosen by the authors have provided the link to $\mu$SR. The short muon lifetime may promote the observation of metastable structures, that are neither present after annealing of hydrogen samples nor formed during hydrogenation under equilibrium conditions. Another link is indicated by channeling data [10] obtained after deuterium-implantation at 30K. The results showed that ~80% of the deuterium atoms occupies near-BC sites while the remaining ~20% is located at near tetrahedral interstitial sites. Redistribution due to annealing sets in at about 140K. Also muon channeling [11] is consistent with a BC-position in pure (float-zone) silicon.

Finding the link between $Mu^*$ and AA9 is an important step for further progress. Below we shall examine the properties of a deep donor (E3') observed by DLTS. A comparison of the AA9-annealing with the E3'-annealing may provide the next step for advance.

## 3. The deep donor E3'.

In DLTS deep centers are observed by recording the thermal emission rates of their trapped carriers. A non-equilibrium situation is created in a diode structure by a sudden increase in the space-charge layer from which the centers subsequently emit electrons or holes. Recently we have carried out DLTS experiments [4] in $p^+n$ diodes after implantation of up to $10^{10}$ H/cm$^2$ ($\sim 3\cdot 10^{14}$ H/cm$^3$) at low temperatures (<50K). The hydrogen E3'-center [12] was examined in detail. The center was found to be a deep donor with activation enthalpy for electron emission $E_c$-$E_t$ = 0.16 eV. It transforms into a DLTS-invisible structure at about 100K (or ~200K) during zero-bias annealing (or reverse-bias annealing), and the decay of the E3'-signal is consistent with first-order processes, $\nu_H = 3.0\cdot 10^{12}\ s^{-1} \times \exp[-0.29eV/k_BT]$ ( or $\nu_H = 1.3\cdot 10^8\ s^{-1} \times \exp[-0.44eV/k_BT]$). In addition E3' is metastable, since the DLTS-signal can be regenerated by forward bias injection of minority carriers at low temperatures. Fig.1 compares the zero-bias annealing for hydrogen and deuterium implanted samples. The value of the pre-exponential factor $\nu_H$ has the correct order of magnitude for a vibrational frequency, as expected for a process governed by a single atomic jump. Furthermore, the difference in the decay rates for hydrogen and deuterium indicates that hydrogen (deuterium) is the jumping entity. In that case, the pre-exponential factors should be related approximately by the harmonic relation $\nu_H = \sqrt{2}\nu_D$ which is used in the least-squares analysis shown. It follows from capacitance-voltage profiling that the conversion of E3' from its ionized state to the DLTS-invisible structure is associated with a shift of two elementary units of charge [4].

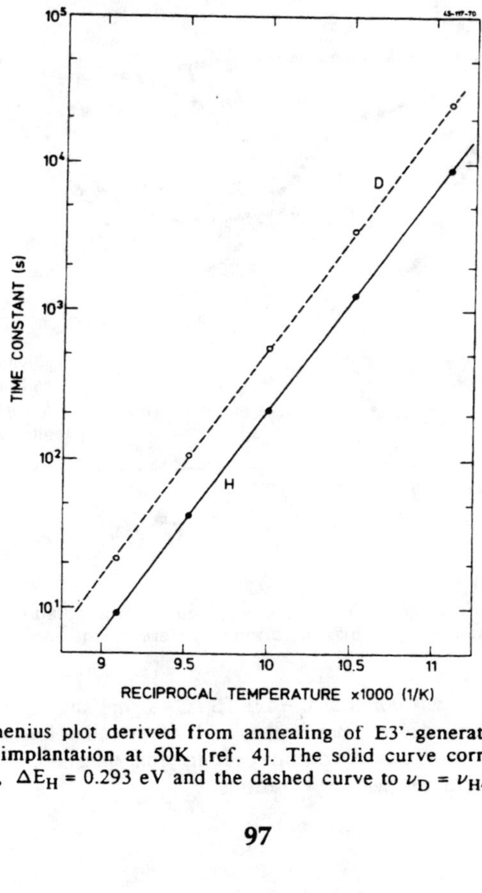


Fig.1. Arrhenius plot derived from annealing of E3'-generated by proton or deuterium implantation at 50K [ref. 4]. The solid curve corresponds to $\nu_H = 3.0\cdot 10^{12}\ s^{-1}$, $\Delta E_H = 0.293$ eV and the dashed curve to $\nu_D = \nu_H/\sqrt{2}$, $\Delta E_D = 0.30$ eV.

A capacitance versus temperature measurement has shown that conversion during the reverse bias-annealing of the ionized E3'-center occurs without any change in charge state. However, as soon as the bias is removed two electrons are captured. Thus in n-type silicon ( $N_D$ about $10^{15}$ $cm^{-3}$ ) the DLTS-invisible component contains one/two additional electrons compared with the non-ionized/ionized deep donor and may therefore reflect that $H^-$ is formed in transformation from $H^0$ and $H^+$. It should be emphasized that a large fraction of the implanted ions (typically 70%) are forming E3'-centers (at 65K) immediately after implantation. Further investigations have shown that the regeneration by injection of minority carriers is incomplete. The injection leads to the formation of two other, metastable hydrogen defects with very similar properties. The injection temperature influences the relative population of the three centers (see fig.2), and for a given temperature the annealing/injection cycle is closed; i.e. repeating the cycle leads to the same center population. After annealing at about 250K neither of the three metastable hydrogen centers can be recovered, implying that their common reservoir disappears by annealing at that temperature.

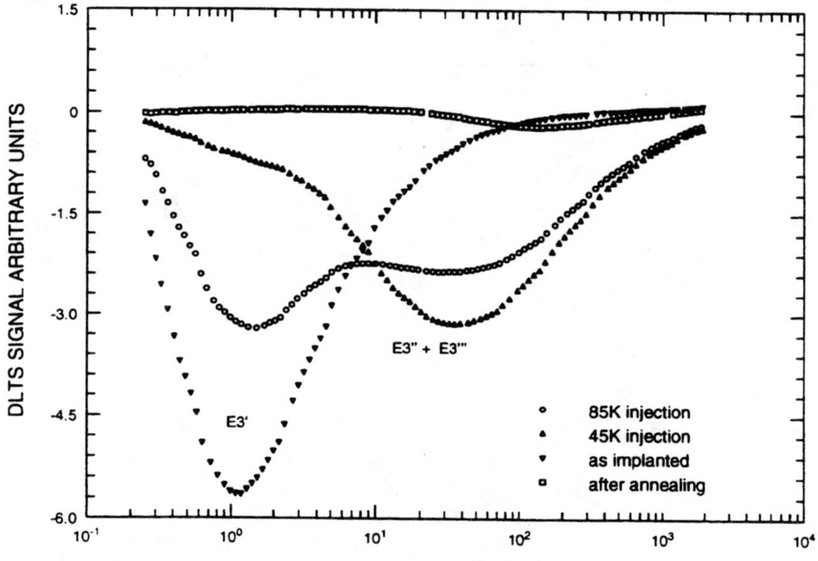


Fig.2. Comparison of DLTS frequency-scan spectra obtained at 65K after implantation and forward-bias injection at different temperatures.

It is tempting to ask whether the DLTS E3'- signal and the ESR AA9-signal may originate from the same defect since both signals are found at low temperature after ion implantation. In the next section we shall examine that question, and relate some of the dynamics features to theoretical expectations.

**4. Support for bond-center assignment of E3'.**

Comparison of the E3' reverse-bias annealing and the annealing in the dark of AA9 [9] shows fair agreement, see fig 3. The "effective" Fermi level is close to midgap in both cases, implying that the annealing proceed under similar conditions. It should be noticed though, that the hydrogen doses differ by at least three orders of magnitudes, which may influence the annealing stage. However, using a 60-layer hydrogen implantation we have recently been able to detect the ESR-AA9 signal at a dose level which was between one and two orders of magnitude lower than in the Gorelkinskii measurement, furthermore our preliminary annealing data is in agreement with those of Gorelkinskii. Thus similar annealing properties support the assumption that E3' and AA9 originate from the same center. Identity of AA9 and E3' would also explain why the AA9-center was observed only during illumination. Work is in progress, aiming at a more quantitative correlation of E3' and AA9 data. Since we have already, from the correlation with muonium data, adopted the bond-center assignment of AA9, we may consequently assume that the E3'-signal marks electron emission from the deep donor state of bond-center hydrogen. We shall now confront this assumption with theory and with present interpretations of other experimental results.

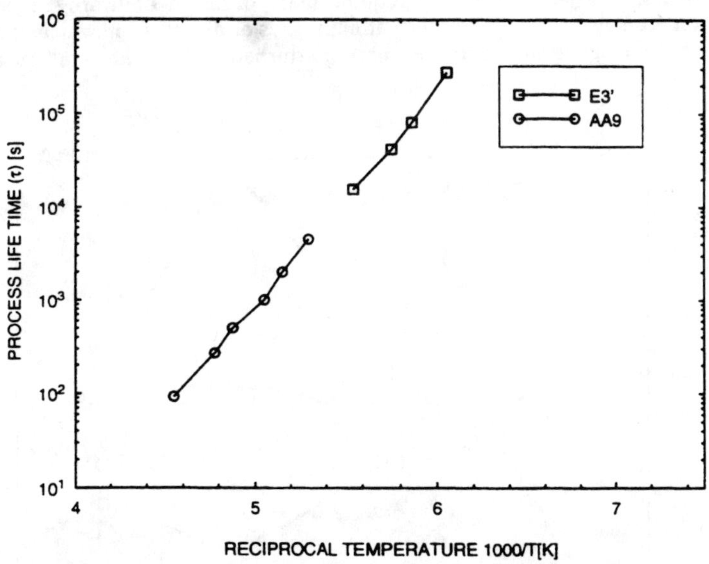


Fig.3. The lifetimes $\tau$ for the annealing of the AA9-center [from ref.9] and for the reverse-bias annealing of the E3'-center [from ref.4] shown as a function of inverse temperature.

Theoretical calculations [13,14,2] have shown that hydrogen in silicon may exist in three charge states $H^+$, $H^0$ and $H^-$, where $H^+/H^0$ and $H^-$ has energy minima at the bond-center site (BC) and at the tetrahedral site (T), respectively (see fig.4). The structural configuration isolated hydrogen in silicon is therefore intimately related to the type of shallow dopant. Hydrogen behaves as an amphoteric impurity. it always counteracts the prevailing doping, i.e. it behaves as an acceptor ($H^-$) in n-type material, and a deep donor ($H^+$) in p-type material. Many experimental findings are consistent with those properties [1].

The ability to passivate both shallow acceptors and shallow donors may be triggered by the amphoteric behavior of hydrogen and involves eventually the formation of neutral hydrogen-acceptor or hydrogen-donor pairs [2]. As mentioned earlier, the observation of AA9 can be explained by the excitation of $H^+$(BC) to $H^0$(BC) in intrinsic and/or damaged material where $H^+$(BC) is of lowest energy. Correspondingly the observation of E3' signifies electron emission from $H^0$ as indicated by the arrow in fig. 4. It is implied that sample preparation by low-temperature ion-implantation leaves hydrogen in the metastable $H^0$(BC) state. The zero-bias annealing stage at ~100K is ascribed to the transformation from $H^0$(BC) to $H^-$(T) controlled by a single, thermally activated jump. The reverse-bias annealing stage at ~200K is believed to reflect diffusion ( because of the low pre-exponential factor) from $H^+$(BC) to a $H^+$(near-sink) from which it relaxes into the equilibrium state $H^-$(T) when electrons become available. The fact that reverse bias annealing followed by injection leads to closed annealing/injection cycles supports the assumption that some kind of intermediate trapping connected with $H^+$ may play a role. Also, if the satellite signals (see fig.2) are assumed to represent perturbed E3' centers, their existence may require that diffusion of hydrogen occurs during injection. This may lead the trapping at impurities or defects from which in turn the E3'-recovery to neighboring perturbed or unperturbed bond-center configurations could occur.

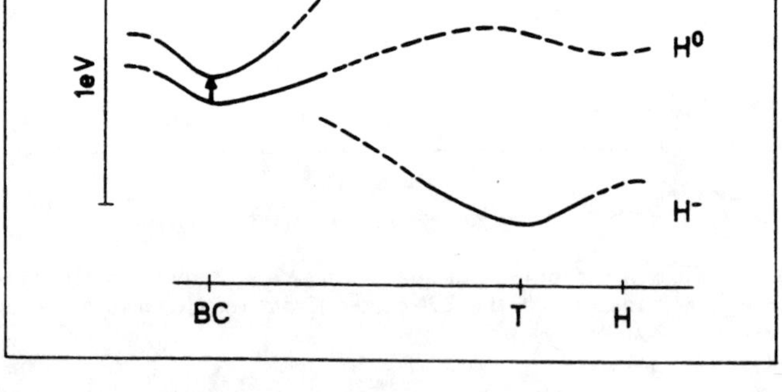


Fig.4. Sketch of the total-energy curves for $H^+$, $H^0$, and $H^-$in n-type silicon, based on Van de Walle et al [ref.13]. The arrow indicate the activation enthalpy for the ionization of $H^0$.

The dynamics of muonium in silicon has much in common with the dynamics of hydrogen described above. We disregard differences in transition temperatures which may be influenced by two opposing elements, quantum enhanced jumping and short observation time. $Mu^{*}$ (BC) is more stable than Mu (near-T) as indicated by the fact that electron irradiation of silicon catalyzes conversion from Mu to $Mu^{*}$. In pure silicon both Mu and $Mu^{*}$ disappear with increasing temperature before the interconversion can be seen. In both cases charge changing reactions to diamagnetic muonium are responsible [3]. Channeling data show that muonium undergoes a transition from T to BC sites in the same temperature range. This supports the conclusion that the diamagnetic center $\mu^{+}$ is positive muonium at the bond-center. Finally we note that tetrahedral-site muon trapping occurs in oxygen rich material [11]. This should be kept in mind in connection with the injection studies of hydrogen discussed above; so far DLTS injection data have been taken only in oxygen-rich samples.

Finally we shall comment on hydrogen diffusion in silicon. From studies of the dissociation of hydrogen-dopant complexes [15] it is inferred that both species $H^{-}$ and $H^{+}$ can be released and diffuses in silicon. Subsequent immobilization can occur by formation of diatomic hydrogen. Molecular hydrogen ($H_2$) ( known to be present in amorphous silicon) is expected from theoretical calculations to form a stable configuration at the interstitial tetrahedral site in crystalline silicon [2]. To date only indirect experimental evidence for $H_2$ has been obtained; direct spectroscopic detection is still lacking. However, a different diatomic hydrogen center ($H_2^{*}$) has been proposed from theoretical work [16,14], and there is now spectroscopic evidence for its existence [17]. In $H_2^{*}$ one hydrogen atom resides at a near-BC position and the other at a adjacent near-T position along the <111> axis. The formation of $H_2^{*}$ is likely to be connected with the diffusion of hydrogen via the BC-site in the presence of near-T site hydrogen, or vice versa.

The most straightforward situation for diffusion is likely to occur at high temperatures, where hydrogen atoms/ions diffuse without interaction with each other or with other impurities. Theoretical calculations [18,19] of diffusion coefficients agree well with experimental data at temperatures above 1000K. It is interesting that ab initio molecular dynamics simulations [18] show that diffusion of $H^{+}$ proceed via two alternating paths in such a way that both regions of high and of low electron density are visited. In some segments of the path the hydrogen ion jumps between near-BC sites, using near-T sites as intermediate ones. In other segments of the trajectory hydrogen is remaining in the low-density region and jumps from one near-antibonding site to another, going through tetrahedral or hexagonal sites. The reason why the proton does not follow the strictly BC-path can be traced back to the large difference in mass between H and Si. When the $H^{+}$ motion is fast the heavy Si ions cannot adiabatically follow the proton. Thus the lattice may not have time to undergo the large relaxation needed to accommodate the $H^{+}$ in BC site. When this happens $H^{+}$ prefers to move through regions of low energy in the undistorted lattice.

The profound role that lattice relaxations play in the case of hydrogen should caution the use of the concept "hydrogen level", e.g. in comparison of levels deduced from DLTS (i.e. from ionization enthalpy), and diffusion (i.e from location of chemical potential).

## 5. Conclusions and some trends for future development.

Many aspects of hydrogen in silicon are now fairly well understood. However, challenging problems remain. Clearly a full understanding of the mechanisms underlying the properties of metastability and diffusion is important since it is intimately related to the coupling between hydrogen and the surrounding atoms, which continuously modifies the potential in which the impurity moves. For this, unambiguous spectroscopic assignments of energy levels and geometrical structures of basic hydrogen related centers in silicon are needed. Such assignments of centers and their interplay via interconversions and defect reactions are still incomplete.

Ion-implantation at low temperature often leave the implanted impurities in non-equilibrium states, that are inaccessible otherwise. However, ion-implantation has the disadvantage of creating damage. This makes non-trivial the comparison of data recorded by detection methods that have very different sensitivity, such as DLTS and ESR discussed in this talk. The inclusion of channeling and infrared absorption is even more difficult. Nevertheless has the interplay between these techniques and also the nuclear techniques played an important role in the past. We believe that further progress can be made. Experiments at much lower local defect concentrations are indeed feasible utilizing deep multilayer implantation at low temperature, followed by in situ applications of the various spectroscopic techniques. The constructive interplay between theory and experiment has been very fruitful in the past and hopefully will remain so.

## References


[1] S.J. Pearton, J.W. Corbett, and T.S. Shi; Appl.Phys. A43, 153 (1987), and references therein.
[2] C.G. Van de Walle; Physica B170, 45 (1991), and references therein.
[3] R.F. Kiefl and T.L.Estle; Physica B170, 33 (1991), and references therein.
[4] B. Holm, K. Bonde Nielsen, and B. Bech Nielsen; Phys.Rev.Lett. 66, 3260 (1991).
[5] S.F.J. Cox and M.C.R. Symons, Chem.Phys.Lett. 126 516 (1986).
[6] T.L. Estle, Hyperfine Interactions 17-19 585 (1984).
N. Sahoo, K.C. Mishra, and T.P. Das; Phys.Rev.Lett. 55 1506 (1985)
[7] R.F. Kiefl, M. Celio, T.L. Estle, S.R. Kreitzman, G.M. Luke, T.M. Riseman, and E.J. Ansaldo; Phys.Rev.Lett. 60 224 (1988)
[8] C.G. Van de Walle, Phys.Rev.Lett. 64 669 (1990)
[9] Yu.V. Gorelkinskii and N.N. Nevinnyi; Sov.Tech.Phys.Lett. 13 45 (1987), and Physica B170, 155 (1991).
[10] B. Bech Nielsen; Phys.Rev. B37 6353 (1988).
[11] B.D. Patterson; Rev.Mod.Phys. 60 69 (1988).
[12] K. Irmscher, H. Klose, and K. Maass; J.Phys. C17 6317 (1984).
[13] C.G. Van de Walle, P.J.H. Denteneer, Y. Bar-Yam, and S.T. Pantelides; Phys.Rev. B39 10791 (1989).
[14] R. Jones; Physica B170, 181 (1991).
[15] N.M. Johnson, C. Doland, F. Ponce, J. Walker, and G. Anderson; Physica B170, 45 (1991), and references therein.
[16] K.J. Chang and D.J. Chadi; Phys.Rev. B40 11644 (1989).
[17] J.D. Holbech, B. Bech Nielsen, and R. Jones; to be published.
[18] F. Buda, G.L. Chiarotti, R. Car, and M. Parrinello; Phys.Rev.Lett.63 294 (1989), Physica B170, 105 (1991)
[19] P.E. Blöchl, C.G. Van de Walle, and S.T. Pantelides; Phys.Rev.Lett. 64 1401 (1990)

Local Order
in Condensed
Matter Physics


ISBN 1-56072-220-7

# ELECTRONIC STRUCTURE OF DEFECTS IN IONIC CRYSTALS

John M. Vail

*Department of Physics, University of Manitoba*
*Winnipeg, MB, R3T 2N2, Canada*

## ABSTRACT

**A method is briefly described for computer simulation of point defects in ionic crystals. Within the unified framework of this method, a wide variety of impurity and color center properties are considered. For impurities these include hole trapping, electron loss, crystal field splitting, spin polarization, and classical and quantum diffusion. Calculated color center properties include localization, distortion and polarization, spin resonance, optical absorption, and impurity effects. Comparison with experiment illustrates the scope of the method, while indicating that it needs to be pursued to higher accuracy in some cases.**

## 1. Introduction

Point defects, along with chemical composition and lattice structure, determine the properties of crystalline ionic insulators. Experimental studies of these materials sometimes need to be supplemented by computer simulation. We shall describe results of a wide variety of such calculations.

First, we briefly describe the model on which our simulations are based. The point defect is considered to be part of a molecular cluster of ions, in which all the electrons are treated quantum mechanically, using the Hartree-Fock method, to which correlation correction can be applied. This cluster is embedded in an infinite static crystal whose ions are modeled classically, as point charges with classical Coulomb- and short-range forces. In this model (the shell model), ions are dipole polarizable, and the model parameters are derived from the experimental harmonic properties of the crystal. The ground state is determined by minimizing the total energy with respect to variations of the electronic structure and nuclear positions in the cluster, and ionic positions and polarizations outside the cluster. This method is embodied in an automated user-friendly program called ICECAP.[1]

## 2. Impurity Processes

### *2.1 Nickel in MgO*

In a study of the charge state stability of nickel (and of copper) in MgO,[2] we have found that $Ni^{+}$ loses an electron to the $Mg^{2+}$ – 3s band, and that $Ni^{3+}$ creates a hole in the

$O^{2-}$ – 2p band. This is illustrated in Table 1, where the Mulliken populations are given for $Ni^{3+}$ and its six nearest oxygen neighbors. We note that the values are close to the total 26 electrons for $Ni^{2+}$, with 14 spin up and 12 spin down, whilst the oxygen spin down manifold lacks one of the thirty electrons it would have in the unperturbed crystal.

**Table 1.** Mulliken populations for the molecular cluster $Ni^{3+}(O^{2-})_6$ in MgO.

| spin | up | down | total |
|---|---|---|---|
| $Ni^{3+}$ | 14.03 | 11.74 | 25.97 |
| $(O^{2-})_6$ | 29.97 | 29.26 | 59.23 |

Our calculations show crystal field splitting of $T_{2g}$ and $E_g$ $Ni^{2+}$ 3d levels of 1.32 eV for the Hartree-Fock eigenstates, and spin polarization energy of 1.69 eV for the $T_{2g}$ level. A disproportionation energy of 10 eV is found, favoring $Ni^{2+}$ over $Ni^{+,3+}$, as follows:

$$Ni^{+} + Ni^{3+} \rightarrow 2\ Ni^{2+} + 10\ \text{eV}.$$

This result is qualitatively supported by the experimental fact that $Ni^{+}$ and $Ni^{3+}$ do not occur in MgO without associated stabilizing defects.

### *2.2 Lithium in NiO*

In NiO, lithium occurs as $Li^{+}$ with an associated hole, for charge neutrality. Experimentally, in pure NiO the top of the valence band is dominated by Ni 3d levels.[3] In that case, holes would be associated with nickel. However, when lithium impurity is added, the compensating holes are found to be associated with oxygen 2p levels.[4] It therefore seems that the role of $Li^{+}$ impurity is to depress the Ni 3d levels locally, exposing oxygen 2p levels at the top of the valence band. This is illustrated in the molecular orbital one-electron energy level spectra calculated[5] for embedded $Ni^{2+}.(O^{2-})_6$ and $Ni^{2+}.(O^{2-})_6.Li^{+}$.hole clusters. In the former case, representing pure NiO, heavy mixing of Ni $T_{2g}$ states with oxygen 2p states is found at the top of the valence band. In the impurity case, the oxygen 2p bands are spread, and straddle the Ni 3d levels, which are depressed. In this case, the top of the valence band is pure oxygen 2p-like, and this band has 18 and 17 electrons in spin up and spin down manifolds respectively. Further analysis shows that the hole localizes on a single oxygen ion, producing a defect with $C_{4v}$ symmetry, with energy 3 eV lower than the $O_h$–symmetric defect would have.

### *2.3 $Cu^{+}$ and $Ag^{+}$ in Alkali Halides*

The classical diffusion processes for these impurities have been studied using short-range repulsive potentials between the impurity and host-crystal ions that were derived quantum mechanically.[6] In the derivation, the total energy of the defect in a quantum cluster is calculated as a function of nearest-neighbor distance, and then classical pairwise potentials for a point-charge impurity are fitted to produce agreement with this energy curve. The resulting classical model is used to calculate the activation energies for diffusion by both vacancy and interstitial mechanisms. These energies are shown in Table 2 for a variety of cases, compared with experimental results where such data exist. The calculated results agree both with the magnitude and ordering of the experimental results. Our calculations indicate that $Ag^+$ diffusion by the vacancy mechanism follows a non-collinear path in RbCl and KCl.

**Table 2. Activation energies ΔE (units eV) for $Cu^+$ and $Ag^+$ by vacancy and interstitial mechanisms in some alkali halides.**

| material | mechanism | ΔE calculated | experiment |
|---|---|---|---|
| KCl:$Cu^+$ | vacancy | 1.19 | 1.1 |
| | interstitial | 0.78 | 0.83 |
| KCl:$Ag^+$ | interstitial | 0.83 | 0.95 |
| RbCl:$Ag^+$ | vacancy* | 0.44 | – |
| KCl:$Ag^+$ | vacancy* | 0.51 | – |
| NaCl:$Ag^+$ | vacancy | 0.93 | – |
| NaF:$Ag^+$ | vacancy | 0.72 | – |

* non-collinear mechanism

## *2.4 Muonium in NaF*

Experimental study of muonium diffusion in KCl[7] exhibits the characteristic temperature dependence of quantum diffusion, wherein as temperature drops, the hop rate undergoes a minimum, as the transport goes from incoherent (random walk) to coherent (Bloch-like). The theory of these processes[8] is expressed in terms of the muonium-phonon coupling constant and the potential due to the crystal, through the transfer matrix. This theory treats the muonium impurity atom ($\mu^+ - e^-$) as a small polaron. We have applied it to muonium diffusion in NaF,[9] calculating coupling and potential from embedded quantum clusters. Three features emerge: (i) the calculated crossover temperature from coherent to incoherent mechanism is plausible; (ii) the theory cannot fit the large rise in hop rate observed experimentally below the crossover temperature; (iii) a large renormalization would be required for the transfer matrix, to account for eight orders of magnitude difference between calculated and observed hop rates, and the mechanism for such renormalization is not known. This work illustrates the importance of carrying theoretical analysis through to the quantitative calculation of properties of real materials.

## 3. Color Centers

### 3.1 *The F Center in NaF*

F-type centers have important optical properties, and play a fundamental role in radiation damage. The simplest of them, the ordinary F center, is exemplified by an electron bound in a fluorine vacancy in NaF. In investigating the ground state of this defect, we found it to be well-localized in the vacancy, not shared by 3s orbitals of nearest neighbor $Na^+$ ions.[10] The latter were only slightly displaced, ~1% of the inter-ionic spacing of the crystal. From the calculated spin densities, isotropic hyperfine constants can be compared with experiment,[11] Table 3. The results show that the calculated ground state is significantly less diffuse than the true ground state.

**Table 3. Isotropic hyperfine constants a for the F center in NaF (units h.Mc, 1 Mc = $10^6 s^{-1}$), for nearest and second neighbors.**

| | calculated | experimental |
|---|---|---|
| nearest ($Na^+$) | 80 | 107 |
| second ($F^-$) | 32 | 97 |

### 3.2 *$F_2^+$ Center in NaF*

This center consists of an electron bound by two adjacent fluorine vacancies: the defect has a net positive charge. It is used in color center lasers.

In the $F_2^+$ center, two $Na^+$ ions are each nearest neighbors of both vacancies. Their coordinates in the unrelaxed crystal are ($\mp$0.5, ±0.5, 0) in units of nearest-neighbor spacing, in a cartesian frame whose origin is midway between the vacancies at (±0.5, ±0.5, 0), and whose axes are parallel to the cubic axes. In Table 4, the relaxed coordinates of these nearest-neighbors, and of the other two inequivalent sets, are given, along with calculated isotropic hyperfine constants.[10] The result agrees with the experimental conclusion[12] that the excess electron is strongly concentrated toward the defect center, rather than symmetrically within the two vacancies.

**Table 4. Calculated isotropic hyperfine constants a (units h.Mc, 1 Mc = $10^6 s^{-1}$), and relaxed nearest-neighbor coordinates (units nearest-neighbor distance of perfect crystal) for the $F_2^+$ center in NaF.**

| coordinates | a |
|---|---|
| $Na^+$ (0.55, 0.55, 0) | 189 |
| $Na^+$ (1.55, 0.49, 0) | 12 |
| $Na^+$ (0.49, 0.49, 1.05) | 31 |

### 3.3 $(F_2^+)^*$ Center in NaF:Mg

This defect produces laser action, whilst maintaining itself stable at room temperature, a merit not shared by other F-type center lasers.

The $(F_2^+)^*$ center is a combination of two defects, an $F_2^+$ center, and an impurity dipole consisting of a $Mg^{2+}$ substitutional cation impurity and a $Na^+$ vacancy ($V'_{Na}$), which compensates for the excess charge of $Mg^{2+}$ relative to the NaF lattice. We have examined a large number of possible configurations of this system,[10] including that proposed by Hofmann et al.,[12] using a model in which interactions of the excess electron with ions are represented by classical potentials derived as in sec. 2.3. We conclude that the ground state configuration has the $F_2^+$ center and $Mg^{2+}$ – $V'_{Na}$ dipole combined in a planar crescent. Relative to the coordinate system of sec. 3.2, the coordinates are $V'_{Na}$ (−0.5, −1.5, 0) and $Mg^{2+}$ (0.5, −2.5, 0). The electron is fully localized in the $F^-$ vacancy at (0.5, 0.5, 0), away from the dipole, which has its negative end ($V'_{Na}$) adjacent to the empty (positively charged) $F^-$ vacancy. This is not an $F_2^+$-like center then, because its electron is localized in a single vacancy: rather it is a complicatedly perturbed F center. This result is in agreement with the experimental deduction of Hofmann et al.[12]

### 3.4 $F^+$ Center in MgO

This defect has a net positive charge because a single excess electron replaces an $O^{2-}$ ion. We have analysed it extensively[13] and found several qualitative and quantitative features. Qualitatively, the ground state is found to have a single electron localized in the vacancy, rather than creating a hole in the second neighbor oxygen manifold and leaving two electrons in the vacancy. Furthermore, in our results the unrelaxed excited state of the defect is also well-localized in the vacancy. Two features that can be compared with experiment have been calculated. One is the ground state spin density at the nearest-neighbor $Mg^{2+}$ nuclei (units $A^{-3}$) 0.171 calculated, 0.274 experimental.[14] The other is the optical excitation energy 7.59 eV calculated, 4.95 eV experimental.[15] Taken with an appreciation of the work done in obtaining the calculated results, the merely qualitative agreement with experiment illustrates the care that must be taken to calculate quantitatively accurate results of this sort.

### 3.5 $(F_2^+)_A$ Center in $CaF_2$:Na

Several activation energies relating to this system have been deduced by Tijero and Jaque.[16] We are now undertaking simulation studies[17] that will estimate these activation energies directly from embedded clusters, rather than by use of derived classical potentials as in sec. 2.3. The $(F_2^+)_A$ center consists of an excess electron shared between two $F^-$ vacancies, perturbed by a nearest neighbor cation substitutional $Na^+$. The net charge of the defect is zero relative to the fluorite crystal. Two formation processes will be studied, starting with a single-vacancy $F_A$ center, where the nearest neighbor impurity $Na^+$ ion is associated with an $F^-$ vacancies ($V_{F-}$) by way of charge compensation. In one process, the $Na^+$ – $V_{F-}$ dipole reorients so that $V_{F-}$ comes to be adjacent to the $F_A$ center, which then spreads out to become an $(F_2^+)_A$ center. In the other process, the $V_{F-}$ is dissociated from the $Na^+$ impurity, and combines with the $F_A$ center other than by dipole reorientation. In similar manner we can estimate the activation energy for $V_{F-}$ diffusion in pure and Na-doped $CaF_2$ in the absence of F centers. Results of these calculations compared with experiment will further test the ability of the method to deal with relatively complicated defect processes.

**Summary**

We have described results, with numerous experimental comparisons, for a wide variety of point defects, host ionic materials, and physical processes. Since all the results derive from a single model and computer program, they provide a strong foundation for assessing the methodology, and hold out the promise of refining it to the point of flexible, reliable applicability.

**Acknowledgements**

I am grateful to my collaborators J. Meng, R. Pandey, T. McMullen, Z. Yang, P. Jena, and A. M. Stoneham, whose work has been described herein. I enjoyed hospitality at Michigan Technological University, Virginia Commonwealth University, and Harwell Laboratory during the course of the work. Financial support came from the Canadian Natural Sciences and Engineering Research Council and the University of Manitoba Research Board.

**References**

1 J. H. Harding, A. H. Harker, P. B. Keegstra, R. Pandey, J. M. Vail, and C. Woodward, Physica B **131** (1985) 151.
2 J. Meng, J. M. Vail, A. M. Stoneham, and P. Jena, Phys. Rev. B **42** (1990) 1156.
3 D. E. Eastman and J. L. Freeouf, Phys. Rev. Lett. **34** (1975) 395.
4 P. Kuiper, G. Kruizinga, J. Ghijsen, G. H. Sawatzky, and H. Verweij, Phys. Rev. Lett. **62** (1989) 221.
5 J. Meng, P. Jena, and J. M. Vail, J. Phys. Cond. Matt. **2** (1990) 10371.
6 J. Meng, R. Pandey, J. M. Vail, and A. B. Kunz, J. Phys: Cond. Matt. **1** (1989) 6049.
7 R. F. Kiefl, R. Kadono, J. H. Brewer, G. M. Luke, H. K. Yen, M. Celio, and J. Ansaldo, Phys. Rev. Lett. **62** (1989) 792.
8 T. McMullen and B. Bergersen, Solid State Commun. **28** (1978) 31.
9 T. McMullen, J. Meng, J. M. Vail, and P. Jena, submitted (1992).
10 Z. Yang and J. M. Vail, unpublished.
11 H. Seidel and H. C. Wolf, in *Physics of Color Centers*, Ed. W. B. Fowler (Academic, New York, 1968), Ch. 8, Table 8-3.
12 D. H. Hofmann, F. Lohse, H. J. Paus, D. Y. Smith, and J.-M. Spaeth, J. Phys. C: Solid State Phys. **18** (1985) 443.
13 R. Pandey and J. M. Vail, J. Phys: Cond. Matt. **1** (1989) 2801.
14 W. P. Unruh and J. M. Culvahouse, Phys. Rev. **154** (1967) 861.

[15] B. Henderson and R. D. King, Phil. Mag. **13** (1966) 1149.
[16] J. M. G. Tijero and F. Jaque, Phys. Rev. B **41** (1990) 3832.
[17] J. M. Vail and Z. Yang, unpublished.

Local Order
in Condensed
Matter Physics


ISBN 1-56072-220-7

# ON THE ORIGIN OF TRANSIENT MAGNETIC FIELDS

F. Hagelberg

*Department of Physics, State University of New York at Albany, Albany, NY, 12222*

**Abstract.** The Transient Field phenomenon has been ascribed to a polarization transfer between the electrons of the ionic projectiles and the surplus of majority spin electrons of the ferromagnetic host over the minority spin electrons. Earlier attempts to explain this crucial process failed to account for the order of magnitude of the experimentally observed Transient Field strengths. A recent model which proposes spin exchange scattering between bound projectile electrons and quasi-free host electrons as mechanism of polarization transfer arrives at the correct orders of magnitude but is in conflict with the weak velocity dependence of the experimental polarization, showing a strongly decreasing behavior with increasing velocity. The new model presented here proposes spin exchange between the ionic shell and localized electrons of the ferromagnet as a more adequate approach to the problem. It is shown that calculations involving hydrogen-like ions explain the size of the experimentally observed polarization effects as well as their velocity dependence for various ion probes traversing thin iron foils.

Transient Magnetic Fields (TF) are observed at the nuclei of swift ions traversing ferromagnetic media [1,2]. Due to their extraordinary size in the 10 megagauss range the TF have made possible the experimental investigation of Nuclear Magnetic Moments of extremely short lived nuclei with lifetimes in the subpicosecond region [3,4]. Thus, using the TF, magnetic properties of high spin states have been explored [5] in nuclear spin precession measurements. Since no successful quantitative description of the TF from first principles has been given so far, g-factor measurements still depend on TF calibrations which are often hard to perform and also decrease the precision of the measured value of the g-factor. Reliable predictions of the TF field strength based on a theoretical explanation of the TF phenomenon could improve considerably upon the quality of nuclear moment studies which employ the TF method. Moreover, an analytic understanding of the TF would allow optimization of the experimental conditions in order to obtain a maximum TF strength corresponding to strongest nuclear spin precession effects.

In the present work we concentrate on the hydrogen-like ion projectiles which makes the atomic physics associated with the problem as simple as possible and thus allows one to attempt first-principle quantitative understanding of the important features involving the valence electrons which play a crucial role in the origin of the TF.

The first experiments with hydrogen like probe ions have clearly demonstrated [3], that the TF is produced by unpaired s-electrons of the ion which are polarized by the interaction with the surplus number of majority electrons in the ferromagnet. Thus, the crucial problem in the understanding of the TF phenomenon consists in the explanation of the polarization transfer mechanism involved. As will be outlined in below, the use of ion probes in a hydrogen like state allows for the determination of the degree of polarization acquired by the K-shell electrons of the projectiles as they pass through the ferromagnetic host. These experiments give the key to any theoretical treatment of the origin of the TF since a quantitative first principles theory has to explain both the magnitude as well as characteristic trends of the ionic polarization with respect to the velocities and atomic numbers of the ion probes as well as the characteristics of the host material.

As the TF-producing mechanism, elastic spin exchange scattering [6] is proposed. This process transfers preferentially the spin orientation of the majority electrons of the ferromagnet into the inner shells of the ion and therefore gives rise to spin polarization of these shells. Any closer examination of the effectiveness of this mechanism has to be based on numerical values of the spin exchange cross section $\sigma^{sf}$ . This quantity has been evaluated before [6] following the formalism developed by Temkin and Lamkin [7] for the scattering of a free electron by a hydrogen like target, taking into account the strongly focussing effect of the high projectile charges [8]. The 1s polarizations deduced from this approach were found to reflect the overall magnitudes of the experimental values reasonably well. However, some characteristic trends of the observed probe ion polarization could not be explained in the framework of this model. Thus, the experimental finding is that the degree of 1s polarization stays constant over a sizable ion velocity range [9], while the theory outlined in Ref.6 leads to the expectation of a strongly decreasing polarization as the ion velocity increases. The two processes which determine the effectiveness of polarization transfer according to the model discussed in [6], spin exchange between ion and target electron as well as host electron density enhancement at the ionic core, are both very markedly attenuated with increasing velocity.

To avoid these contradictions with experiment, one has to reinvestigate the spin exchange approach. In the first place, the unrealistic assumption of a free electron representing the contribution of the host to the scattering process has to be replaced by a more appropriate description of the polarized electron states in the ferromagnet. As is well known, these states are localized to a very high degree at least, if not fully [10], in all ferromagnetic metals which have been used so far in TF experiments, these are Fe, Co, Ni and Gd. It is thus necessary for theory to take the localized nature of these electrons properly into account.

The present work concentrates on the TF produced by the passage of light ions ranging from carbon (Z = 6) to sulphur (Z = 16) through Fe foils. The purpose of the research reported here is to demonstrate that the magnitude of the observed polarization effects is readily explained by elastic spin exchange scattering between bound projectile electrons and the localized electrons of the ferromagnet. The model introduced in this work is also shown to be in much better agreement with experimental trends than is the case for any of the foregoing theoretical attempts. Before giving details on this theory, we want to summarize the main empirical findings which our model tries to account for.

The experimental determination of the degree of ion polarization is based on a microscopic description of the TF [11]:

$$B_{TF}(v,Z,host) = \sum_{ns} q_{ns}(v, Z, host)\, p_{ns}(v, Z, host)\, B_{ns}(Z) \qquad (1)$$

In this formula, $B_{ns}(Z)$ refers to the Fermi Contact Field produced by an unpaired electron in an ns-orbital at the nuclear site of the ion. The "single electron factor" $q_{ns}$ indicates the fraction of ions with half-filled ns orbital and $p_{ns}$ their degree of polarization. Both quantities depend on the elementary parameters of the experiment, i.e. the ion's atomic number Z, its velocity v as well as on properties of the ferromagnetic host.

The expression (1) simplifies considerably when the TF is associated with H-like ions. In this case, the sum over all ns orbitals is reduced to the contribution from the K shell:

$$B_{TF}(v, Z, host) = q_{1s}(v, Z, host)\, p_{1s}(v, Z, host)\, B_{1s}(Z) \qquad (2)$$

The Fermi Contact Field $B_{1s}(Z)$ can be evaluated analytically [12], and the single electron factor $q_{1s}$ is known from measurements for a wide range of different ion sorts and ion velocities in Fe and Gd [13,14]. Thus, using equation (2), it is possible to extract the degree of polarization $p_{1s}$ from the measured TF strength $B_{TF}$, the latter quantity being experimentally determined by observation of the nuclear spin precession angles of the probe ions as they traverse thin ferromagnetic layers [3]. The precession angle is measured using the technique of Perturbed Angular Particle Gamma Correlations [15].

Equation (2) is strictly valid only at high enough ion velocities $v \geq Zv_0$ ( the "1s Bohr velocity"). In this velocity regime, the projectiles traverse the solid essentially in a hydrogen-like state or as bare nuclei which do not contribute to the field strength. The microscopic description of the TF given by formula (2) can still be applied in a situation of smaller ion velocities, corresponding to more complex electron configurations of the projectile, provided the single electron factors $q_{ns}$ of higher ns orbitals are small as compared to $q_{1s}$

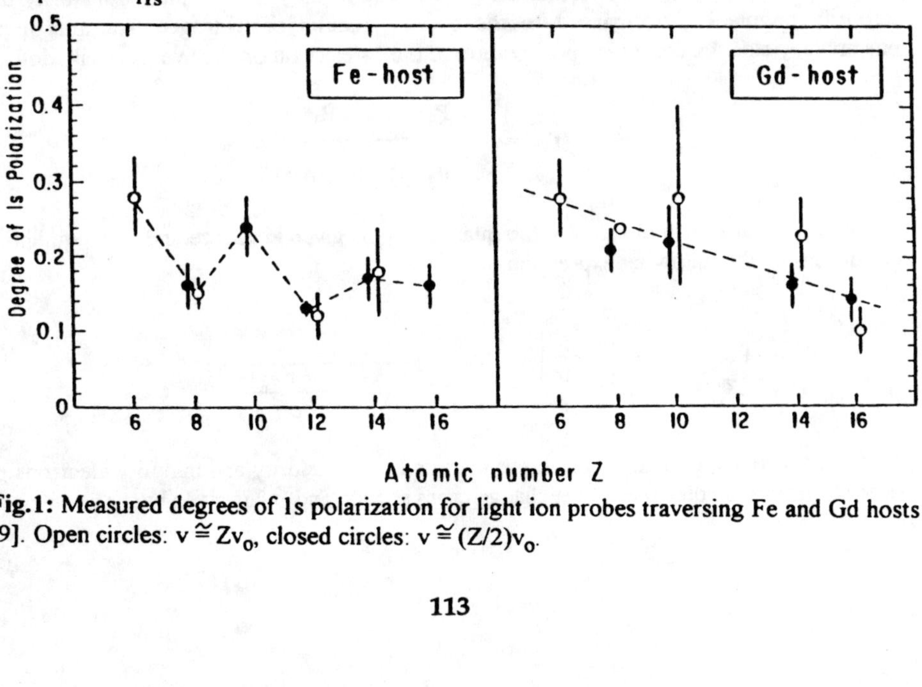


**Fig.1:** Measured degrees of 1s polarization for light ion probes traversing Fe and Gd hosts [9]. Open circles: $v \cong Zv_0$, closed circles: $v \cong (Z/2)v_0$.

Fig. 1 summarizes the results of a series of experiments where highly charged hydrogen-like ions ($6 \leq Z < 16$) were observed in Fe and Gd host. The ion probe K-shell polarization as obtained from Eq.(2) is shown versus the atomic number of the ion.. From this systematic experimental study, three features emerge that any theoretical treatment of the polarization transfer problem would have to explain:

1. In both metals, high degrees of 1s-polarization in the range of 10 - 30% are found.

2. There appears to be no significant difference between polarization data taken at $v \cong Zv_0$ and at $v \cong Z v_0/2$.

3. In both cases, the degree of 1s polarization seems to drop with increasing atomic number of the ion probe. This effect, however, is more pronounced in Gd than in Fe where a weakly decreasing tendency of the polarization with atomic number is accompanied by a marked oscillating behavior.

The apparent velocity independence of the polarization data poses a special challenge to theoretical explanation. Intuitively, one expects a decrease of polarization transfer efficiency as the ion velocity rises and , consequently, the ion-atom interaction time drops.

In the following paragraphs, we introduce a model which considers spin exchange scattering by localized moments of the host atoms as the crucial polarization transfer mechanism.

As a hydrogen like ion ensemble traverses a ferromagnetic medium, it acquires polarization via spin exchange scattering between the electron bound by the ion nucleus and polarized host electrons. With the elapse of time, the ion population $P_1\downarrow$ with the majority spin orientation (electron spin "down") will increase over the ion population $P_1\uparrow$ with minority spin orientation (electron spin "up"). The surplus of majority over minority population, normalized to the overall fraction of hydrogen like ions in the ensemble, yields the degree of polarization of the 1s electron on the hydrogen-like ion:

$$p_{1s}(t) = \frac{P_1\downarrow(t) - P_1\uparrow(t)}{P_1\downarrow(t) + P_1\uparrow(t)} \tag{3}$$

As has been shown in Ref. 16, the quantity $p_{1s}$ is given in charge and spin equilibrium conditions by the following expression:

$$p_{1s} = \frac{n\downarrow - n\uparrow}{n\downarrow + n\uparrow}\left(1 - \frac{1}{1 + \frac{(n\downarrow + n\uparrow)\sigma^{sf}}{\sigma^{q}}}\right) \tag{4}$$

In this formula, $n\downarrow$ and $n\uparrow$ represent the number of majority and minority electrons per host atom, $\sigma^{sf}$ denotes the spin exchange cross section and $\sigma^q$ is given by:

$$\sigma^q = \sigma^{cap} + \sigma^{loss} \tag{5}$$

The numerical evaluation of expression (4) requires the knowledge of all contributing microscopic quantities : $n\downarrow$, $n\uparrow$, $\sigma^q$ and $\sigma^{sf}$.

The results which will be discussed below were obtained adopting for $n\downarrow/\uparrow$, the average numbers of unpaired electron spins with down and up polarization, respectively, per Fe atom: $n\downarrow = 2.6$ and $n\uparrow = 0.4$, corresponding to the effective magnetization of 2.2 Bohr magnetons per atom [17]. These values refer to an undisturbed situation while the highly charged ion projectiles are likely to induce some ionization of a host target atom thus enhancing the total number of unpaired electron spins per atom, $n\downarrow + n\uparrow$. This quantity, however, has only a very slight influence on the equilibrium degree of 1s polarization as given by formula (4).

The spin quenching cross section $\sigma^q$ has been experimentally observed for the systems O in Fe [13] and for S in Fe [14] at velocities close to $v = Z\, v_0$. In the remaining cases of C, Ne, Mg and Si in Fe, where no experimental cross sections are available so far, a scaling procedure based on Oppenheimer-Brinkmann-Kramers and Binary encounter approximations [6] has been employed to obtain realistic estimates of the capture and loss cross sections involved.

In the following, we want to give an account of the basic approach adopted in the evaluation of the quantity $\sigma^{sf}$.

Our treatment of the spin exchange process between the 1s projectile electron and a 3d electron of an Fe host atom follows the lines of the Impact Parameter Approach [18]. In view of the high velocities involved, the probe ion can be assumed to move in a straight path with a well defined impact parameter p and ion velocity v. Fig.2 indicates the scattering geometry adopted.

The elementary exchange process is described as a two center interaction of two electrons, one being attached to the probe ion and the other to a host atom core. The influence of the remaining host atom valence electrons on the 3d electron during the scattering process is neglected in the present context. With these simplifying constraints, the problem becomes tractable without demanding prohibitive computational efforts.

The system investigated consists of a 1s electron bound to a bare probe ion nucleus with charge Z as well as a 3d electron attached to an Fe atom core with effective charge $Z_{Fe}^{eff}$. The time evolution of this system is governed by the time dependent Schrodinger equation:

$$(H_{el} - i\frac{\partial}{\partial t})\, \Psi_T(\mathbf{r}_1, \mathbf{r}_2, \mathbf{R}, t) = 0 \qquad (6)$$

where the trial function $\Psi_T$ depends on electron coordinates $\mathbf{r}_1$, $\mathbf{r}_2$ while $\mathbf{R}$ is the vector of internuclear separation. The electronic Hamiltonian $H_{el}$ is given by:

$$H_{el} = H_1 + H_2 + H_{int} \qquad (7)$$

The operator $H_1$ represents the Hamiltonian of an electron of a free hydrogen-like ion, while $H_2$ describes the Hamiltonian of a 3d electron of the Fe atom. The Hamiltonian $H_{int}$ which describes the Coulomb interaction between target atom and probe ion is given by

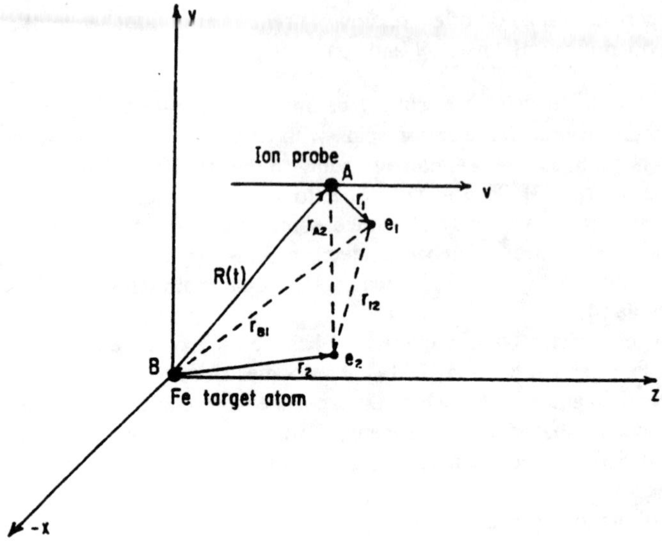


**Fig.2:** Geometry of the two center interaction between a hydrogen like ion probe ("center A") and a Fe target atom ("center B").

$$H_{int} = -Z/r_{A2} -. Z^{eff}{}_{Fe}/r_{B1} + 1/r_{12} + Z\,Z^{eff}{}_{Fe}/R \qquad (8)$$

The potentials $Z/r_{A2}$ and $Z_{eff}{}^{FE}/r_{B1}$ refer to the Coulomb attraction of the 3d target electron by the screened projectile ("center A") and of the 1s projectile electron by the core of the Fe atom ("center B"). As illustrated in Fig.2, $r_{A2}$ is the distance between the ion nucleus and the 3d electron attached to the Fe core, $r_{B1}$ is defined correspondingly. The third term in (8) denotes the Coulomb interaction between the two electrons involved , their distance being given by $r_{12}$. The fourth term refers to the repulsion between the Fe core characterized by the effective charge $Z^{eff}{}_{Fe}$ and the nuclear charge of the ion, Z.

As a trial function $\Psi_T$ we choose:

$$\Psi_T(\mathbf{r_1}, \mathbf{r_2}, \mathbf{R}, t) = c_1(t)\, \Phi^A{}_{1s}(\mathbf{r_1})\; \Phi^B{}_{3d}(\mathbf{r_2}) + c_2(t)\; \Phi^A{}_{1s}(\mathbf{r_2})\; \Phi^B{}_{3d}(\mathbf{r_1}) \qquad (9)$$

This function is of the Heitler-London type [19], i.e. we neglect the "ionic solutions" which involve the capture of a 3d electron from the target atom into the 1s orbital of the projectile or vice versa. This choice is justified by the extremely small likelihood of the capture processes in question, the corresponding cross sections being in the order of $\sigma^{cap} \cong 10^{-21}$ [cm$^2$] and lower [20].

The exchange cross section $\sigma^{sf}$ is obtained by an integration of the spin exchange probability $P^{sf}(p,v)$ over both the impact parameter and the azimuthal angle $\varphi$:

$$\sigma^{sf} = \int dp \; d\varphi \; p \; P^{sf}(p, v) \qquad (10)$$

For computational details on the derivation of $P^{sf}(p,v)$, see Ref. 16.

A comparison between measured and calculated degrees of 1s polarization for various ion species with $6 \leq Z \leq 16$ in Fe is made in Table I. The theoretical values quoted refer to an ion velocity of $v = Z v_o$, i.e. the 1s Bohr velocity. At velocities $v \geq Zv_o$ the ion probes are predominantly in a hydrogen-like state. Since the spin exchange formalism outlined above is derived under the assumption of a hydrogen-like situation, it should apply best in the range of velocities equal to or greater than $Zv_o$. It must be noted, that a number of the experimental velocities indicated in Table I are somewhat lower than the 1s Bohr velocity, which can correspond to a more complex electron configuration of the probe ions than the hydrogen-like one. However, as has been already pointed out, no velocity dependence of polarization effects has been found empirically (see Fig.1) in the region $(Z/2)\, v_o \leq v \leq Zv_o$ for ions traversing iron foils. This observation justifies a comparison between theoretical values referring to the 1s Bohr velocity and experimental data taken at velocities slightly lower than $v = Zv_o$.

A successful theory of the polarization exchange mechanism has to explain in the first place the order of magnitude of the observed polarization effects. Two other significant points of comparison between experimental and theoretical degrees of 1s polarization are their trends with ion velocity and atomic number.

| **Ion sort** | $(v/v_o)^{exp}$ | $\sigma^q$ [$10^{-17}$ cm$^2$] | $\sigma^{sf}$[$10^{-18}$ cm$^2$] | $p_{1s}^{cal}$ | $p_{1s}^{exp}$ |
|---|---|---|---|---|---|
| C | 6 | 3.2(11) | 2.51 | 0.14(6) | 0.28(5) |
| O | 6.9 | 2.8(6) | 2.35 | 0.15(3) | 0.14(2) |
| Ne | 6.8 | 1.9(7) | 1.79 | 0.17(7) | 0.25(5) |
| Mg | 8.7 | 0.8(3) | 1.27 | 0.25(5) | 0.14(1) |
| Si | 10.8 | 0.4(1) | 0.89 | 0.31(4) | 0.19(6) |
| S | 7.8 | 0.4(1) | 0.62 | 0.25(3) | 0.16(3) |

**Table 1:** Comparison of experimental and calculated degrees of 1s polarization. The theoretical quantities are evaluated at an ion velocity $v = Zv_o$.

Turning first to the magnitudes of calculated and measured values, one can notice in Table I that our calculations yield spin exchange cross sections in the order of $\sigma^{sf} \approx 10^{-18}$ [cm$^2$] which lead to degrees of polarization in the range of 15% to 30%, i.e. the magnitude found in experiment. The calculated degrees of polarization are quoted with error limits which stem from our use of measured capture and loss cross sections in the evaluation of $p_{1s}^{cal}$. The error limits of these quantities thus reflect typical uncertainties of experimental values of $\sigma^{cap}$ and $\sigma^{loss}$. Within these uncertainties, there is an overlap of experimental and theoretical findings in a number of cases. For the heavier projectiles

Mg, Si and S, however, the calculation overestimates the experimental results by factors of 1.5 to 1.8. This feature might be ascribed to the high charges of the respective ion probes which cause excitation and ionization effects in the target atom and therefore could diminish the polarization transfer efficiency. A more refined model based on the approach presented in this work will have to examine the impact of these effects on the spin exchange process.

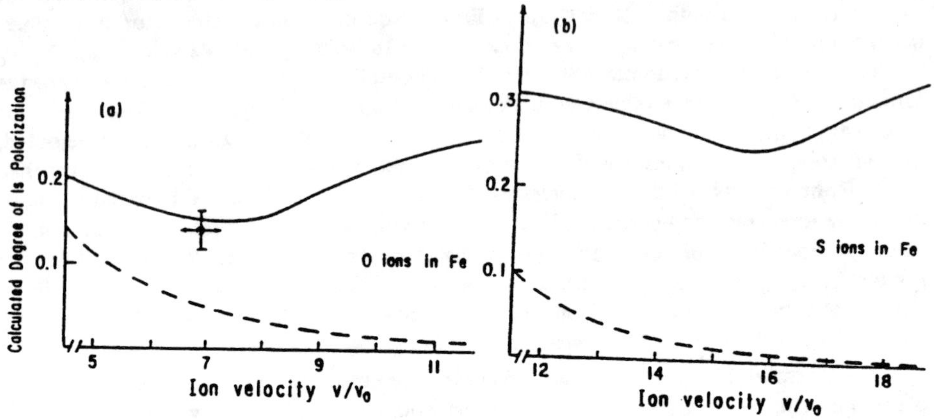


**Fig.3:** Calculated degrees of 1s polarization for O ions (a) and S ions (b) both traversing Fe are shown as functions of ion velocity in the regime $v \approx Zv_0$. Solid lines: results based on the approach presented in this work. Dashed lines: calculations using the model introduced in Ref.6. An experimental degree of polarization for fast O ions in Fe is shown also.

As to the velocity dependence of the degree of 1s polarization, a direct comparison of our model with experimental data is difficult to perform in the present state. The low velocity data indicated in Fig.1 refer to velocities $v \approx (Z/2)\,v_0$, i.e. to a region where the model does not apply since the hydrogen-like electron configuration is not realized. The velocity dependence predicted for the region $v \approx Zv_0$ is plotted in Fig.3 for the case of O and S ions. As can be seen from this figure, the calculated degrees of polarization pass through a minimum close to the 1s Bohr velocity. This observation is explained by the fact that for $v \leq Zv_0$, the polarization quenching cross section $\sigma^q$ decreases more strongly with increasing velocity than the spin exchange cross section $\sigma^{sf}$, while for $v \geq Zv_0$, $\sigma^q$ decreases more rapidly than $\sigma^{sf}$. This velocity behavior is a particularly interesting aspect of the model discussed here since any theoretical approach proposed formerly to account for the polarization transfer implies a markedly decreasing trend with velocity, which is in contradiction to experiment. While all data obtained so far for ions traversing iron foils suggest velocity independence of the degree of 1s polarization, a rise of this quantity with velocity has been observed in Gd host (see Fig.1). Especially, for Ne ions in Gd at $v = 12.5v_0$, measurement yielded a degree of polarization $p_{1s} = 0.32(13)$ [21] which is distinctly larger than the value $p_{1s} = 0.25(5)$ found experimentally at $v = 6.8\,v_0$. It would thus be interesting to test the validity of our present model by application to the computationally more complicated case of Gd as host material.

Already, it can be said that the very moderate velocity dependence implied by the approach presented here seems to be much more compatible with the experimental observations than the strongly decreasing velocity dependence found in all foregoing models. In Fig.3, comparison is made between calculations based on our present theory and values resulting from the approach introduced in Ref. 6 , where spin exchange between bound projectile electrons and quasi-free host electrons is proposed as the mechanism of polarization transfer. According to the prediction of this latter model, the 1s degree of polarization decreases monotonically as the velocity increases and attains extremely small values $p_{1s} < 0.01$ for $v > Zv_0$, as can be seen in the figure. Obviously, this behavior is not confirmed by trends and magnitudes of the experimental data.

For a more detailed comparison with the data and especially for a more quantitative understanding of the trends they show as a function of the projectile atomic number and velocity, one has to generalize the theoretical approach introduced above to apply also to the situation of velocities smaller than the 1s Bohr velocity. Thus, the model has to include the case of projectile electron configurations more complicated than the hydrogen-like one on which we concentrate in this work. Also, a more refined treatment of the electronic structure of the host target atom will contribute to make our model more quantitative. In particular, the assumption made here that the exchange process involves on the side of the target atom a single electron which does not interact with other valence electrons of the atomic shell has to be replaced by a more accurate approach which takes the multiplet nature of the polarized electron states properly into account during the spin exchange process. As a further step of generalization, one should examine to what extent the polarization transfer efficiency is influenced by target atom excitation and ionization due to the highly charged projectile. Incorporation of these features into the presently existing theory will allow one to decide if the nature of the observed effects can all be explained satisfactorily on the basis of an ion-atom interaction model.

## References


[1] P.Borchers, J.D.Bronson, D.E.Murnick and L.Grodzins, Phys.Rev.Lett.17, 1900 (1966).

[2] N.Benczer - Koller et al., Ann.Rev.Nucl.Sci.30, 53 (1980).

[3] G.J.Kumbartzki, K.Hagemeyer, W.Knaur, G.Krosing, R.Kuhnen, V.Mertens and K.-H.Speidel, Hyp.Int.7, 253 (1979).

[4] N.Benczer-Koller, G.Lenner, R.Tanczyn, A.Pakov, G.Kumbartzki, A.Pique, D.Barker, D.Berdichevsky and L.Zamick, Phys.Rev.C40, 77 (1989).

[5] M.Hass, N.Benczer-Koller, G.Kumbartzki, T.Lauritsen, T.L.Khoo, I.Ahmad, M.P.Carpenter, R.V.F.Janssens, E.F.Moore, F.L.H.Wolfs, Ph.Benet and K.Beard, Phys.Rev.C44, 1397 (1991)

[6] F.Hagelberg, K.-H.Speidel, P.N.Tandon, Z.Phys.D17, 17 (1990).

[7] A.Temkin and J.C.Lamkin, Phys.Rev.121, 788 (1961).

[8] W.Brandt, in: Atomic Physics in Nuclear Experiments, Ann.Israel Phys.Soc.1, 441 (1977).

[9] K.-H.Speidel, Hyp.Int., in press.
[10] M.B.Stearns, Phys.Rev.B8, 4383 (1973)....
[11] K.-H.Speidel, F.Hagelberg, M.Knopp, W.Trolenberg, W.Neuburger, J.Gerber, S.S.Hanna, H.Dekhissi, P.N.Tandon, Z.Phys.D1, 363 (1989).
[12] T.P.Das, Relativistic Quantum Theory of Electrons, Harper & Row Publishers, New York (1973); T.P.Das, Hyp.Int.3, 149 (1987).
[13] K.Dybdal and N.Rud, Physica Scripta 34, 561 (1987).
[14] L.C.Tribedi, K.G.Prasad and P.N.Tandon, submitted for publication in: Z.Phys.D
[15] H.Frauenfelder and R.M.Steffen, in: "Alpha, Beta and Gamma-Ray Spectroscopy", E.Karlsson, E.Mathias and K.Siegbahn ed., North Holland Publishing Co., Amsterdam 1964.
[16] F.Hagelberg, T.P.Das, K.-H.Speidel, accepted for publication in Phys.Rev.C.
[17] M.B.Stearns in: Landolt-Bornstein, New Series, Magnetic Properties of Matter, Subvolume a, Springer Verlag (1986).
[18] M.R.C. McDowell and J.P.Coleman, "Introduction to the Theory of Ion-Atom Collisions", North Holland Publishing Company; Amsterdam, London (1970).
[19] H.Eyring, J.Walter, G.E.Kimball, "Quantum Chemistry", New York, J.Wiley & Sons (1944).
[20] G.Lapicki and F.D.McDaniel, Phys.RevA22, 1896 (1980).
[21] K.-H.Speidel, J.Cub, U.Reuter, F.Passek, H.-J.Wollersheim, N.Martin, P.Egelhof, H.Emling, W.Henning, R.S.Simon, R.Schmidt, H.-J.Simonis, N.Gollwitzer, Z.Phys.A339, 265 (1991).

Local Order
in Condensed
Matter Physics


ISBN 1-56072-220-7

# ELECTRON-PHONON INTERACTION, PHONON RESPONSE AND RAMAN SCATTERING IN A CORRELATED SUPERCONDUCTOR[1]

Surjyo N. Behera[2] and Peter Entel

*Theoretische Tieftemperaturphysik*
*Universität-GH-Duisburg, 47048 Duisburg, Germany*
*and*

Debanand Sa[3]

*Institute of Physics, Bhubaneswar-751 005, Orissa, India*

**Abstract**

**A brief summary of some of the properties of the copper oxide superconductors is presented with emphasis on the anomalous behaviour of the Raman scattering data. The temperature dependence of the Raman scattering cross section is calculated from first principles assuming that the normal state of these systems is a correlated metal and that superconductivity is described by the correlation bag model. It was demonstrated that the correlated metallic state exhibits the marginal Fermi liquid behaviour. A new mechanism of charge carrier-phonon interaction involving the decay of the phonon into a quasiparticle pair, due to the breaking of the local bonds in the normal state, is postulated and the phonon response of the system is calculated. The temperature dependence of the phonon frequencies and their widths arising due to their self energies are shown to be strongly influenced by this new mechanism of interaction. The calculated Raman spectrum reproduces the observed constant background intensity due to scattering by charge carriers. It also predicts the softening of the phonon frequencies and broadening of their peaks in the normal state, in qualitative agreement with the recent measurements.**

---

[1]**Transcript of the talk presented by one of us (S.N.B.) at the *International Symposium on Local Order in Condensed Matter Physics* held at Jekyll Island, Georgia, U.S.A. during June 14-17, 1993.**

[2]**Permanent address: Institute of Physics, Bhubaneswar-751 005, Orissa, India**

[3]**e-mail: debanand@iopb.ernet.in**

## 1. INTRODUCTION

Since the discovery of high temperature superconductivity by Bednorz and Müller [1] in $La_{2-x}Ba_xCuO_4$ (the so called 214 system with a transition temperature ($T_c$) of $\sim$ 35 K), several other families of copper oxides and other systems have been discovered, and the transition temperature has been raised to a highest of 134 for the system $HgBa_2Ca_2Cu_3O_{8+x}$ (Hg-1223). To make a brief mention of these systems, among the copper oxides, there are the 123 system $YBa_2Cu_3O_{7-\delta}$ with a $T_c$ of 90 K, the Bismuth 2212 and 2223 systems $Bi_2Sr_2CaCu_2O_8$, $Bi_2Sr_2Ca_2Cu_3O_{10}$ with $T_c$'s of 80 and 110 K and finally the families of thallium 2223 i.e. $Tl_2Ca_2Ba_2$ $Cu_3O_{10}$ with the $T_c$ of 125 K which were discovered in quick succession. All the above mentioned cuprates are grouped as the hole-doped systems as the charge carriers in these are holes. The discovery of these cuprate supercondocturs broke the liquid nitrogen barrier, since they have $T_c$'s above the liquid nitrogen temperature. This development was followed by the discovery of yet another system $(Nd_{1-x}Ce_x)_2CuO_4$ with a $T_c$ of 30 K, which is an electron doped superconductor. For some time it was thought that the presence of copper in the system was a prerequisite for producing a high $T_c$ superconductor, till superconductivity was found in the system $Ba_{1-x}K_xBiO_3$ with a $T_c$ of $\sim$30 K. Obviously this is a non-copper system and its properties differ considerably from those of the cuprates. The latest addition to the list of high temperature superconductors are the so called bucky ball superconductors or the alkali doped fullerides with $T_c$'s of 18 K for $K_3C_{60}$, 28 K for $Rb_3C_{60}$ and 33 K for $Cs_2RbC_{60}$. Again these systems have very different properties than the cuprates. Here we shall be concerened only with the cuprates.

The basic properties of all the cuprates can be briefly summarized by going through their phase diagrams, which depict the different phases of the material as a function of the dopant concentration (or equivalently the oxygen content). Since all the cuprates have similar phase diagrams, we discuss only the one for the 214 system which is depicted in Fig. 1. In the figure the different transition temperatures are plotted as a function of the dopant concentration $x$. The pure material $La_2CuO_4$ ($x = 0$) undergoes a structural change from the tetragonal to the orthorhombic phase at a rather high temperature. On lowering the temperature it becomes an antiferromagnet below the Néel temperature ($T_N$) of 350 K; and it is an insulator. At the outset it is even difficult to understand why the pure material is an insulator. On matching the valence of the stoichiometric compound it is clear that copper exists in the $Cu^{2+}$ state and has an atomic structure of $3d^9$. Since the $d$-shell can accomodate 10 electrons, there is a partialy filled $d$ level in the system, and this will give rise to partially filled band when the solid is formed. Therefore, band theory predicts this system to be a metal. Even if we try to refine the picture by taking into account the crystal field splitting of the $d$ level and so on, one still ends up with a half filled valence band, which should be a metal. The only way that the system could become an insulator is by invoking strong electron correlations which will produce antiferromagnetism (AF) in the system, and thereby split the half-filled band into two bands due to the doubling of the periodicity. This suggests that we have a strongly correlated system at hand, which was first recognized by Anderson [2]. On doping the system, the Néel temperature decreases very rapidly, and the long range AF order vanishes, when the dopant concentration is of the order of an atomic per cent. On further doping there results a disordered magnetic state, sometimes denoted the spin-glass state, and the system continues to be an insulator. Of course doping creates holes which are the charge carriers in the insulator, simply because the divalent alkaline earth metal atom (Ba or Sr) replaces the trivalent La atom. When the dopant concentration increases to about 5 at % the system undergoes a transition from the insulating to the metallic state and on lowering the temperature superconductivity appears in the system. The insulator-metal transition is again believed to be a characteristic feature associated with strongly correlated charge carriers in the system. The SC transition temperature at first increases with $x$, reaches a saturation, then decreases, finally vanishing for $x \approx 0.3$ as can be seen from Fig. 1. Beyond this dopant concentration, the system is known to be a good Fermi liquid like metal. On the other hand the range of the dopant concentratin $0.05 \leq x \leq 0.3$, where SC has been observed, the normal state corresponds to a bad metal. To distinguish between the good and the bad metallic states, the behaviour of the resistance as a function of the temperature in the three different regions marked 1, 2, 3 in the phase diagram is shown schematically in the inset of fig. 1. In the good

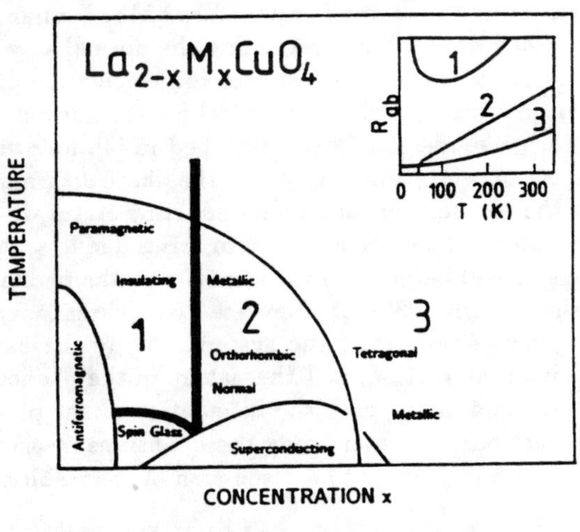


Figure 1: The schematic phase diagram for the $La_{2-x}Sr_xCuO_4$ system. The transition temperatures are plotted as function of the dopant concentration ($x$). The inset shows the temperature dependence of the resistance in the three regions marked 1, 2 and 3. For a discussion see the text.

metallic region (3) the resistance shows the expected linear temperature dependence at high temperatures due to scattering by phonons, the temperature independent behaviour at low temperatures corresponding to the residual resistance and $T^2$ dependence in the intermediate region corresponding to the scattering by Coulomb interaction. In region (1) the temperature dependence of the resistance is typically that of a semiconductor. In contrast in the region of bad metal (2) the resistance shows a linear temperature dependence from high down to the lowest temperatures. This linear temperature dependence happens to be a characteristic feature of all the cuprate systems and is believed to be caused by a mechanism of scattering intrinsic to these systems. A proper understanding of this anomalous temperature dependence of the resistance in the metallic state is recognized to be the key feature to a proper understanding of the mechanism of superconductivity in these systems. This follows from a rather universal thumb rule that bad metals are good superconductors. In the conventional superconductors the resistance in the metallic state is caused by phonons and so also is the mechanism for superconductivity. So it is natural to believe that in the cuprates, whatever causes the resistance in the bad metallic state, should also be responsible for the mechanism of pairing of charge carriers and hence superconductivity. It is not only the resistance but there are many other properties which show anomalous behaviour in the bad metallic state. One such property is the Raman scattering cross section, a close examination of which reveals the nature of the bad metallic state.

In section 2 we discuss the interesting feature of the Raman data and point out the inferences that one can draw from it regarding the normal as well the superconducting state. An interpretation of the data requires the calculation of the Raman scattering cross section. This in turn necessitates a model for the normal as well as the superconducting states. Such a model has been developed in [3], following the evolution of the system with dopant concentration as given by the phase diagram. The essential features of the models for the normal and the superconducting states will be discussed in section 3. Since in these systems the Raman spectrum arises due to scattering by charge cariers as well as phonons; it will be important to know how the phonons couple to the charge carriers. It was shwon earlier [3] that there are two different ways in which the phonons interact with the charge carriers in the system. Again the basic mechanisms for the charge carrier-phonon interactions and the nature of the phonons in these systems will be presented in section 4, along with the calculation of the phonon response and their spectral density functions, the later being the quantities measured in a Raman experiment. Finally section 5 is devoted to a discussion of the results.

## 2. THE RAMAN DATA

Raman scattering [4] from the copper oxide superconductors has shown several unusual features which provided much usful information about the normal as well as the superconducting states. A summary of the essential features is given below:

(i) In the the normal state, a rather large but constant background intensity has been observed at all frequencies up to the highest measured frequency as can be seen from Fig. 2, which depicts the spectrum for the Bismuth (2212) [5] system. This is a general feature of all the copper oxide superconductors. This background intensity has been attributed to the scattering by the charge carriers in the system and it dictates a particular form for the frequency and temperature dependence of the polarizability of the medium as can be seen from the arguments given below: The Raman intensity is proportional to a thermal factor times the imaginary part of the polarizability at zero wave vector $\Pi(\mathbf{q}=0,\omega)$ and is given by

$$I \approx \left(1 - e^{-\beta\omega}\right)^{-1} \operatorname{Im}\Pi(\mathbf{q}=0,\omega). \tag{1}$$

The high and low temperature behaviour of the thermal factor is

$$\left(1 - e^{-\beta\omega}\right)^{-1} \approx \begin{cases} 1 & \text{for} \quad \omega \gg T \\ T/\omega & \text{for} \quad \omega \ll T \end{cases} \tag{2}$$

Since the intensity is a constant independent of frequency and temperature it follows from eqns. (1) and (2) that the imaginary part of the polarizability must have the following form

$$\operatorname{Im}\Pi(\mathbf{q}=0,\omega) = \begin{cases} S_0 & \text{for} \quad \omega \gg T \\ S_0(\omega/T) & \text{for} \quad \omega \ll T \end{cases} \tag{3}$$

This particular frequency and temperature dependence of the polarizability is the basis of the phenomenological marginal Fermi liquid theory for the cuprates proposed by Varma et al. [6]. Recently more careful analysis of the scattering intensity by charge carriers in the case of the (214) [7] and the (123) [8] systems has revealed that the intensity background persists even in the insulating as well as the highly doped good metallic states.

(ii) Usually at low frequencies the phonon peaks ride over the constant intensity background of the Raman spectrum. The number of phonon peaks seen may differ form one Cuprate to the other. In the (123) system [4] three prominent phonons having frequencies of 116, 340 and 504 $cm^{-1}$ have been observed (see Fig. 3) and identified to be associated with the out of phase vibrations of the Ba, in plane Oxygen and the apex Oxygen atoms, respectively. Some of these show the Fano line shape indicating a strong coupling to the charge carriers in the system.

(iii) On lowering the temperature below the superconducting transition temperature the 116 $cm^{-1}$ phonon sharpens while the 340 $cm^{-1}$ phonon shows anomalous softening while its width increases. All these features again indicate a strong coupling to the charge carriers in the superconducting state [9].

(iv) In the superconducting state the constant background intensity is suppressed at low frequencies while it develops a broad peak with its maximum around 490 $cm^{-1}$. The suppression of the intensity in the low frequency region is interpreted as being due to the appearance of the superconducting gap. However, the slow linear rise of the

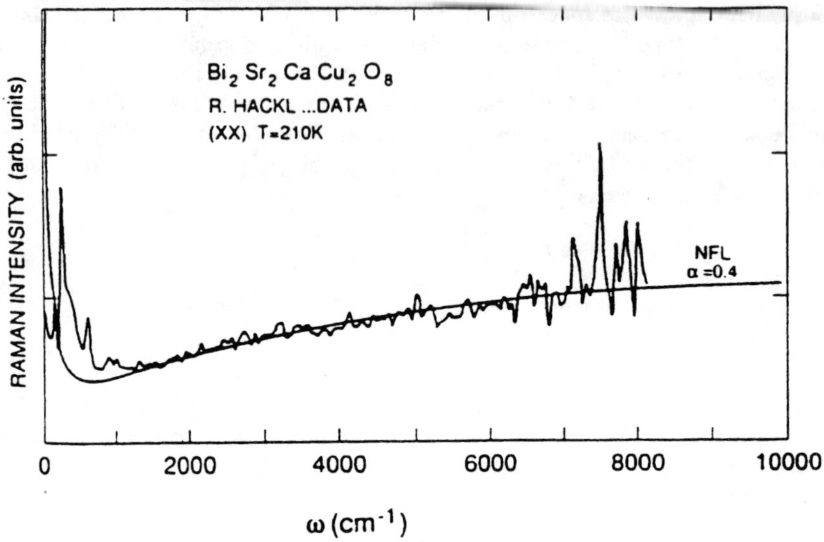


Figure 2: Raman spectrum for $Bi_2Sr_2CaCu_2O_8$ at 210 K, taken from Ref. [5].

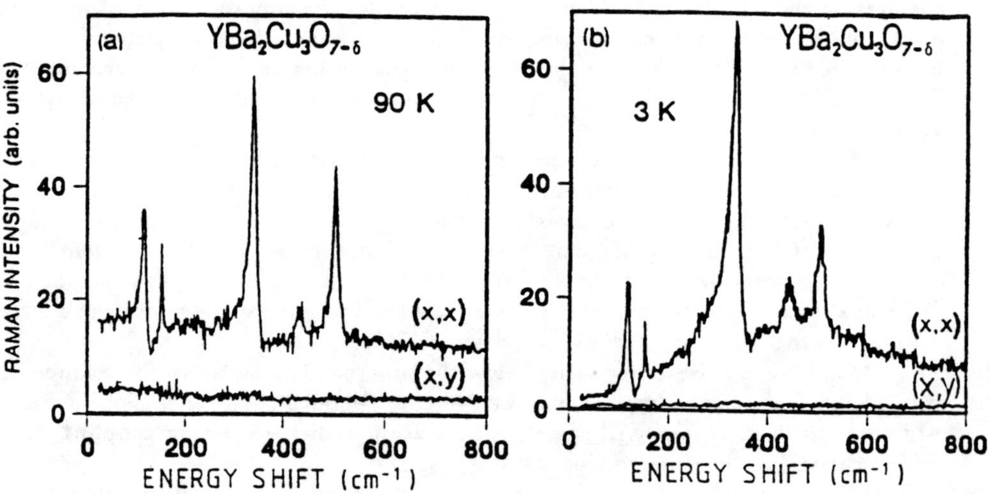


Figure 3: Raman spectrum of $YBa_2Cu_3O_{7-\delta}$, (a) at 90 K and (b) at 3 K from Ref. [4].

intensity in this region also indicates the presence of a finite density of states inside the superconducting gap.

(v) The broad peaking of the background intensity in the SC state is a puzzling feature. While some peaking is expected from the redistribution of the density of states in the vicinity of the SC gap, the occurrence of the peak maximum at a rather high frequency and its invariance with samples of varying $T_c$ point to the fact that it may be a normal state feature.

(vi) Recently the shift in the phonon frequency and change in their width has been studied more carefully as a function of the oxygen content of the sample [10]; which shows the softening of the high frequency mode at a temperature much higher than $T_c$ in a low $T_c$ sample with low oxygen content. This is indicative of the presence of a gap even in the normal state of the system. A theoretical analysis [11] of all these puzzling features will throw light on the nature of the normal as well as the SC states of the cuprates.

In a metal normally Raman scattering proceeds as follows: The incident photon excites an electron-hole pair, which finally recombines to emit the scattered photon. Simultaneously, the electron or hole can emit or absorb a phonon because of the electron-phonon interaction as depicted in Fig. 4a [12]. However, only the zero momentum ($\mathbf{q} = 0$) phonons will take part in the process because of the conservation of momentum. The $\mathbf{q} = 0$ phonon can further acquire a self energy due to its coupling to the quasiparticles of the ground state of the system as shown in Fig. 4b. Thus the phonon response of the system is capable of providing informaton about the ground state, because of which Raman scattering can be successfully used to probe both the normal as well as the SC states as demonstrated in 1980 [13] for the first time. The Raman intensity is proportional to the sum the imaginary parts of the polarizabilities and the phonon spectral density functions, and is given by

$$I \approx \left(1 - e^{-\beta\omega}\right)^{-1} \left[ \sum_i \operatorname{Im} \chi^i(\omega) + \sum_i \operatorname{Im} \mathcal{D}^i(0,\omega) \right] , \tag{4}$$

where $\chi^i(\omega)$'s are the different response functions of the system and the phonon spectral density functions are the imaginary parts of the phonon Green's functions $\mathcal{D}^i(0,\omega)$. Obviously the evaluation of the response functions $\chi^i(\omega)$ as well as the phonon self energy in the normal and the SC state requires proper models for these states.

## 3. THE CORRELATED METALLIC STATE AND SUPERCONDUCTIVITY

The metallic state in the cuprates evolves from that of the AF-insulator on increasing the dopant concentration. It was pointed out in the introduction that the properties of the pure system can be understood only when the strong electron correlations are taken into account. Since these are highly anisotropic systems with the action taking place

In the $CuO_2$ planes a realistic model for the cuprates should involve the orbitals of one Cu and two O-atoms which constitute the unit cell in this plane. The Cu $d$-orbitals are expected to exhibit strong correlations. If one takes only those orbitals which lie in the $CuO_2$ plane it will lead to a three or a five band model [14]. However, for simplcity we consider only the one band Hubbard model, as the one appropriate to take care of strong correlations in the cuprates. The Hamiltonian for the Hubbard model is [15]

$$\mathcal{H} = \sum_{ij,\sigma} t_{ij} c^{\dagger}_{i\sigma} c_{j\sigma} + U \sum_{i} n_{i\uparrow} n_{i\downarrow} , \tag{5}$$

where the symbols have their usual meaning. In the limit of the intra-atomic Coulomb repulsion $U \rightarrow \infty$, the Hubbard Hamiltonian transforms into the Heisenberg antiferromagnetic exchange interaction [16], given by

$$\mathcal{H} = J \sum_{\langle ij \rangle} \left( \mathbf{S}_i \mathbf{S}_j - \frac{1}{4} n_i n_j \right) , \tag{6}$$

where the kinetic exchange integral $J = 2t^2/U$. This Hamiltonian describes the AF-insulating ground state of the pure system. On doping, the long range AF order disappears, giving way to a disordered magnetic state, which could be the insulating resonating valence bond (RVB) state described by the Hamiltonian [17]

$$\mathcal{H} = -J \sum_{\langle ij \rangle} b^{\dagger}_{ij} b_{ij} , \tag{7}$$

where

$$b^{\dagger}_{ij} = \frac{1}{\sqrt{2}} \left( c^{\dagger}_{i\uparrow} c^{\dagger}_{j\downarrow} - c^{\dagger}_{i\downarrow} c^{\dagger}_{j\uparrow} \right) \tag{8}$$

is the valence bond operator. This is a spin liquid like state consisting of nearest neighbour spin pairs. In the mean field approximation the Hamiltonian of eqn. (7) can be diagonalized and be described by a non-vanishing value of the valence bond order parameter $\Delta = \langle b^{\dagger}_{ij} \rangle$. At this stage because of the presence of a small number of holes in the system there can be charge and spin seperation and the elementary excitations will be the holons and the spinons [18]. However, because of the disorder the states are localized and there is no possibility for the charge transport in the system. On further increasing the dopant concentration the RVB metallic state results, which is described by the mean field Hamiltonian

$$\mathcal{H}_0 = -t' \sum_{ij,\sigma} c^{\dagger}_{i\sigma} c_{j\sigma} - J\Delta \sum_{\langle i,j \rangle} (b_{ij} + b^{\dagger}_{ij}) , \tag{9}$$

$t'$ being a modified hopping integral proportional to the dopant concentration. The presence of mobile holes in this state will self consistently generate strong fluctuations of the valence bond because of the possibility of virtually breaking and making the bonds as the holes move. Therefore, the holes are expected to couple strongly with the fluctuations over the mean field. The fluctuations in the valence bond operators can be written as

$$b_{ij}^{\dagger} = \Delta + \tilde{b}_{ij}^{\dagger}, \tag{10}$$

where $\tilde{b}_{ij}^{\dagger}$ is the fluctuating part of the operator. The interaction Hamiltonian is obtained by substituting eqn. (10) into eqn. (7) and is given by

$$\mathcal{H}_I = -J\sum_{\langle ij\rangle}\left(b_{ij}^{\dagger}\tilde{b}_{ij} + \tilde{b}_{ij}^{\dagger}b_{ij}\right). \tag{11}$$

The quanta of the fluctuations over the mean field are just the collective modes of the system describing the oscillations of the amplitude and the phase of the order parameter. Hence the operators corresponding to the fluctuations can be decomposed into bosonic anihilation and creation operators while the original valence bond operators are expressed in terms of the hole anihilation and creation opeators using eqn. (8). Then the interaction Hamiltonian of eqn. (11) describes the interaction of quasiparticle pairs with the bosonic degrees of freedom. This interaction given by eqn. (11) together with the mean field Hamiltonian of eqn. (9) will describe the *correlated metallic* state as was first pointed out in [3]. The other characteristics of the correlated metal are the following: (i) The presence of valence bonds in the metallic state opens up a gap over regions of the Fermi surface. With increasing dopant concentration the gap region of the Fermi surface decreases. (ii) The elementary excitations of the system are regular Fermions carrying both spin and charge. (iii) The interaction of eqn. (11) will renormalize the quasiparticles of the system, which will acquire self energy due to the fluctuations. The propagator for the quanta of the fluctations is the pair susceptibility function. This latter quantity when calculated using the mean field Hamiltonian (9) is given by [11]

$$\chi^{p-p}(0,\omega) = \sum_{\mathbf{k}} \frac{4(\varepsilon_{\mathbf{k}} - \mu)^2}{E_{\mathbf{k}}} \frac{\tanh(\beta E_{\mathbf{k}}/2)}{(\omega^2 - 4E_{\mathbf{k}}^2)}, \tag{12}$$

where

$$E_{\mathbf{k}} = \left[(\varepsilon_{\mathbf{k}} - \mu)^2 + \Delta_R^2\right]^{1/2}. \tag{13}$$

$\Delta_R = -J\Delta$ being the gap parameter of the metallic state. In the region of $(\omega/2\Delta_R)^2 > 1$, the frequency and temperature dependence of the imaginary part of $\chi^{p-p}(0,\omega)$ is given by

$$\operatorname{Im}\chi^{p-p}(0,\omega) = -\frac{\pi N(0)}{4}\frac{[(\omega/2\Delta_R)^2 - 1]^{1/2}}{(\omega/2\Delta_R)}\begin{cases}(\omega/4T) & \text{for} \quad \omega < 4T\\ 1 & \text{for} \quad \omega > 4T\end{cases} \tag{14}$$

where $N(0)$ is the density of states at the Fermi energy. In the limit of vanishing normal state gap ($\Delta_R \rightarrow 0$) eqn. (14) reduces to

$$\mathrm{Im}\,\chi^{p-p}(0,\omega) = -\frac{\pi N(0)}{4}\begin{cases} (\omega/4T) & \text{for} \quad \omega < 4T \\ 1 & \text{for} \quad \omega > 4T \end{cases} \tag{15}$$

which is identical in form to eqn.(3) and is the Varma-polarizability. Therefore, the fluctuations over the mean field renormalize the quasiparticles so that the marginal Fermi liquid behaviour is generated. Because of this the theory of the correlated metallic state is capable of explaining many of the anomalous properties of the normal state of the cuprates. However, it should be pointed out that the marginal Fermi liquid like behaviour is also generated in the nested Fermi liquid theory [19] as well as in theories with strong electron-phonon coupling [20] and those with Fermi liquids involving two-dimensional plasmon exchange [21].

The density response function (particle-hole polarizability) for the correlated metallic state can also be calculated [3,11]. This will contribute to the phonon response as well as to the scattering by the charge carriers. In the case of a normal Fermi liquid like metal, the density response function is given by the Lindhard function. For the calculation of the Raman intensity we have to evaluate the imaginary part of the Lindhard function at $\mathbf{q} = 0$; and this quantity simply goes to zero. Therefore, there is no scattering by charge carriers in normal metals. However, in the model of the correlated metallic state, there will be a finite contribution to the density response function at $\mathbf{q} = 0$, because of the presence of local pairs in the system. This contribution will be proprtional to the gap ($\Delta_R$) in the normal state and is given by [3,11]

$$\chi^{p-h}(0,\omega) = \Delta_R^2 \sum_{\mathbf{k}} \frac{\tanh(\beta E_{\mathbf{k}}/2)}{E_{\mathbf{k}}(\omega^2 - 4E_{\mathbf{k}}^2)} . \tag{16}$$

The imaginary part of $\chi^{p-h}(0,\omega)$ will contribute to the light scattering by charge carriers. The temperature dependence of this quantity for different ranges of frequency are given below. For $(\omega/2\Delta_R)^2 > 1$

$$\mathrm{Im}\,\chi^{p-h}(0,\omega) = -\frac{\pi N(0)}{4}\frac{1}{(\omega/2\Delta_R)}\frac{1}{[(\omega/2\Delta_R)^2 - 1]^{1/2}}\begin{cases} (\omega/4T) & \text{for} \quad \omega < 4T \\ 1 & \text{for} \quad \omega > 4T \end{cases} \tag{17}$$

and for $(\omega/2\Delta_R)^2 < 1$

$$\mathrm{Im}\,\chi^{p-h}(0,\omega) = \frac{N(0)}{4}\frac{1}{[1-(\omega/2\Delta_R)^2]^{1/2}} \cdot \ln\left|\frac{[1-(\omega/2\Delta_R)^2]^{1/2} - 1}{[1-(\omega/2\Delta_R)^2]^{1/2} + 1}\right| \begin{cases} (2\Delta_R/4T) & \text{for} \quad \omega < 4T \\ 1 & \text{for} \quad \omega > 4T \end{cases} \tag{18}$$

Unlike the pair polarizability (eqn. (14)), the imaginary part of the density response function vanishes in the limit of finite $\Delta_R$, there will be a contribution to the Raman intensity due to charge carrier scattering coming from the density response function. Besides eqn. (16) will also contribute to the phonon response.

Within this model of the correlated metallic state superconducting pairing of the quasiparticles is generated by an effective interaction mediated by the exchange of the collective modes of the system; as has been shown in detail in [3]. Such a model for superconductivity is called the correlation bag model which is essentially dictated by the thumb rule mentioned in the introduction. The particle-hole and the particle-particle (or pair) polarizabilities in the superconducting state can also be calculated [3,11] to compute the Raman intensities.

## 4. THE CHARGE CARRIER-PHONON INTERACTION AND THE PHONON RESPONSE

It was mentioned in the introduction that the Raman intensities are determined by scattering due to charge carriers as well as the phonons. The phonon contribution is proportional to its spectral density function as given by eqn. (4), and its evaluation requires the calculation of the phonon self energy. The phonon response of the system will therefore depend on the nature of the charge carrier-phonon interaction as well as the ground state of the charge cariers in the system. Usually the displacements of the ions ($\mathbf{u}_i$) from the equilibrium position modulates the hopping integral $t_{ij}$ in eqn. (5) which gives rise to the normal electron-phonon interaction

$$\mathcal{H}_{c-p}^{(1)} = g \sum_{\langle i,j \rangle} c_{i\sigma}^{\dagger} c_{j\sigma} (\mathbf{u}_i - \mathbf{u}_j) , \tag{19}$$

where a phonon decays by creating an electron-hole pair as shown in Fig. 5a. The phonon self energy due to this process comes from the density response function given by eqn. (16). On the other hand it was argued in [3] that the displacement of ions can modulate the kinetic exchange integral $J_{ij}$ as well, which will give rise to the AF magnon-phonon interaction in the undoped insulating system. However, on doping the system, this interaction evolves with the increase in the dopant concentration as the ground state of the system changes. In the disordered magnetic state, since the ground state is an RVB insulator, this interaction will correspond spinon-phonon interaction. On further increasing the dopant concentration one ends up in the correlated metallic state, where the charge carrier-phonon interaction due to the modulation of $J$ will essentially amount to the phonon creating or anihilating a pair as shown in Fig. 5b, and is given by

$$\mathcal{H}_{c-p}^{(2)} = g' \Delta \sum_{\langle i,j \rangle} \left[ b_{ij}^{\dagger} (\mathbf{u}_i - \mathbf{u}_j) + \text{h.c.} \right] . \tag{20}$$

The contribution of $\mathcal{H}_{c-p}^{(2)}$ to the phonon self energy, involves the pair correlation function as given by eqn. (12). In the correlated metallic state, both the mechanisms of

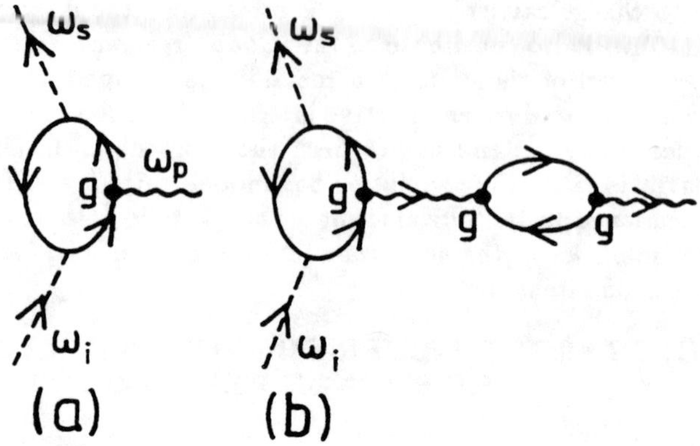


Figure 4: The process by which Raman scattering proceeds in a metal; (a) one phonon process, (b) probing the ground state through the phonon self energy.

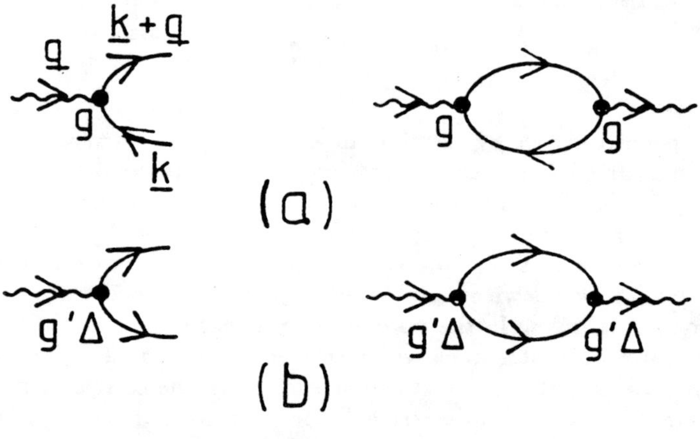


Figure 5: Process contributing to the charge carrier-phonon interaction: (a) the normal electron-phonon interaction and its contribution to phonon self energy in the lowest order, (b) the phonon interaction involving a local pair and its contribution to phonon self energy.

interactions of eqns. (19) and (20) will be in operation, since the system is described by the Hamiltonian of eqn. (9). Furthermore, it is worth pointing out that the interaction of eqn. (20) vanishes if $\Delta = 0$, i.e. if there are no local pairs present in the ground state; therefore, this interaction is characteristic of the correlated metallic state alone.

It was pointed out in [3] that there are propagating as well as localized phonons present in the system. The localized phonons again owe its origin to the evolution of the correlated metallic state from the Néel ordered state. The presence of Néel ordereing will double the periodicity of the system which will thereby produce new phonon branches due to the halving of the Brillouin zone. On doping the system, the long range order vanishes and a disordered magnetic state results, which in turn is expected to localize those phonon branches which originated due to the Néel order. This is schematically depicted in Fig. 6. It is therefore expected that the localized phonons will interact with the charge carriers through the mechanism of eqn. (20) while the propagating phonon-charge carrier interaction is dominated by eqn. (19).

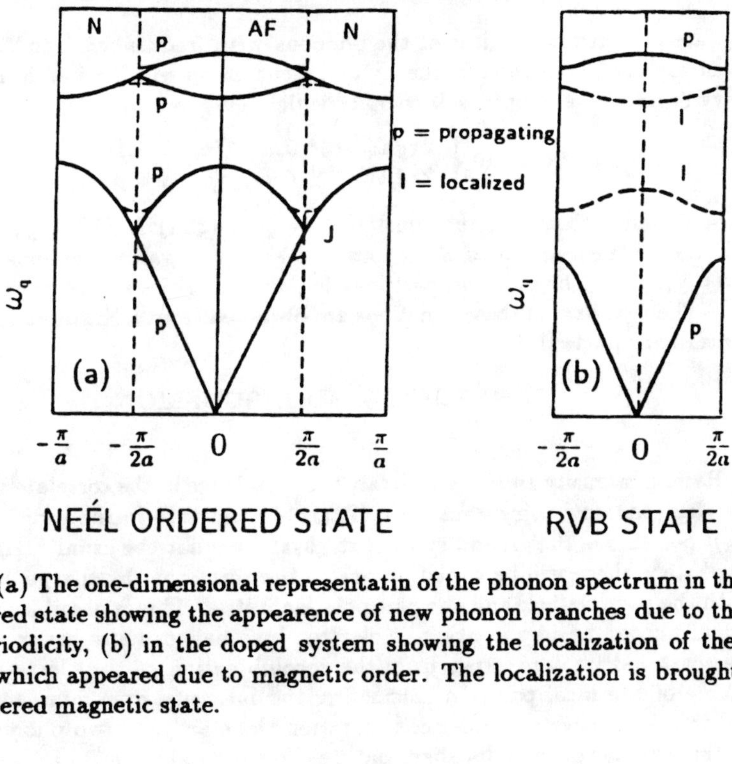


Figure 6: (a) The one-dimensional representatin of the phonon spectrum in the undoped Néel ordered state showing the appearence of new phonon branches due to the doubling of the periodicity, (b) in the doped system showing the localization of these phonon branches which appeared due to magnetic order. The localization is brought about by the disordered magnetic state.

The Raman spectrum of the system will be dominated both by the propagating and the localized phonons and requires the calculation of their propagators which are given by

$$\mathcal{D}_{\mathbf{q}}^{(i)}(\omega) = (\omega_{\mathbf{q}}^{(i)}/\pi)\left[\omega^2 - \omega_{\mathbf{q}}^{(i)2} - \Pi^{(i)}(\mathbf{q},\omega)\right]^{-1}, \tag{21}$$

where $i = l, p$ denote the localized and the propagating modes, $\Pi^{(i)}(\mathbf{q},\omega)$ being their self energies arising out of the interactions $\mathcal{H}_{c-p}^{(i)}$ with $i = 2$ and 1, respectively. From the knowledge of the polarizabilities the spectral density functions for the different phonons can be calculated, which are given by

$$\begin{aligned} \mathcal{S}^{(i)}(0,\omega) &= -2\mathrm{Im}\mathcal{D}_{\mathbf{q}=0}^{(i)}(\omega + i\eta^i) \\ &= \frac{-2\omega^i(2\omega\eta^i + \mathrm{Im}\Pi^{(i)}(0,\omega)}{\pi\left(\left[\omega^2 - \omega^{i2} - \mathrm{Re}\Pi^{(i)}(0,\omega)\right]^2 + \left[2\omega\eta^i + \mathrm{Im}\Pi^{(i)}(0,\omega)\right]^2\right)} \end{aligned} \tag{22}$$

where $\eta^i$ are the intrinsic widths of the phonons with frequency $\omega^i$ and a temperature dependent shift of the phonon frequency and a change in width arises from the real and imaginary parts of the phonon self energies which are given by

$$\Pi^{(i)}(0,\omega) = \begin{cases} 2\pi\tilde{g}_l\Omega_0^2\,\tilde{\chi}^{p-p}(0,\omega) & \text{for } i = l \\ 2\pi\tilde{g}_p\omega_0^2\,\tilde{\chi}^{p-h}(0,\omega) & \text{for } i = p \end{cases} \tag{23}$$

where the dimensionless coupling constants are $\tilde{g}_l = (g'\Delta)^2N(0)/\Omega_0$, $\tilde{g}_p = g^2N(0)/\omega_0$; $\Omega_0$ and $\omega_0$ being the frequencies of the localized and propagating phonons and $\tilde{\chi}(0,\omega) = \chi(0,\omega)/(\pi N(0)/4)$. The pair and partile-hole polarizabilities are as given by eqns. (12) and (16). The real part of these functions are obtained from a Kramers-Kronig analysis of the imaginary parts [11].

## 5. RESULTS AND DISCUSSION

The Raman intensity and its temperature dependence in the correlated metallic state can be calculated now using eqns. (4), (22), (14) and (17) in the range of frequencies $(\omega/2\Delta_R)^2 > 1$. Since the ground state of the system is not the usual Fermi liquid but a correlated state there will be contributions to the scattering by charge carriers coming from both the particle-hole and the pair polarizabilities. This is possible simply because an incident photon besides exciting an electron-hole pair can also break up a local pair and the scattered photon comes from the recombination of the electron and hole or the making of the local pair. In calculating the intensity from eqn. (4) it is further assumed that in the system under consideration there are only two phonons, one of the propagating and the other of localized nature. So the total intensity arises from the sum of the imaginary part of the two polarizabilities and the spectral density functions of the two phonons. Ther are several parameters which enter the calculation, e.g., the two dimensionless coupling constants for the phonons, $\tilde{g}_l$ and $\tilde{g}_p$, their intrinsic

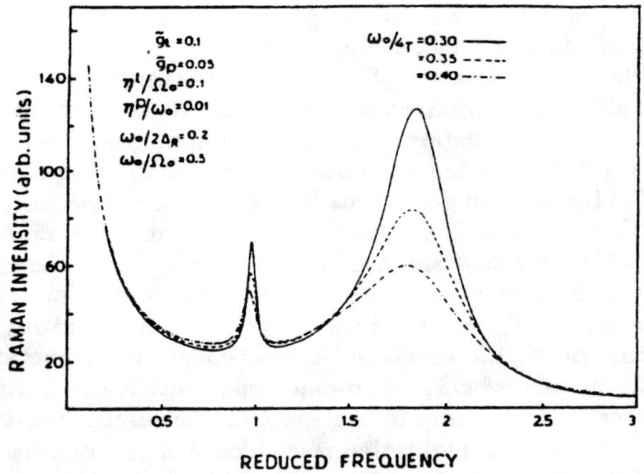


**Figure 7: Raman intensity as a function of the reduced frequency ($\omega/\omega_0$) plotted for different values of the parameter ($\omega_0/4T$) in the high temperature region $\omega > 4T$. The values of the other parameters entering the calculation are $\tilde{g}^l = 0.1$, $\eta^l/\Omega_0 = 0.1$, $\tilde{g}^p = 0.05$, $\eta^p/\omega_0 = 0.01$, $\omega_0/2\Delta_R = 0.2$ and $\omega_0/\Omega_0 = 0.5$.**

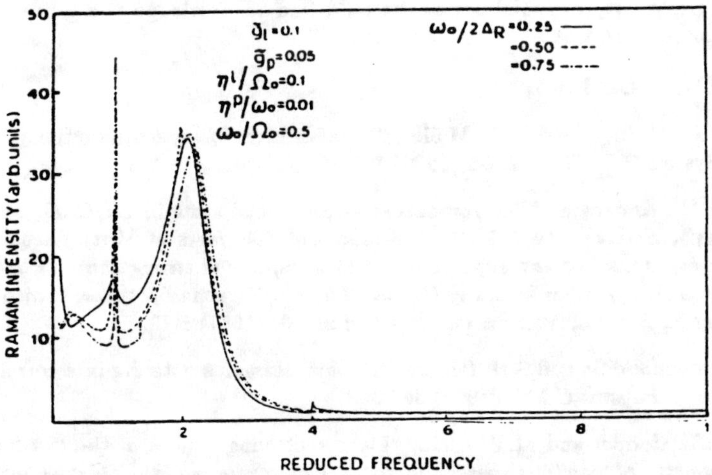


**Figure 8: Raman intensity as function of reduced frequency plotted in the low temperature region $\omega > 4T$ for different values $\omega_0/2\Delta_R$. All the other parameters are the same as in Fig. 7.**

widths $\eta^l/\Omega_0$ and $\eta^p/\omega_0$, the ratio of the phonon frequency to the normal state gap $\omega_0/2\Delta_R$; the ratio of the two phonon frequencies $\omega_0/\Omega_0$ and of course the temperature $\omega_0/4T$. Figure 7 shows the variation of the Raman intensity with temperature for high temperatures $\omega < 4T$ in the normal state, plotted as a function of the reduced frequency $\omega/\omega_0$ for a given set of parameters. The characteristic features of this plot are: (i) there is a temperature independent constant background intensity extending up to high frequencies and, (ii) the softening and broadening of both the propagating and localized phonon peaks with decreasing temperature, in the normal state. In fact, as pointed out in the introduction, recent Raman measurements [10] does show the softening and broadening of some phonons in the normal state. Figure 8 shows the same spectrum at low temperatures $\omega > 4T$, for the case of varying normal state gaps. Of course in the low temperature phase the spectrum is independent of temperature, and one can notice the constant intensity background extending up to a much higher frequency. In addition one can see a sharp jump in the spectrum corresponding to the value of the normal state gap. There is a qualitative resemblence of the calculated spectrum with the measured Raman spectrum for some cuprate superconductors. The spectrum in the superconducting state can also be calculated. The details of this calculations are presented in [11].


## ACKNOWLEDGEMENT

One of us (S.N.B.) would like to thank Prof. P. Entel for hospitality during his stay at the University of Duisburg and would like acknowledge the grant from SFB 166 for the financial support.


## REFERENCES


1. J.G. Bednorz and K.A. Müller, "Possible high $T_c$ superconductivity in the Ba-La-Cu-O System", *Z. Phys. B* **64** (1986) 189-194.

2. P.W. Anderson, "The resonating valence bond state in $La_2CuO_4$ and superconductivity", *Science* **235** (1987) 1196-1198 and "50 years of Mott phenomenon; insulators, magnets, solids and superconductors as aspects of strong coupling theory", in *Frontiers and Boarderlines in Many Particle Physics*, Varenna Lectures, Enrico Fermi School, Ed. by R. Broglia, Amsterdam, North Holland, (1987) 1-40.

3. Debanand Sa and S.N. Behera, "Phonon-Raman scattering in a correlated superconductor", *Physica C* **203** (1992) 347-368.

4. S.L. Cooper and M.V. Klein, "Light scattering studies of the low frequency excitation spectra of high $T_c$ superconductors", *Comments in Cond. Mat. Phys.* **15** (1990) 99-124 and C. Thomsen and M. Cardona, "Raman scattering in high $T_c$ superconductors", in *The Physical Properties of High Temperature Superconductors*, Vol. I, Ed. by D.M. Ginsberg, Singapore, World Scientific, (1989) 409-507.

5. T. Stauffer, R. Hackl and Müller, "Raman study of the electronic continuum in oxide superconductors", *Solid State Commun.* **75** (1990) 975-980.

6. C.M. Varma, P.B. Littlewood, S. Schmitt-Rink, E. Abrahams and A.E. Ruckenstein, "Phenomenology of the normal state of Cu-O high temperature superconductors", *Phys. Rev. Lett.* **66**, (1989) 1996-1999 and C.M. Varma, "Phenomenological constraints on theories for high temperature superconductivity", *Int. J. Mod. Phys. B* **3** (1989) 2083-2118.

7. S. Sugai, T. Ido, H. Takagi, S. Uchida, M. Sato and S. Samoto, "Electronic Raman scattering from the hole-spin composite states in $La_{2-x}Sr_xCuO_4$", *Solid State Commun.* **76** (1990) 365-370.

8. A.A. Maxsimov, A.V. Puchkov, I.I. Tartakovskii, D. Reznik, M.V. Klein, W.C. Lee and D.M. Ginsberg, "Temperature dependence of electron Raman scattering in superconducting and insulating $YBa_2Cu_3O_{6+x}$", *JETP Lett.* **56** (1992) 570-574.

9. R. Zehyer and G. Zwicknagl, "Superconductivity induced phonon self energy effects in high $T_c$ superconductors", *Z. Phys. B* **78** (1990) 175-190.

10. E. Altendorf, J.C. Irwin, R. Liang and W.N. Hardy, "The superconducting gap in pure crystals of $YBa_2Cu_3O_7$ from phonon-Raman spectra", *Solid State Commun.* **80** (1991) 627-630 and E. Altendorf, X.K. Chen, J.C. Irwin, R. Liang and W.N. Hardy "Temperature dependence of the 340, 440, and 500 $cm^{-1}$ Raman modes of $YBa_2Cu_3O_y$ for $6.7 \leq y \leq 7.0$", Simon Fraser Univsity Preprint, (1992) pp.47.

11. Debanand Sa and S.N. Behera,"Raman scattering in a correlated superconductor II: A finite temperature study", IP/BBSR/93-41, Submitted to *Physica C*,, (1993) pp. 62.

12. M.V. Klein, "Raman studies of phonon anomalies in transition metal compounds", in *Light Scattering in Solids*, Vol. 3, Ed. by M. Cardona and G. Guntherodt, Berlin, Springer, (1982) 121-172.

13. R. Sooryakumar and M.V. Klein, "Raman scattering by superconducting gap excitations and their coupling to charge density waves", *Phys. Rev. Lett.* **45** (1980) 660-662.

14. P. Entel, J. Zielinski, M. Matschke and S.G. Mishra, "Theoretical models for high temperature superconductors", in *Electronic Correlation and Disorder Effects in Metals*, Ed. by S.N. Behera, Singapore, World Scientific, (1991) 146-185 and P. Entel and J. Zielinski, "Many body derivation of quasiparticle bands and effective interaction of fermions in $CuO_2$ layers", *Phys. Rev. B* **42** (1990) 307-324.

15. J. Hubbard, "Electron correlations in narrow energy bands", *Proc. Roy. Soc. London A* **276** (1963) pp.40.

16. K.A. Chao, J. Spalek and A.M. Oles, "Kinetic exchange interaction in a narrow S-band", *J. Phys. C* **10** (1977) L271-L276.

17. E. Baskaran, Z. Zou and P.W. Anderson, "The resonating valence bond state and high $T_c$ superconductivity-A mean field theory", *Solid State Commun.* **63** (1987) 973-976.

18. S. Kivelson, D. Rokhasar and J. Sethna, "Topology of the resonating valence bond state: Solitons and high $T_c$ superconductivity", *Phys. Rev. B* **35** (1987) 8865-8868.

19. A. Virosztek and J. Ruvalds, "Nested Fermi liquid theory", *Phys. Rev. B* **42** (1990) 4064-4072.

20. P. Entel, "Strong coupling theory of superconductivity", talk presented in the present conference (1993).

21. S. Bose, "Plasmon exchange in multilayered systems and application to high $T_c$ superconductors", talk presented in the present conference (1993).

Local Order
in Condensed
Matter Physics


ISBN 1-56072-220-7

# STRONG COULPING THEORY OF HIGH TEMPERATURE SUPERCONDUCTIVITY

Peter Entel

*Theoretische Tieftemperaturphysik*
*Universität-GH-Guisburg, 47048 Duisburg, Germany*
*and*

Janusz Zielinski and Rafal Grabowski

*Institute of Physics, Silesian University, 40-007 Katowice, Poland*

ABSTRACT

Superconductivity in the cuprates and fullerenes will be discussed on the basis of strong electron-phonon coupling arising from large long-range Madelung-like contributions. A first principles calculation yields $\lambda = 1.37$ for $La_{1.85}M_{0.15}O_4$ ($M=$ Sr, Ba) [H. Krakauer, W. E. Pickett, and R. E. Cohen, Phys. Rev. B **47**, 1002 (1993)] which is sufficiently large to account for moderate transition temperatures between 30 and 40 K. For the case of $YBa_2Cu_3O_7$ $\lambda$ was estimated to be of the order of 2.36 [R. Zeyher, Z. Phys. B **80**, 187 (1990)]. This value allows transition temperatures of 90 to 100 K provided the Coulomb pseudopotential $\mu^*$ is not too large. By using an Eliashberg-like scheme the difference between superconductivity induced by electron-phonon interactions and by charge and spin fluctuations due to strong local Coulomb correlations is critically discussed.

PACS numbers: 74.20.-z, 74.65.+n, 74.30.Gn

## 1. Introduction

Since the discovery of superconductivity in heavy fermion systems [1] in 1979, in perovskite-derived systems [2] in 1986 and in the alkaline-doped fullerenes [3,4] in 1991, intense research work is still going on to elucidate the pairing mechanisms which are responsible for superconductivity in the structurally quite different intermetallic compounds.

In heavy fermion (HF) systems ($m^* \sim \gamma/n^{1/3} \sim 300\ m_0$) the Kondo interaction competes with the Ruderman-Kittel-Kasuya-Yosida (RKKY) interaction. Whereas the first interaction tends to form a nonmagnetic singlet state below a characteristic temperature $T^*$ (= several Kelvins), the second interaction tends to stabilize the local $f$

moments, leading to an inherent tendency of the heavy-Fermi liquid to become unstable against either a superconducting, an itinerant magnetic phase transition or both. Characteristic for those heavy fermion systems which become superconducting, are the low superconducting transition temperatures of order 1 K and tiny itinerant magnetic moments of order $10^{-2}$ $\mu_B$. The ratio of Néel and superconducting transition temperature is of the order of 10 in the case that both phase transitions exist. The origin of Cooper-pair formation is still at debate. It is sometimes assumed that the attractive interaction originates from the strong volume dependence of the characteristic temperature $T^*$ [5]. The two competing interactions are often dealt with by using the Anderson lattice hamiltonian or an extended version of it. For a compact presentation of the most recent experimental and theoretical results we refer to Ref. [6].

While all HF compounds are metallic, the cuprates can be either semiconductors, metals or superconductors. For example, $YBa_2Cu_3O_6$ is an antiferromagnetic semiconductor with a Néel temperature of 508 K while $YBa_2Cu_3O_7$ is a superconductor with a superconducting transition temperature of 92 K. Superconductivity above 130 K was recently found in the $HgBa_2Ca_2Cu_3O_{1+x}$ and $HgBa_2CaCu_2O_{6+x}$ systems [7]. State of the art band structure calculations based on the local density approximation (LDA) for this prototypical 1-2-3 system $YBa_2Cu_3O_{7-\delta}$ yield conduction bands which are ~9 eV wide and which arise predominantly from strongly hybridized Cu d-O p states. If no constraints are used to fix the $Cu^{2+}$ valence state for an oxygen deficiency $\delta = 1$, then LDA calculations are unable to describe antiferromagnetism and the metal-insulator transition in high temperature superconductors. Because of this and because of the strong Coulomb repulsion on the Cu site (which is of the order of 10 eV between two holes with opposite spin) new electronic models like an effective one-band [8] or three-band Hubbard model [9] were introduced. The discussion of the strong coupling limit of these models has led to new concepts like Anderson's concept of a quantum liquid of spin singlets [10] or Varma's marginal Fermi liquid theory for high temperature superconductivity [11].

These purely electronic models for high $T_c$ materials can account for their magnetic properties. On the other hand, it was early recognized that the coupling of electrons to lattice vibrations can be rather strong due to long-range contributions to the electron-phonon interaction [12]. The aim of this short contribution is to show, in how far phonons can be relevant for high $T_c$ superconductivity. Investigations are limited to the prototypical 1-2-3 system $YBa_2Cu_3O_{7-\delta}$.

The recently discovered superconducting phase of fcc $C_{60}$ doped with K or Rb, is remarkably similar to the superconducting phase of high $T_c$ compounds. The basic ingredient of high $T_c$ materials and of the fullerenes, i.e. the element oxygen and carbon, respectively, stems from the strong electronegative and nonmetallic elements of the periodic table [13]. In high $T_c$ materials it is the impact of larger local Coulomb energies as compared to the charge transfer energy which allows *oxygen to become metallic and superconducting*. In the fullerenes it is the electropositive element K or Rb which forces *carbon to become metallic and superconducting*. Therefore, if strong electronegativity is to guarantee strong electron-phonon interaction in the metallic state, then materials (still to be found) containing metallic hydrogen, nitrogen or fluor should also have very high superconducting transition temperatures.

In addition, the relative light mass of the oxygen and the carbon atom gives rise to high frequency phonon modes; typically up to ~80 meV. This wide distribution of modes together with large $\lambda$ values ($1 < \lambda < 3$) could easily explain high temperature superconductivity.

The electron-doped superconductor $Nd_{2-x}Ce_xCuO_4$ is of particular interest [14]. Here it is important that over 50% vibrational amplitude is localized on a specific kind of atoms. This and the small overlap of the partial phonon density of states of Nd (centered below 20 meV), of Cu (between 20 and 40 meV), of O(1) (between 30 and 75 meV), and of O(2) (between 40 and 75 meV) is responsible for a very small oxygen isotope exponent $\alpha_O = 0.035$ while $\alpha_{Cu} = 0.148$ is relatively large. A measurement of $\alpha_{Cu}$ would show whether phonons may be providing the predominant pairing mechanism in $Nd_{2-x}Ce_xCuO_4$.

## 2. Strong Electron-Phonon Coupling

Varma and Weber have shown that strong electron-phonon interaction in transition metal compounds leads to renormalization effects which can be taken into account by using for the dynamical matrix the relation $\mathcal{D} = \mathcal{D}_1 + \mathcal{D}_2$ [15]. Here, $\mathcal{D}_1$ arises essentially from short range forces (short range ion-ion interactions and first order perturbation contributions due to changes of the single particle energies with ionic displacements), while $\mathcal{D}_2$ includes long range forces and represents the second order perturbation contribution:

$$\mathcal{D}_2(\kappa\alpha, \kappa'\beta, \mathbf{q}) = \frac{1}{N} \sum_{\mathbf{k},\mu,\nu} \frac{f_{\mathbf{k},\mu} - f_{\mathbf{k+q},\nu}}{\varepsilon_{\mathbf{k},\mu} - \varepsilon_{\mathbf{k+q},\nu}} g^{\kappa\alpha}_{\mathbf{k},\mu,\mathbf{k+q},\nu}\, g^{\kappa'\beta}_{\mathbf{k+q},\nu,\mathbf{k},\mu}\,, \tag{1}$$

where $\kappa$ is a site index (of the sublattice atoms), $\alpha$ a cartesian coordinate, and $\mu$ a band index; $\varepsilon_{\mathbf{k},\mu}$ is the energy of the $\mu$th band at $\mathbf{k}$; and $f_{\mathbf{k},\mu}$ is the Fermi distribution function. The matrix element $g^{\kappa\alpha}_{\mathbf{k},\mu,\mathbf{k}',\nu}$ of the electron-phonon interaction is given by (with the notation used in Ref. [16])

$$\begin{aligned} g^{\bar{\kappa}\alpha}_{\mathbf{k},\mu,\mathbf{k}',\nu} = & \\ & \sum_{\kappa,m_1} \sum_{\kappa',m_2} A^*(\mathbf{k}\mu|\kappa m_1) \cdot \\ & [\gamma_\alpha(\kappa m_1, \kappa' m_2, \mathbf{k}')\,\delta_{\kappa,\bar{\kappa}} - \gamma_\alpha(\kappa m_1, \kappa' m_2, \mathbf{k})\,\delta_{\kappa',\bar{\kappa}}]\, A(\mathbf{k}'\nu|\kappa' m_2)\,, \end{aligned} \tag{2}$$

$$\gamma_\alpha(\kappa m_1, \kappa' m_2, \mathbf{k}) = \sum_{l_1 - l_2} \left\langle \phi_{\kappa l_1 m_1} \middle| \frac{\partial \mathcal{H}}{\partial u_\alpha(\kappa l_1, \kappa' l_2)} \middle| \phi_{\kappa' l_2 m_2} \right\rangle e^{-i\mathbf{k}\cdot\mathbf{R}(\kappa l_1, \kappa' l_2)}\,. \tag{3}$$

Here, $\phi_{\kappa l m}$ is the $m$th orbital of the atomic wave function of sublattice atom $\kappa$ in cell $l$; $\mathbf{R}(\kappa l_1, \kappa' l_2) = \mathbf{R}(\kappa l_1) - \mathbf{R}(\kappa' l_2)$; $A(\mathbf{k}\mu|\kappa m)$ is the $(\kappa m)$ component of $\mu$th eigenstate at the point k; $\mathcal{H}$ is the Hamiltonian and $u_\alpha(\kappa l)$ is the displacement of the atom at the site $\kappa$ of the $l$th cell. The quantities (1-3) have successfully been computed for a series of compounds by using a tight-binding Hamiltonian whose matrix elements are found from fits to first principles band structure calculations. The original work [15,16] contains a non-orthogonal tight-binding formulation which leads to correction terms in (3) coming from the overlap of wave functions of adjacent atoms.

Moreover, with the help of (1-3) one can evaluate the spectral density $\alpha^2F(\omega)$ and the electron-phonon coupling parameter $\lambda$ (or the coupling function $\lambda_{m-n}$ which enters the Eliashberg equations), once the phonon spectrum $\omega_{\mathbf{q}\lambda}$ has been obtained by diagonalizing $\mathcal{D}$. To this one has first to evaluate the imaginary part of $\mathcal{D}_2$ which is given by a Fermi surface integral, and the contribution to the phonon linewidth $\gamma_{\mathbf{q}\lambda}$ originating from electron-phonon coupling

$$\Gamma(\kappa\alpha, \kappa'\beta, \mathbf{q}) = \sum_{\mu\nu}\sum_{\mathbf{k}'} \int \frac{dS_{\mathbf{k}\mu}}{|\mathbf{v}_{\mathbf{k}\mu}|} \int \frac{dS_{\mathbf{k}'\nu}}{|\mathbf{v}_{\mathbf{k}'\nu}|} g^{\kappa\alpha}_{\mathbf{k},\mu,\mathbf{k}',\nu} g^{\kappa'\beta}_{\mathbf{k}',\nu,\mathbf{k},\mu}\, \delta(\mathbf{k}-\mathbf{k}'-\mathbf{q}), \tag{4}$$

$$\gamma_{\mathbf{q}\lambda} = \frac{\pi\hbar}{2}\sum_{\kappa\kappa'}\sum_{\alpha\beta} \frac{1}{\sqrt{m_\kappa m_{\kappa'}}} e(\kappa\alpha, \mathbf{q}\lambda)\, e(\kappa'\beta, -\mathbf{q}\lambda)\, \Gamma(\kappa\alpha, \kappa'\beta, \mathbf{q}), \tag{5}$$

where $v^{\alpha}_{\mathbf{k}\mu} = \partial\varepsilon_{\mathbf{k}\mu}/\partial k_\alpha$ is the electron velocity and $e(\kappa\alpha, \mathbf{q}\lambda)$ is the phonon polarization vector associated with the phonon frequency $\omega_{\mathbf{q}\lambda}$. The spectral function, the coupling function $\lambda_{m-n}$ and the effective mass-renormalization parameter $\lambda$ are then given by

$$\alpha^2F(\omega) = \frac{1}{2\pi\hbar N(\varepsilon_F)} \sum_{\mathbf{q}\lambda} \frac{\gamma_{\mathbf{q}\lambda}}{\omega_{\mathbf{q}\lambda}} \delta(\omega - \omega_{\mathbf{q}\lambda}), \tag{6}$$

$$\lambda_{m-n} = 2\int_0^\infty d\omega \frac{\omega\alpha^2F(\omega)}{\omega^2 + (\omega_m - \omega_n)^2}, \tag{7}$$

where $N(\varepsilon_F)$ is the density of states for one spin direction at the Fermi energy, and $\omega_n = \pi k_B T(2n-1)$ is the $n$th Matsubara frequency. $\lambda$ can be obtained from $\lambda_{m-n}$ by setting $m = n$.

With the spectral function as input one must then solve the Eliashberg equations

$$\Delta_n Z_n = \pi T \sum_m [\lambda_m n - \mu^*_N] \frac{\Delta_m}{\sqrt{\omega_m^2 + \Delta_m^2}}, \tag{8}$$

$$Z_n = 1 + \frac{\pi T}{\omega_n} \sum_m \lambda_m n \frac{\omega_m}{\sqrt{\omega_m^2 + \Delta_m^2}}, \tag{9}$$

where the pseudopotential comes with a Matsubara cut-off frequency $\omega_N$.

A thorough numerical investigation of the Eliashberg equations shows that there is in principle no limitation for $T_c$ [17]. Even if $\lambda$ in the cuprates and fullerenes is only slightly enhanced as compared to usual superconductors, there is still an increase possible due to the wide phonon spectrum, provided the holes (electrons) of metallic oxygen or carbon couple to the high-frequency modes, and provided that $\mu^*$ is not too large.

Weber calculated the spectral function $\alpha^2F(\omega)$ of $La_{2-x}(Ba,Sr)_xCuO_4$ and of $YBa_2Cu_3O_7$ according to the equations (1-6) [18,19]. In both cases he finds very strong coupling of specific O phonons to the conduction electrons ($\lambda \geq 2$). He concludes that this strong coupling (which is comparable to the coupling strength in A15 compounds) can not alone explain the high $T_c$ values of 35-40 K in $La_{2-x}(Ba,Sr)_xCuO_4$, but that

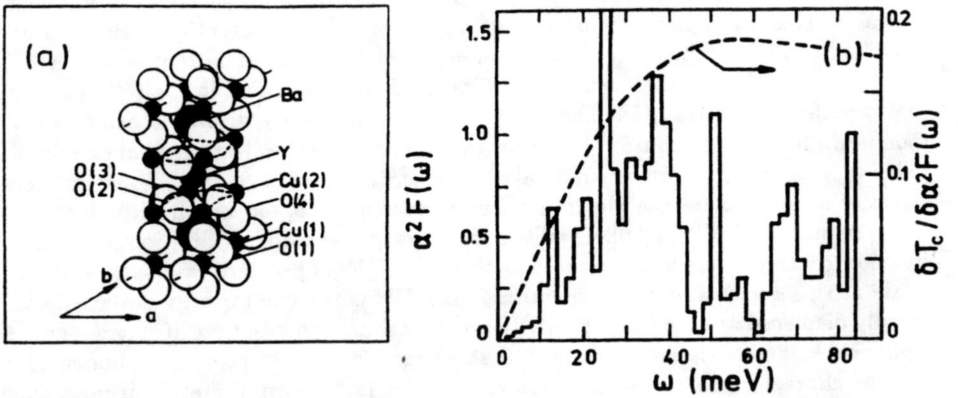


Figure 1: (a) Crystal structure of $YBa_2Cu_3O_7$, (b) spectral function $\alpha^2F(\omega)$ arising from the long-range electron-phonon interaction and functional derivative of the superconducting transition temperature.

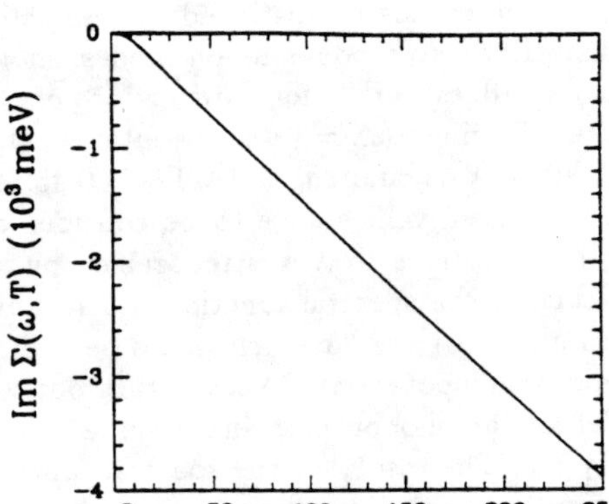


Figure 2: Temperature dependence of the imaginary part of the self-energy for $\omega = 1$ meV (with $\alpha^2F(\omega)$ for $YBa_2Cu_3O_7$ as input).

$\alpha^2 F(\omega)$ extending to much higher frequencies than in a typical transition-metal compound, is responsible for the high $T_c$. For the case of $YBa_2Cu_3O_{7-\delta}$ he finds smaller $T_c$ values than 92 K and concludes that in this compound conventional electron phonon coupling can not account for the high superconducting transition temperatures.

On the other hand, it was early recognized that screening in high $T_c$ systems may be imperfect and that Madelung terms arising from rigidly displaced atoms causing a non negligible charge transfer, can substantially enhance the electron-phonon coupling $\lambda$. This type of contribution was not taken into account in [18,19]. A different microscopic approach to calculate the electron-phonon coupling is based on the Rigid Muffin-Tin Approximation (RMTA) [20]. In its simplest version the main contibution to the electron-phonon matrix element is the on-site term $\langle \kappa\, l | \nabla V | \kappa\, l' \rangle$. Here $|\kappa\, l\rangle$ is a Wannier-like state at site $\kappa$ with angular momentum $l$, and $\nabla V$ is the band energy potential due to a rigidly displaced atom $\kappa$ leading only to non-vanishing contributions if $\Delta = l - l' = \pm 1$ (if spherical symmetry is assumed). So the short-range part of the electron-phonon coupling can be characterized by the selection rule $\Delta = \pm 1$. In normal metals displacements of atoms are well screened out except very near the displacement. This is the physical basis of Weber's method and of the RMTA approach to calculate the electron-phonon interaction. But the cuprates are strongly ionic compounds (as well as the fullerenes) and the long-range electron-phonon interaction arising from the $\Delta = 0$ contribution may become important.

Jarlborg performed self-consistent Linear Muffin Tin Orbital (LMTO) band calculations for the normal and distorted 1-2-3 system [21]. He used the usual RMTA approach to calculate the $\Delta = 1$ contribution to $\lambda$, while frozen displacement calculations were used to calculate the $\Delta = 0$ contribution as $\int d^3r\, [V_0(\mathbf{r}) - V(\mathbf{r})]/\Delta R$, where $V$ and $V_0$ are the self-consistently obtained potentials for distorted and undistorted lattices. The calculations for various O distortions, to model phonon modes, show that Madelung shifts and charge transfer are not small, especially for $\hat{z}$-distortions of O(4) sites (compare Fig. 1a). Jarlborg obtained the following $\lambda(\Delta = 0)$ contributions: O(1)-$\hat{z}$: 0.2, O(2)-$\hat{z}$: 0.2, O(3)-$\hat{z}$: 0.2, O(4)-$\hat{z}$: 0.9, while $\lambda$ calculated in RMTA is 0.45. He argues that due to the approximations involved, these values have to be considered as lower limits. The importance of Madelung contributions to $\lambda$ was recognized by several authors [22,23]. In particular, Zeyher calculated the spectral function $\alpha^2 F(\omega)$ arising from the $\Delta = 0$ contribution [23]. He finds $\lambda(\Delta = 0) = 2.36$ which would be sufficient to explain high $T_c$ values, even if the Coulomb pseudopotential $\mu^*$ should turn out to be large. Zeyher used a phenomenological model for the phonon part and assumed that the long-range forces are screened essentially by $\varepsilon_\infty$. The result for the spectral function $\alpha^2 F(\omega)$ is shown in Fig. 1b. We use these results to calculate various normal and superconducting state properties.

### 3. Tunneling and Optical Properties of High Temperature Superconductors

Strong electron-phonon interaction can give rise to marginal Fermi liquid effects. This is shown in Fig. 2 which displays the temperature dependence of the imaginary part of the self-energy (of the Green's function) for the case of $YBa_2Cu_3O_7$ which is approximately given by

$$\operatorname{Im}\Sigma(\omega) = -\frac{\pi}{2}\int dz\,\alpha^2F(z)\left[2\coth\frac{z}{2T} - \tanh\frac{\omega+z}{2T} + \tanh\frac{\omega-z}{2T}\right]. \tag{10}$$

One notes that $\Sigma(\omega)$ varies linear with temperature down to low temperatures but that there is a crossover to Fermi-liquid behaviour at very low temperatures ($T < 10$ K). This means that strong electron-phonon interaction leads to a low-energy scale; below a cut-off energy of a few Kelvins (similar to the case of HF systems) the systems behaves as a common Fermi liquid; above this cut-off the systems behaves as a marginal Fermi liquid. It is beyond the scope of this paper to elucidate the origin of this low temperature scale. The curve was calculated for $\omega = 1$ meV. It might be that marginal Fermi liquid terms arising, for example, from charge/spin fluctuations mix up with the contributions from electron-phonon coupling. This would hinder to define a unique mechanism for pairing in high temperature superconductors.

The spectral function $\alpha^2F(\omega)$ presented in Fig. 1a has been used to study various properties, especially the tunneling and optical properties of $YBa_2Cu_3O_7$ by solving the Eliashberg equations numerically. Results are displayed in Fig. 3. Fig. 3a shows the behaviour of the isotope exponent when the Coulomb pseudopotential is increased. Unfortunately no detailed calculation of the pseudopotential exists for HTSC. A recent estimation of the Coulomb pseudopotential in $C_{60}$ and doped $C_{60}$ indicates that $\mu$ and $\mu^*$ can be of the same magnitude (0.4) [24]. Fig. 3b, c and d shows the energy dependence of the energy-gap, of the mass renormalization parameter, and of the tunneling density of states (without impurities), respectively. A comparison to experiment is difficult. The reconstruction of the tunneling density of states (TDOS) from the differential current as function of the voltage yields a maximum of the TDOS at $\omega > 40$ meV. This leads to very large ratios $2\Delta/k_BT_c \sim 8$ which is probably too large. This implies that the true gap is masked in the tunneling data. (a) Optical conductivity as function of frequency of $YBa_2Cu_3O_7$ [25], (b) calculated optical conductivity for different temperatures.

More easy is the comparison to experiment with respect to optical properties like the optical conductivity. Fig. 4a shows experimental results for $YBa_2Cu_3O_7$ [25] while Fig. 4b shows the theoretical results (for convenience we used a scaling factor $x$; $x \cdot \alpha^2F(\omega)$). It is obvious from the figure that strong-electron phonon coupling can as well account for the optical data as results obtained with the marginal Fermi liquid theory [26].

## 4. Strong Electron-Electron Interaction

Strong electron-electron interactions can be delt with exactly by diagonalizing small system sizes and using simplified model Hamiltonians like the Hubbard model or by applying the Quantum Monte Carlo Method. In spite of finite size scaling there is no definite answer to the question whether the Hubbard model can have a superconducting ground state or not. Some success is connected with phenomenological models emphasizing spin/charge fluctuations. Such a phenomenological model is the marginal Fermi liquid model of Varma et al. [10]. The Eliashberg equations can be generalized to incorporate this aspect. This was done by Nicol and Carbotte [26]. Many physical properties

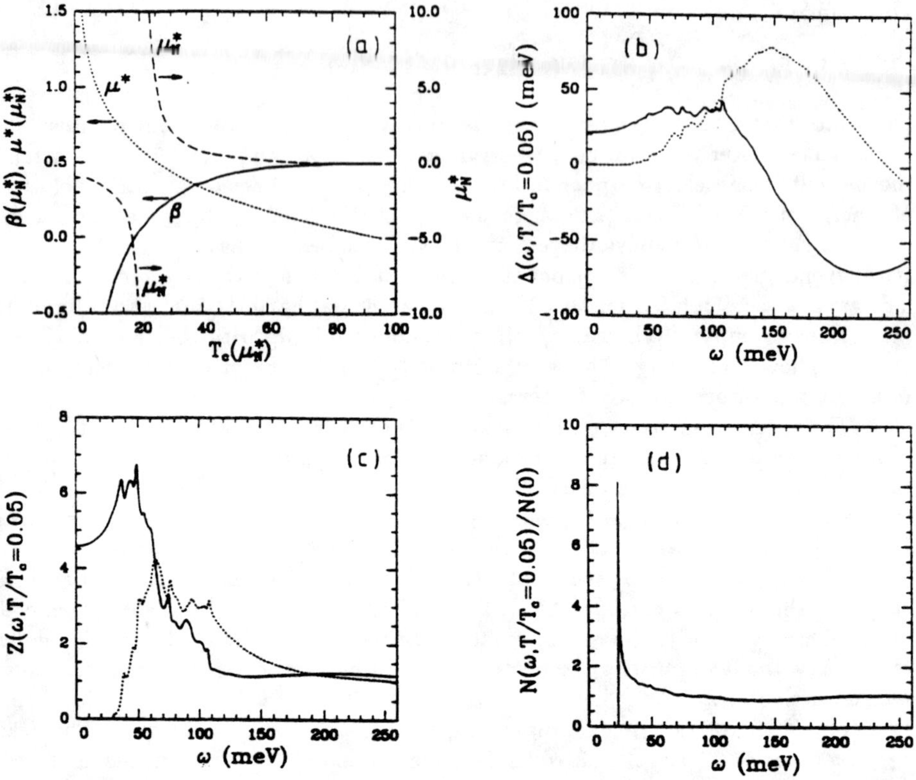


Figure 3: (a) Variation of isotope exponent with the Coulomb pseudopotential, (b) energy dependence of the real (solid line) and imaginary (dashed line) part of the energy-gap function, (c) energy dependence of the real (solid line) and imaginary (dashed line) part of the mass renormalization parameter, (d) energy dependence of the tunneling density of states.

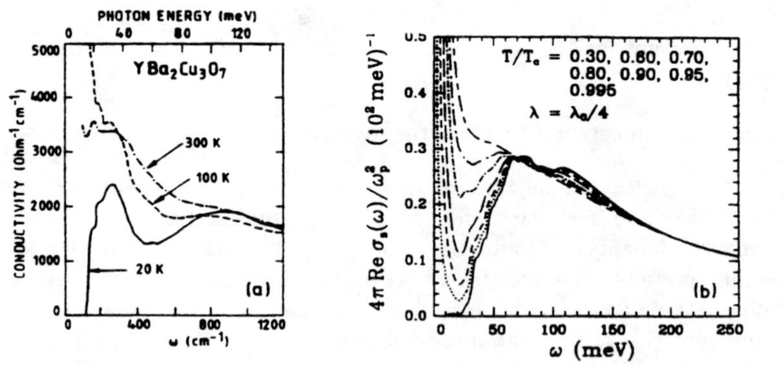


Figure 4: (a) Optical conductivity as function of frequency of $YBa_2Cu_3O_7$, and (b) calculated optical conductivity for different temperatures.

which can be calculated from the Eliashberg equations look very much the same if one compares results obtained for the case of strong electron-phonon coupling and for the case of marginal Fermi liquid theory. However, there is one feature where one obtains different results. This feature is connected with the quasiparticel damping rate as function of $T$ which in contrast to the BCS case shows an exponential decrease for $T < T_c$ (see Fig. 5).

The problem to deal simultaneously with strong electron-phonon and strong electron-electron interactions is difficult and will be discussed only briefly. One difficulty is connected with the observation that - if strong nonlocal electron-phonon interaction is responsible for HTSC - then these systems can not be treated as an array of two-dimensional $CuO_2$ layers. The reason is that Cooper pairing occurs then in the three-dimensional network of oxygen ions. Two-dimensional features become only important for the case of small doping when antiferromagnetic order sets in. Therefore, the difficulty one faces is twofold, (i) with decrease in doping a crossover occurs from the metallic nearly correlation free state becoming superconducting at low temperatures to a highly correlated insulating state which is unstable with respect to antiferromagnetic order, and (ii) with decrease in doping the system changes from three-dimensional to nearly two-dimensional behaviour.

Neglecting the dimensional crossover, we had recently proposed a five-states model for mobile carriers in a $CuO_2$ layer which feel strong local correlations at the Cu sites and less strong correlations at the O sites. In fact, it was assumed that Cu sites do not participate much in transport and superconductivity. Therefore, strong electron-phonon interaction involves only carriers at the O sites. A thorough discussion of this Hamiltonian and its physical properties can be found in Ref. [27,28]. The formulation of generalized Eliashberg equations for this model including interlayer coupling is in progress.

In order to give at least one example of the influence of local correlations on superconductivity we show in Fig. 6 the results for $\lambda$ and $T_c$ for the $U \to \infty$ limit of the single-band mean-field-slave-boson Hubbard model coupled to an Einstein oscillator. The increase of $\lambda$ and of $T_c$ is mainly an effect of the increase of the DOS with band narrowing when the band filling is close to one. For the evaluation of $T_c$ we have used Kresin's formula [29]. The result in Fig. 6 is for isotropic s-wave pairing. The more interesting case of d-wave pairing which is much at debate at present [30], is under consideration.

## 5. Conclusions

As conclusions we would like to present a few arguments which show that - besides other interactions - strong electron-phonon coupling is important for HTSC.

- The temperature dependence of the resistivity is linear at high $T$ and nonlinear at low T for $K_xBa_{1-x}BiO_3$, $Ne_{2-x}Ce_xCuO_4$ and $YBa_2Cu_4O_8$. Zeyher has shown that the temperature dependence of the resistivity originating from the electron-phonon interaction is in excellent agreement with experiment [31]. He also discusses a severe constraint for any nonphonon-scattering mechanism for the resistivity.

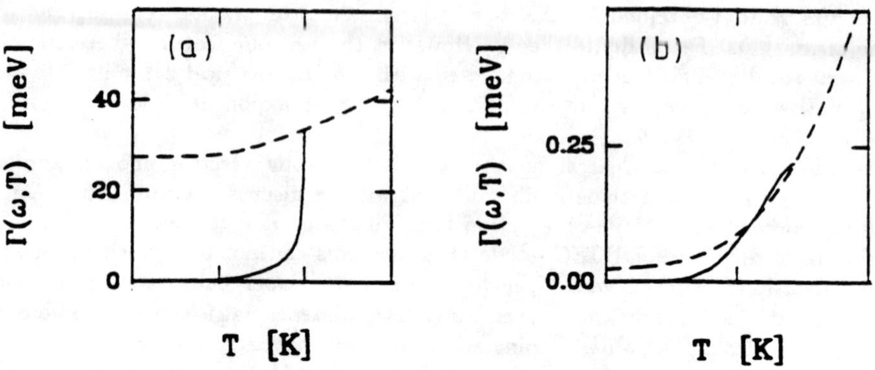


Figure 5: The quasiparticle damping rate $\Gamma(\omega, T)$ in the marginal Fermi-liquid model (a) and for Pb (b) for $\omega = \Delta_0$. The dashed curve is the normal state and the solid curve is the superconducting state in both cases.

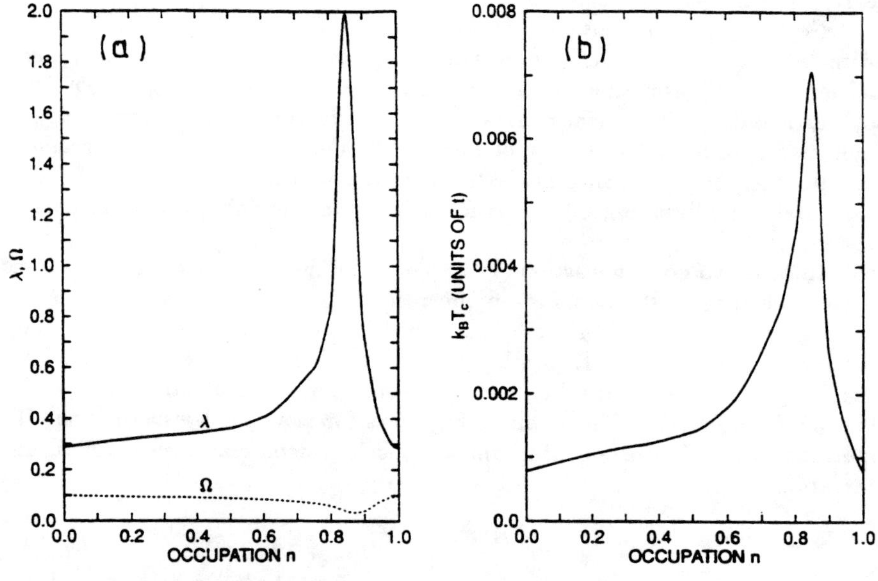


Figure 6: (a) The interaction $\lambda$ and mean phonon frequency $\Omega$, and (b) the superconducting transition temperature as function of band filling.

- The isotope effect is rather large ($\alpha = 0.36$) for $La_{2-x}Sr_xCuO_4$ ($T_c = 36$ K).
- From tunneling follows that the ratio $2\Delta_0/k_BT_c$ is larger than the BCS value 3.52.
- There is some agreement between the tunneling spectral function and the phonon density of states.
- With respect to the optical properties there is some similarity between the optical properties of HTSC and $K_xBa_{1-x}BiO_3$.

As has been stated in the introduction and in section 1, there have been only a few attempts to put the possibility of strong electron-phonon coupling on a thorough ground. First principles calculation of the frozen phonon type are clearly needed to further underline the importance of long-range electron-phonon interaction in HTSC. In this context we would like to draw attention to a very recent evaluation of the electron-phonon matrix interaction in $La_{2-x}(Ba,Sr)_xCuO_4$ by Krakauer, Picket and Cohen [32]. Their results of self-consistent linearized-augmented-plane-wave calculations within the local-density-functional approximation for the interaction strength of selected phonons yield (a Brillouin-zone average) $\lambda = 1.3$. Supercell calculations were used to obtain results for $x = 0.15$. A further important task is to calculate the effective Coulomb interaction and the Coulomb pseudopotential for HTSC. The first serious attempt by Gunnarsson and Zwicknagl to do this shows that one faces a real problem here, since the results depend crucially on the way how high energy-band states are integrated out in order to obtain the renormalized Coulomb parameter [24].

**Acknowledgements**

It is a pleasure to thank Prof. Jules Carbotte and Dr. Elizabeth Nicol for the hospitality during my Sabbatical stay at the McMaster University in Hamilton, Canada.

## References

1. F. Steglich, J. Aarts, C. D. Bredl, W. Lieke, D. Meschede and J. Schäfer, *Phys. Rev. Lett.* **43** (1979) 1892.
2. J. G. Bednorz and K. A. Müller, *Z. Phys. B - Condensed Matter* **64** (1986) 189.
3. A. F. Hebard, H. J. Rossinsky, R. C. Haddon, D. W. Murphy, S. H. Glarum, T. T. M. Palstra, A. P. Ramirez and A. R.Kortan, *Nature* **350** (1991) 600.
4. S. P. Kelty, C. C. Chen and C. M. Lieber, *Nature* **352** (1991) 223.
5. P.Fulde, J. Keller and G. Zwicknagl, *Solid State Physics* **41**, edited by H. Ehrenreich and D. Turnbull (Academic Press, New York, 1988) p. 1.
6. Proceedings of the *International Conference on Magnetism 1991*, *J. Magn. Magn. Mat.* **108** (1992).
7. A. Schilling, M. Cantoni, J. D. Guo, and H. R. Ott, Nature **363**, 56 (1993).

8. F.C. Zhang and T.M. Rice, *Phys. Rev. B* **37**, 3759 (1988).

9. V.J. Emery, *Phys. Rev. Lett.* **58**, 2794 (1987).

10. P.W. Anderson, G. Baskaran, Z. Zou and T. Hisu, *Phys. Rev. Lett.* **58**, 2790 (1987).

11. C.M. Varma, P.B. Littlewood, S. Schmitt-Rink, E. Abrahams and A.E. Ruckenstein, *Phys. Rev. Lett.* **63**, 1996 (1989).

12. R. Zeyher, *Z. Phys. B-Condensed Matter* **80**, 187 (1990).

13. F.J. DiSalvo, *Science* **247**, 649 (1990).

14. H. Chen and J. Callaway, *Phys. Rev. B* **46**, 14 321 (1992).

15. C.M. Varma and W. Weber, *Phys. Rev. Lett.* **39**, 1094 1977.

16. W. Weber, in *Electronic Stucture of Complex Systems*, edited by P. Phariseau and W.M. Temmermann, NATO ASI Series B: Physics, Vol 113 (Plenum, New York, 1984), p. 345.

17. J.P. Carbotte, *Rev. Mod. Phys.* **62**, 1027 (1990).

18. W. Weber, *Phys. Rev. Lett.* **58**, 1371 (1987).

19. W. Weber and L.F. Mattheis, *Phys. Rev. B* **37**, 599 (1988).

20. G. Gaspari and B. Gyorffy, *Phys. Rev. Lett.* **28**, 801 (1972).

21. T. Jarlborg, *Solid State Commun.* **67**, 297 (1988).

22. R.E. Cohen, W.E. Pickett, L.L. Boyer and H. Krakauer, *Phys. Rev. Lett.* **60**, 817 (1988). S. Barisic, K. Kupcic and I. Batistic, *Int. J. Mod. Phys. B* **3**, 2051 (1989).

23. R. Zeyher, *Z. Phys. B-Condensed Matter* **80**, 187 (1990).

24. O. Gunnarsson and G. Zwicknagl, *Phys. Rev. Lett.* **69**, 957 (1992).

25. K. Kamara, S.L. Herr, C.D. Porter, N. Tache, D.B. Tanner, S. Etemad, T. Venkatesan, E. Chase, A.Inam, X.D. Wu, M.S. Hedge and B. Dutta, *Phys. Rev. Lett.* **64**, 84 (1990).

26. E.J. Nicol and J.P. Carbotte, in *Studies of High Temperature Superconductivity*, edited by A. Narlikar, Vol. 11 (Nova Science, New York, 1993).

27. P. Entel and J. Zielinski, *Phys. Rev. B* **42**, 307 (1990).

28. P. Entel, S.N. Behera, J. Zielinski and E. Kaufmann, *Int. J. Mod. Phys. B* **5**, 271 (1991).

29. V. Z. Kresin, Phys. Lett. A **122**, 434 (1987).

30. D. Thelen, D. Pines, and Jian Ping Lu, Phys. Rev. B **47**, 9151 (1993).

31. R. Zeyher, *Phys. Rev. B* **44**, 10 404 (1991).

32. H. Krakauer, W.E. Pickett and R.E. Cohen, *Phys. Rev. B* **47**, 1002 (1993).

Local Order
in Condensed
Matter Physics


ISBN 1-56072-220-7

# PLASMON EXCHANGE IN MULTILAYERED SYSTEMS AND APPLICATION TO HIGH-$T_C$ SUPERCONDUCTORS

S. M. Bose

*Department of Physics and Atmospheric Science*
*Drexel University, Philadelphia, PA 19104, U.S.A.*
*and*

P. Longe

*Institut de Physique, B5, Université de Liège*
*Sart-Tilman, B-4000 Liège, Belgium*

In this paper we present a study of the multilayered high-$T_c$ cuprate superconductors based on a model where the plasmon mediated effective interaction provides an attraction between the charge carriers[1,2]. We first calculate analytic expressions for the effective interaction between the charge carriers in the CuO layers. We then calculate the critical temperature $T_c$ and show how it can provide an explanation for the rise in $T_c$, with a saturation effect, with the increase in the number of CuO layers per cell, as is observed in the thallium[3] and bismuth[4] based compounds. Our model is then used to calculate the loss function, which is then utilized to interpret the infrared absorption spectra of $La_{1.85}Sr_{0.15}CuO_4$ and the electron energy loss spectra of $Bi_2Sr_2CaCu_2O_8$, which provide good spectra[5-8] as well as good set of physical parameters[9].

We assume that the metallic layers in the multilayered compound can be considered as two-dimensional electron gases (2DEG) and that all these layers have the same 2D density $n_s$ of charge carriers (henceforth designated as 'electrons'). These electrons interact with each other within the same layer, as well as from layer to layer, via an effective Coulomb interaction which we treat in the framework of the random phase approximation (RPA).

The bare Coulomb interaction between two electrons located in layers of ordinates $z_0$ and z is given by

$$v(q,z-z_0) = v_0(q)\exp(-q|z-z_0|), \tag{1}$$

where **q** represents a 2D momentum parallel to the layers and where

$$v_0(q) = 2\pi e^2/\varepsilon_0 q. \tag{2}$$

Since the electrons constitute a many-body system, their Coulomb interaction can be replaced by a frequency dependent effective interaction, including plasmon exchanges in various layers. This effective interaction for a system with L (=1,2,3,4) 2DEG layers/cell has been shown to take the form[10]

$$V_L(q,\omega)/v_0(q) = 1 + (1/\pi)\, S\, w_{\kappa\kappa'}/(\beta - \beta_{\kappa\kappa'}), \tag{3}$$

with the following notations,

$$S\ldots = \sum_{\kappa=1}^{L} \int_0^{\pi} d\kappa' \ldots$$

$$\beta = \omega^2/\sigma q$$

$$\beta_{\kappa\kappa'} = [\omega_\kappa(\kappa',q)]^2/\sigma q$$

$$\sigma = 2\pi e^2 n_s/\varepsilon_0 m,$$

a form which can be identified with the more prevalent form

$$V_L(q,\omega) = v_0(q) + S\, 2\omega_\kappa(\kappa',q)|M_\kappa(\kappa',q)|^2/[\omega^2 - (\omega_\kappa(\kappa',q))^2], \tag{4}$$

used by several authors[11,12] for the phonon coupling. Here, however, $M_\kappa$ is the electron-plasmon coupling matrix. In Ref.10, we have calculated the $\omega_\kappa(\kappa',q)$, $\beta_{\kappa\kappa'}$ and $w_{\kappa\kappa'}$ appearing in (3) and (4) for L=1, 2, 3 and 4. The poles of (3) or (4) give the dispersion relations for the L plasmon bands.

In Ref.10, we have shown that for all bands, except for the one labeled by $\kappa=1$, one has $\omega_\kappa(0,q) \approx \omega_\kappa(\pi,q)$. Moreover, since $\beta_{\kappa\kappa'}$ is proportional to q (for small q and $\kappa \neq 1$), these bands appear as very narrow acoustic bands. Comparatively the band with $\kappa=1$ is a 'pseudo-optical' band which presents a larger width.

Using the effective interaction given by (3) or (4), we now calculate the $T_c$ using the Eliashberg theory[11]. Because of the complexity of the original theory, some approximate analytic forms have emerged[12,13] and here we use the expressions recently given by Kresin[1]. He showed that $T_c$ can be obtained from

$$T_c = 0.25 \langle \omega \rangle \ [\exp(2/\lambda_{eff}) - 1\ ]^{-1/2}, \tag{5}$$

where $\langle\omega\rangle$ is the average plasmon frequency and $\lambda_{eff}$ is the effective interaction strength which depends on the attractive strength $\lambda$ and the effective Coulomb repulsion $\mu^*$ Our problem is then to calculate $\langle\omega\rangle$, $\lambda$ and $\mu^*$. Using the effective interaction (Eq. (4)) in the standard expressions[12,13] for $\lambda$ and $\lambda\langle\omega^2\rangle$ we find that[14]

$$\begin{aligned}\langle\omega\rangle &= (\lambda\langle\omega^2\rangle/\lambda)^{1/2} \\ &= [(2\pi e^2/\varepsilon_0 m n_s L/(La+b))]^{1/2}.\end{aligned} \tag{6}$$

Here a is the distance between adjacent layers within the same cell and b is the smallest distance between the layers of two adjacent cells.

In order to calculate $T_c$ we have also to calculate $\lambda_{eff}$ which requires the knowledge of both $\lambda$ and $\mu^*$, which are both L independent. However, it is difficult to calculate $\mu^*$ since it depends logarithmically on a screening boson frequency which is not known. We can still calculate the variation of $T_c$ with L, using the following argument. Since both $\lambda$ and $\mu^*$ are independent of L, for all L we can write from (5)

$$T_{cL} = C \ \omega_L, \tag{7}$$

where $C=0.25[\exp(2/\lambda_{eff}) - 1\ ]^{-1/2}$. To estimate the value of the constant C valid for all values of L we use the experimental value of $T_{c2}$=105 K and the calculated value of 0.088 eV for $\langle\omega\rangle_2$ in (7). Once C is determined by such a fit, we can then proceed to calculate $T_c$'s for the other three values of L from (7) by using the calculated values of $\langle\omega\rangle_L$'s. The results of our calculation are shown and compared with experimental results in Table I.

**Table 1.** The calculated values for $<\omega>_L$ as obtained from (6) and $T_{cL}$ as obtained from (7) are shown for L=1, 2, 3 and 4. The last column shows the experimental $T_c$'s in the Bi-based compounds.

| L | $<\omega>_L$/eV | $T_{cL}$/K (Calc.) | $T_{cL}$/K (Exp.) |
|---|---|---|---|
| 1 | 0.071 | 84 | 80 |
| 2 | 0.088 | 105 (exact) | 105 |
| 3 | 0.098 | 117 | 125 |
| 4 | 0.104 | 124 | - - - |

We can thus conclude that the plasmon exchange model for the interaction between charge carriers can explain the rise in $T_c$ with the number of CuO layers/cell in the Tl- and Bi-based compounds. Furthermore, there is a saturation effect indicated by $T_{c4}-T_{c3}<T_{c3}-T_{c2}<T_{c2}-T_{c1}$.

In the past several years, the loss function of several cuprate superconductors has been obtained by various authors from the measurements of their infrared [IR] absorption[5,6], and the inelastic scattering cross sections of electrons[7,8,15]. Theoretical studies of these phenomena have been limited. The aim of this calculation is to emphasize the possible effect of the plasmon excitation in a multilayered system on the IR and EEL spectra.

The dielectric function $\varepsilon(q,\omega)$ appearing in the loss function $\mathrm{Im}(-1/\varepsilon)$ must be considered to be a tensor because of the anisotropy of the layered electron system. Thus in our calculation of the IR absorption[16], the usual scalar loss function has to be replaced by a more general loss function $\mathrm{Im}(-\mathbf{u}.\varepsilon^{-1}.\mathbf{u})$ where $\varepsilon^{-1}$ is the inverse dielectric tensor and **u** is the unit IR polarization vector. The IR loss function can be written as

$$\mathrm{Im}(-\mathbf{u}.\varepsilon^{-1}.\mathbf{u}) = v(\mathbf{Q})/\varepsilon_0\,(\pi/\Omega)\sum_f |<f|\mathbf{Q}|0>|^2\,\delta(E_f-E_0-\omega), \quad (8)$$

where **Q** is the photon wave vector associated with the momentum transfer

($Q=\omega/c_0$ and $u.Q=0$, $c_0$ being the velocity of light), $v(Q)=4\pi e^2/\varepsilon_0 Q^2$ is the Coulomb potential, $\varepsilon_0$ being the background dielectric function and $\Omega$ the volume of the sample. Also $|0\rangle$ and $|f\rangle$ represent the initial (ground) and final states of the electron system. The operator O has the form

$$O = \sum_{n=1}^{N} \exp(iQx_n)\; u.p_n/mc_0, \tag{9}$$

where the $x_n$'s and $p_n$'s are the positions and momenta of the N electrons.

Expression in (8) can be expanded diagrammatically[17], up to the first order in plasmon excitation, and the contributions of these diagrams can be written as[16]

$$\begin{aligned} \mathrm{Im}(-u.\varepsilon^{-1}.u) = (\pi/2\varepsilon_0)\ (\omega_p/\omega)^4\ \omega \sin^2\alpha \sin^2\beta\ \delta(\omega - \omega_p \cos\alpha) \\ + (1/\pi\varepsilon_0)\ (\omega_p/\omega)^4 \cos^2\beta\, A(\omega) \end{aligned} \tag{10}$$

where $\alpha$ and $\beta$ are the angles between the layer planes and the wave vector, and between the layer planes and the polarization vector, respectively. Note that the first term on the right hand of (10) gives zero contribution for normal incidence ($\beta=0$). It corresponds to the excitation of a plasmon without pair excitation in the final state, which does not appear in an isotropic 3D system[17]. The second term of (10) is the contribution which is non-zero for all incidence angles. Function $A(\omega)$ is a computed function[16] proportional to $\omega^5$ for $\omega \ll \omega_p$.

Equation (10) has been evaluated numerically for a good set of values for the parameters such as the interlayer distance c, the 2D density ns of electrons, their effective mass m obtained by Kresin and Wolf[9] for La-Sr-Cu-O. It is found that a reasonable agreement with the experimental peak of Kim *et al*[6], in both the position and the intensity, is obtained if we choose $\varepsilon_0=1.7$. Our IR absorption spectrum also gives a fairly accurate description of the experimental spectrum over the entire frequency range, even though at very low frequencies our loss function has a linear dependence on $\omega$ and not the quadratic behavior reported by Bozovic[5] and Kim *et al.*[6] Furthermore, our theory gives us a method for choosing an appropriate value for $\varepsilon_0$ from a concurrent fit of the position and intensity of the peak with experiment.

Next we relate the loss function of our model to its EEL spectra[18]. This EEL loss function $\mathrm{Im}(-1/\varepsilon)$ is given by (8), with the modification that the operator O has to be replaced by

$$O = \sum_{n=1}^{N} \exp(iQx_n), \tag{11}$$

In this case $Q = (q, q_z)$ is related to the momentum transferred by the incident electron to the multilayered system, and ω represents the electron energy loss. The EEL process is in a sense, simpler than the IR absorption considered above. Here the plasmon excitation can occur without the assistance of an electron-hole pair production which was demanded in the IR case by the symmetry requirement.

In a first order calculation, the loss function related to the EEL spectra is given by

$$Im(-1/\varepsilon) = (\pi\,\omega/2\varepsilon_0)\,\delta(\omega - \omega(Q)), \tag{12}$$

which is a scalar since only the diagonal tensor components appear here.

In the plasmon region, for L=1 we obtain the following dispersion relation

$$\omega(Q) = \omega_p[(cq/2)f(Q)]^{1/2}, \tag{13}$$

where

$$f(Q) = \sinh cq/(\cosh cq - \cos cq_z)$$

In a Q-expansion, (13) takes the form

$$\omega = \omega_p + \alpha\, Q^2/m_e, \tag{14}$$

where

$$\alpha = 3\pi/(2M^2A^2\Omega_p) + \Omega_p C^2/48, \tag{15}$$

with $\Omega_p = \omega_p/Ry$, $M = m/m_e$, $A = a/a_B$ and $C = c/a_B$ ($a_B$ being the Bohr radius and $Ry = (2m_e a_B^2)^{-1}$, the Rydberg unit). Here a and c are the lattice constants in the layer plane and the interlayer distance, respectively. Note that we assume that we have one electron per unit cell as it is the case for $Bi_2Sr_2CaCu_2O_8$, the compound investigated here.

A dispersion relation like (14) can be detected in the EEL spectra by observing the position of the δ-peak of (12) as a function of the energy loss $\omega$ and the momentum transfer **Q**. This momentum is perpendicular to the incident electron beam. Since generally this beam is normal to the layer planes, **Q** is parallel to them, i.e. $q_z=0$. It is this geometric situation that we have assumed in calculating (15), and that we will assume henceforth. This kind of EEL experiment was performed by Nücker *et al.*[7] (referred as NI) for Bi-Sr-Ca-Cu-O. However these authors have neglected the multilayered structure of this compound in the interpretation of their spectra. Using a 3D bulk model they obtained

$$\alpha = (3/5)(a_B k_F)^2 / M^2 \Omega_p, \qquad (16)$$

where $k_F = (3\pi^2 n_v)^{1/3}$ is the 3D Fermi momentum and where one has

$$\omega_p^2 = 4\pi e^2 n_v / m. \qquad (17)$$

Their observed spectra essentially yield the values of $\omega_p$ and $\alpha$. They found $\omega_p=1$eV and $\alpha=0.6$. These experimental values, together with a=3.8Å and c=7.725Å, were inserted in (16) and (17) which yielded the numbers of the first line of Table 2, namely their effective mass m, the Fermi energy $k_F^2/2m$ and the background dielectric constant $\varepsilon_0$, which do not seem to be very realistic. These numbers can be compared with the numbers of the second line where (15) has been used. This second set of

**Table 2.** Here the α's are the input parameters as given in NI (first two models) and in NII (last two models). The other parameters (m, $E_F$ and $\varepsilon_0$) have been derived using these α's together with a=3.8Å, c=7.725Å and $\omega_p$=1eV.

| Model | α | | $m/m_e$ | $E_F$/eV | $\varepsilon_0$ |
|---|---|---|---|---|---|
| Eq.(24) | 0.6 | | 1.25 | 1.25 | 9.85 |
| Eq.(23) | 0.6 | | 2.13 | 0.78 | 5.80 |
| Eq.(26) | 0.6(110) | 0.4(110) | 2.62 | 1.00 | 4.73 |
| Eq.(26) | 0.6(110) | 0.4(110) | 2.70 | 0.80 | 4.58 |

numbers are quite acceptable and look more realistic than in the first line, especially since the Fermi energy becomes smaller than $\omega_p$.

In a second paper by Nücker *et al.*[8] (referred as NII), the effect of the lattice structure within the layers has been investigated. Here, because of the (a,b)-anisotropy, the Fermi curve which was treated as a circle in (15), has to be replaced by a square. If the sides of the first Brillouin zone are defined by $\pm k_x = \pm k_y = \pi/a$, the sides of the Fermi square are $\pm k_x \pm k_y = \pi/2a$. Here one has one electron (or hole) per unit cell and the first Brillouin zone is thus half filled.

In this new investigation in NII, the value of $\alpha$ in (14) depends on the orientation of **q** with regard to the a- or b-axis and varies between two limits which are $\alpha_{100}$ and $\alpha_{110}$. These two limits were detected by the authors of NII who propose $\alpha_{100}$=0.63 and $\alpha_{110}$=0.35. We have extended our multilayered model to include this anisotropy and have found two relations that $\alpha_{100}$ and $\alpha_{110}$ have to satisfy. We find that these equations are satisfactorily satisfied if together with $\omega_p$=1eV and the above mentioned lattice constants, we use slightly modified $\alpha_{100}$=0.6, as in NI, and $\alpha_{110}$=0.4, a value measured by Wang, Feng and Ritter[19]. Using these adapted values of the $\alpha$'s, we have calculated the numbers of the last two lines of Table 2. These two lines correspond to two different electron energy functions that we have used[18], but these numbers are quite similar however. We also note that they are not much different from those of the second line where the lattice is neglected inside the layers.

This work has been partially supported by the NATO Research Grant CRG 910966. The work of P.L. has also been supported by the Fonds National de la Recherche Scientifique, Belgium.

## REFERENCES

1. V. Z. Kresin, Phys. Lett. A 1 2 2, 434 (1987); Phys. Rev. B 3 5, 8716 (1987); Solid State Commun. 6 3, 725 (1987); V. Z. Kresin and H. Morawitz, Phys. Rev. B 3 7, 7854 (1988).
2. Ashkenazi, C. G. Kuper, and R. Tyk, Solid State Commun. 6 3, 1145 (1987); J. Ruvalds, Phys. Rev. B 5 3, 8869 (1987).
3. M. K. Wu et at., Phys. Rev. Lett. 5 8, 908 (1987); Z. Z. Sheng et al., Phys. Rev. Lett. 6 0, 937 (1988); Z.Z. Sheng and A. M. Hermann,

Nature 3 3 2, 55 (1988); Nature 3 3 2, 138 (1988); R. M. Hazen et al.,Phys. Rev. Lett. 6 0, 1657 (1988).
4. H. Maeda et al., Jpn. J. Appl. Phys. Pt. 2 2 7, 209 (1988); C. W. Chu et al., Phys. Rev. Lett. 6 0, 941 (1988).
5. I. Bozovic, Phys. Rev. B 4 2, 1969 (1990); I. Bozovic, J. H. Kim, J. J. S. Harris Jr., et al., Phys. Rev. B 4 3, 1169 (1991), Phys. Rev. B 4 6,1182 (1992).
6. J. H. Kim, I. Bozovic, C. B. Eom, T. H. Geballe,and J. S. Harris Jr., Physica C 1 7 4, 435 (1991).
7. N. Nücker, H. Romberg, S. Nakai, B. Scheerer, J. Fink, Y. F. Yan, and Z. X. Zhao, Phys. Rev. B 3 9, 12379 (1989).
8. N. Nücker, U. Eckern, J. Fink, and P. Müller, Phys. Rev. B 4 4, 7155, (1991).
9. V. Z. Kresin and S. A. Wolf, Phys. Rev. B 4 1, 4278 (1990).
10. S. M. Bose and P. Longe, J. Phys.: Condens. Matter 4, 1799 (1992).
11. G. M. Eliashberg, Zh. Eksp. i Teor. Fiz. 3 8, 966 (1960); 3 9, 1437 (1960) (English transls.: Soviet Phys.-JETP 1 1, 696 (1960); 1 2, 1000 (1961)).
12. W. L. McMillan, Phys. Rev. 1 6 7, 331 (1968).
13. P. B. Allen and R. C. Dynes, Phys. Rev. B 1 2, 905 (1975); P. B. Allen, in 'Dynamical Properties of Solids' vol. 3, 95, ed. G. K. Horton and A. A. Maradudin (Amsterdam, North Holland, 1980).
14. P. Longe and S. M. Bose, J. Phys.: Condens. Matter 4, 1811 (1992).
15. C. Tarrio and S. E. Schnatterly, Phys. Rev. B 3 8, 921 (1988).
16. S. M. Bose and P. Longe, Phys. Rev. B 4 8, 6632 (1993).
17. N. Tzoar and A. Klein, Phys. Rev. 1 2 4, 1297 (1961).
18. P. Longe and S. M. Bose, Phys. Rev. B 4 7, 11 611 (1993).
19. Y. Y. Wang, G. Feng, and A. L. Ritter, Phys. Rev. B 4 2, 420 (1990).

Local Order
in Condensed
Matter Physics


ISBN 1-56072-220-7

# DIFFUSION AND CLUSTERING OF IMPURITIES ON SURFACES

G. Schatz

*Department of Physics, University of Konstanz, D-7750 Konstanz, Germany*

ABSTRACT

**Recent applications of the perturbed γγ-angular correlation (PAC) technique to free surfaces have revealed numerous interesting results. The microscopic sensitivity of the method permits to study binding, diffusion and clustering of impurities on surfaces.**
**For metal surfaces two problems will be adressed: the roughening transition on vicinal copper surfaces and, secondly, the two-dimensional clustering of indium atoms on Cu(100). In the case of semiconductors adsorption sites of indium on silicon and their thermal stability will be discussed.**

## 1. Introduction

In recent years, the use of hyperfine techniques has been established as an advantageous complementary approach to study surface, interface and thin film phenomena by local probes [1,2,3]. In particular, the perturbed-angular-correlation (PAC) method has proven to be quite powerful, since the rather small number of radioactive probe nuclei needed for the experiment does not significantly disturb the system under consideration. Using the electric field gradient as a characteristic measure for the lattice site occupied by the probe atoms, the technique has been used to study problems like surface diffusion, surface roughening, surface- and thin-film magnetism as well as interface compound formation on microscopic scales. Recent results on free semiconductor surfaces and metal/semiconductor interfaces showed that the technique is by no means restricted to metals and introduced the method into a new area of fundamental and technological interest.

In this paper the discussion shall be concentrated on the aspect of mobility and adsorption-site changes of impurities, as well as impurity clustering on metal surfaces.

## 2. Indium Impurities on Metal Surfaces

### 2.1. Adsorption Sites on Metal Surfaces

The diffusion behaviour of adsorbed species on metal surfaces has been subject to numerous experimental studies in the past, mainly due to its importance for a better understanding of processes like crystal growth, epitaxy and catalysis.

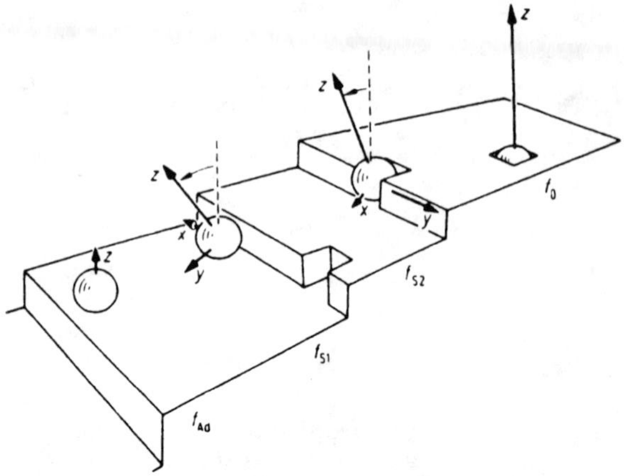


Fig. 1: Different adsortion sites of indium impurities on fcc metal surfaces. The principal axes of the electric-field-gradient tensor is indicated by arrows.

Yet, only a limited number of experimental techniques is capable of following diffusive motions of dilute adsorbates on a microscopic scale. Here, the high sensitivity of the PAC technique together with its short-range character is of outstanding advantage. First results on the diffusion of isolated $^{111}$In probes on stepped Cu(111)[4], Ag(111) and Ag(100) surfaces[5] have already been achieved some years ago; meanwhile, a large variety of stepped surfaces has been studied and a comprehensive set of data has accumulated. For these experiments, about $10^{-4}$ monolayers of the radio-

Table 1: Activation energies for different microscopic diffusion paths on stepped metal surfaces as determined from PAC experiments. The data have been obtained assuming Arrhenius-like diffusion behaviour.

| | adatomic diffusion | diffusion along steps | diffusion within terrace | diffusion into bulk | Ref. |
|---|---|---|---|---|---|
| Ag(100) | 0.31(3) eV | 0.50(3) eV | 0.64(3) eV | > 1.55 eV | 5 |
| Cu(100) | 0.28(5) eV | 0.45(5) eV | 0.65(4) eV | 2.05(5) eV | 9 |
| Ni(100) | 0.31(3) eV | – | 0.64(6) eV | – | 9 |
| Pd(100) | 0.45(4) eV | – | 1.04(6) eV | > 2.22 eV | 9 |
| Ag(111) | – | 0.52(3) eV | 0.69(3) eV | 1.61(3) eV | 5 |
| Cu(111) | – | 0.41(3) eV | 0.64(3) eV | 2.20(4) eV | 4 |
| Pd(111) | 0.23(3) eV | 0.46(7) eV | 0.76(8) eV | – | 19 |
| Ag(110) | – | – | < 0.50 eV | – | 9 |
| Au(110) | – | – | – | 1.88(3) eV | 18 |

active PAC probe $^{111}In$ are deposited onto a stepped metal surface at 77 K and PAC spectra are taken in the course of an isochronal annealing sequence. With increasing annealing temperature, different adsorption sites are populated, each being characterized by a well-defined electric-field-gradient tensor. In particular, four different adsorption sites have been identified (Fig. 1). After residing on adatomic positions on top of a low-index terrace after deposition at 77 K, the probes diffuse towards adatomic step sites, are then incorporated into the steps and finally populate substitutional lattice sites within the topmost low index layer. The site population is monitored as a function of annealing temperature and from the conversion temperatures activation energies can be estimated. Table 1 summarizes the diffusion parameters achieved in this way.

### *2.2. Roughening of Copper Surfaces*

While the experiments mentioned above have mainly been concerned with surfaces of low step density, the study of highly stepped surfaces only recently led to quite interesting results [6]. In particular, Cu(115) and Cu(117) surfaces have been investigated. These surfaces consist of narrow terraces of (100) orientation separated by monoatomic steps. The typical width between two steps amounts to only 3 - 5 atomic rows. It has been known from various scattering experiments [7] that the Cu and Ni surfaces of (11m) orientation (with m odd) undergo a reversible structural phase transition at some critical temperature $T_R$, where the regular step structure becomes thermodynamically unstable. An increasing degree of disorder is observed above $T_R$ resulting in atomically rough surfaces at sufficiently high temperatures. The critical temperatures $T_R$ are decreasing with increasing m.

In Fig. 2 PAC spectra and corresponding Fourier-transforms for a Cu(115) surface labelled with $10^{-4}$ monolayers $^{111}In$ are shown for different measurement temperatures. Other than on the flat surfaces studied so far, on Cu(115) both substitutional step sites and substitutional terrace sites are populated in thermal equilibrium, characterized by two well-known electric field gradients. This can be seen in the top part of the Figure. The narrow line widths in the Fourier spectra indicate unique probe sites with no structural disorder in the near neighbourhood of the probes. As a matter of fact, this situation is observed up to temperatures well above the critical temperature $T_R$ (Cu(115): $T_R$ = 380 K [7]). Only about 150 K above $T_R$ the field-gradient distributions start broadening finally leading to the situation shown in the bottom part of Fig. 2. This transformation is completely reversible. Fig. 3 summarizes the observed behaviour showing the relative width of the field-gradient distributions for the substitutional step sites on both Cu(115) and Cu(117) as a function of temperature. While no changes are observed in the vicinity of $T_R$, only at much higher temperatures structural disorder is observed leading to broad field-gradient distributions. The onset of disorder occurs at a higher temperature on Cu(115) as compared with Cu(117) in agreement with the higher critical temperature found for this surface [7].

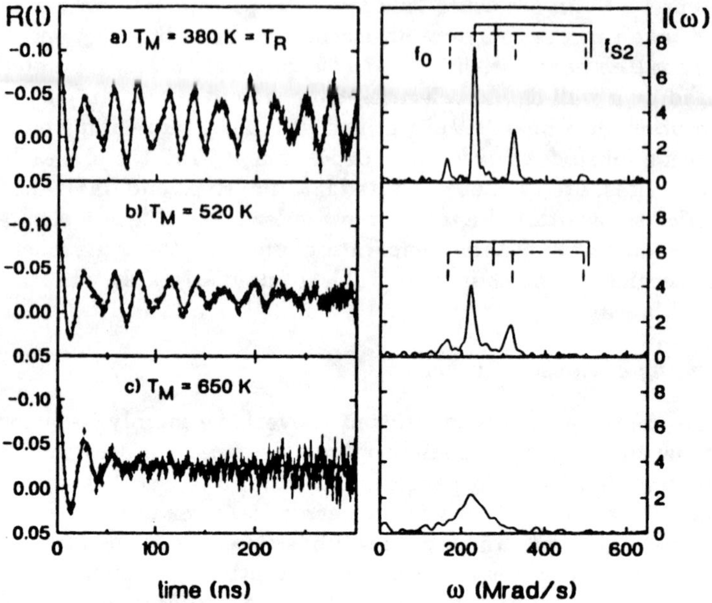


Fig. 2: PAC spectra and corresponding Fourier-transforms for $^{111}In$ on Cu(115) for different measurement temperatures. The solid line represents a least-squares fit to the data. In the Fourier-spectra corresponding lines are indicated; some lines are missing due to the chosen sample-detector geometry.

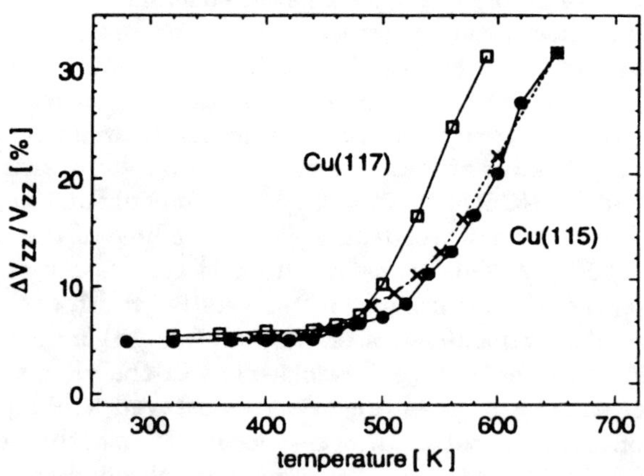


Fig. 3: Relative width of the field-gradient distribution for $^{111}In$ on substitutional step sites on Cu(115) and Cu(117) respectively. The solid lines are guides to the eye.

No quantitative theoretical model is yet available, however a qualitative understanding of the above results is quite straightforward. While the scattering techniques used so far are sensitive to structural disorder on a length scale of some 5 nm , the short range character of the electric field gradient gives insight to structural information in the near-neighbour vicinity of the probe atoms. The PAC technique therefore senses changes in the structural order on a much smaller length scale, i.e. about 0.5 nm. Hence, we conclude that while the long range step structure is distroyed at $T_R$ as observed in scattering experiments, the local order within the steps is preserved and much higher temperatures are needed to finally lead to disorder on an atomic scale. It is worth mentioning that no other experimental technique is capable of studying thermal roughening on an atomic length scale; this even holds for scanning tunneling microscopy, which in principle offers the necessary spatial resolution. Yet, this technique is restricted to processes which are static on a time scale of some milliseconds.

### *2.3. Clustering on Copper Surfaces*

So far experiments have been discussed, where PAC-isotopes in extremely low concentrations have been used to study surface phenomena and therefore $^{111}$In-interactions could be neglected. If the indium concentration comes in the range of a monolayer, indium-indium interactions become important and clustering effects may occur.

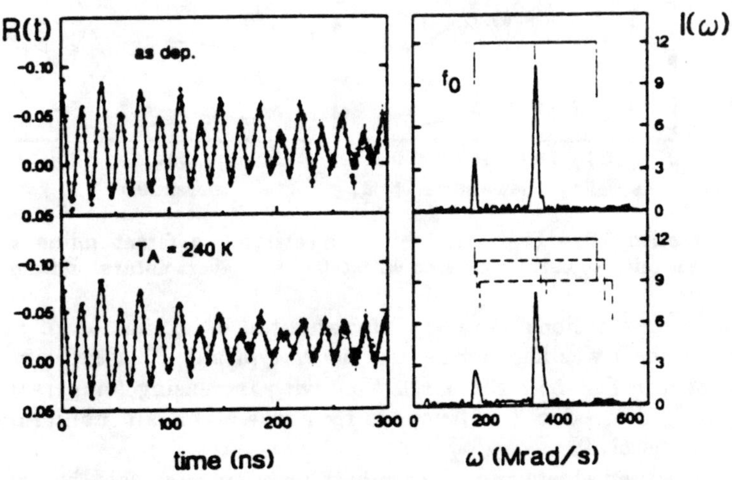


Fig. 4: PAC spectra and corresponding Fourier-transforms for $^{111}$In on Cu(100) covered with 0.04 monolayers indium after preparation (upper part) and after annealing to 240 K (lower part). The solid line represents a least-squares fit to the data. In the Fourier-spectra corresponding lines are indicated; some lines are missing due to the chosen sample-detector geometry.

A first experiment, where in addition to the $^{111}$In-probes about 0.07 monolayers natural indium was deposited on a Cu(100) surface, was performed some years ago [8]. More recently experiments of this kind were performed with varying indium coverages [9]. A Cu(100) surface was prepared carefully and $^{111}$In-probes have been positioned into the surface layer, as we have discribed in chapter 2.1. After that, natural indium with different concentrations was deposited. Fig. 4 shows PAC-spectra for 0.04 monolayers indium on Cu(100) right after deposition at 78 K and after annealing to 240 K, respectively.

The PAC spectrum for the system as prepared exhibits a weakly damped modulation, which can be described with exactly the same field-gradient parameters as for the free copper surface labelled with $^{111}$In. This electric field gradient corresponds to PAC probes incorporated in the first copper layer (fraction $f_0$). After annealing to 240 K a splitting into three sets of transition frequencies occurs. The annealing behaviour of the different fractions related to the three field-gradient situations is depicted in Fig. 5 (left part). At around 240 K the undisturbed terrace site $f_0$ converts into fraction $f_0'$, which we identify with indium-indium pairs. A third somewhat smaller fraction $f_0''$ also appears within this temperature range, this fraction we might identify with small indium clusters.

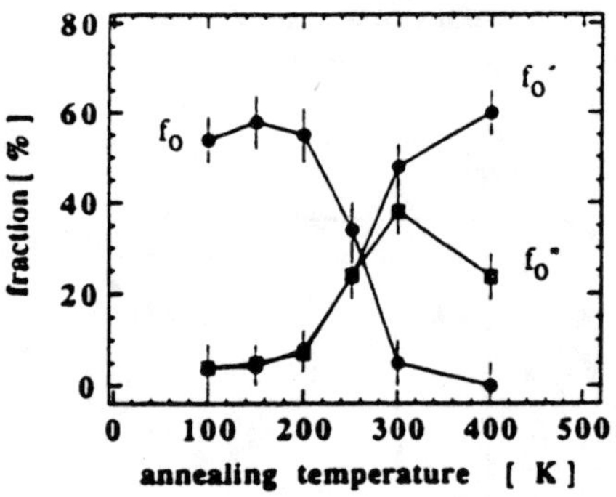


Fig. 5: Fractions of $^{111}$In probes, which are related to different indium configurations, on a Cu(100) surface covered with 0.04 monolayers natural indium.

The $^{111}$In fractions were also investigated as a function of measurement temperature, which was kept for several hours typically. This experimental dependence is shown in Fig. 5 (right part). Now with increasing temperature the fractions $f_0'$ and $f_0''$ are reduced in favour of $f_0$; above 500 K all indium-indium configurations are dissolved.

The observed clustering of adsorbats on a cubic surface has led to several theoretical investigations of such a lattice gas [10,11,12]. In these models an attractive nearest-neighbour indium-indium interaction is assumed and the thermal equilibrium of the different indium configuarations as a function of temperature and indium concentration has been determined. The agreement with the experimental observation is quite satisfying.

## 3. Indium Impurities on Semiconductor Surfaces

While metal surfaces have been extensively studied by PAC throughout the last couple of years, only recently first results on semiconductor surfaces have been reported [13,14]. The adsorption of In on the low-index planes of Si has been widely investigated by a variety of surface sensitive techniques, mainly because of its outstanding technological importance. The In/Si(111) and In/Si(100) interfaces are therefore well-characterized and serve as ideal test systems for first PAC experiments.

### *3.1. Si(111)7x7*

In Fig. 6 PAC spectra and corresponding Fourier-transforms for $^{111}In$ on Si(111)7x7 are shown for two different sample/detector geometries. The probes have been deposited onto the clean reconstructed surface at room temperature and the sample has been annealed to 550 K for 10 min prior to the PAC experiment. Two different electric-field-gradient situations ($f_1$, $f_2$) are observed indicating two different however similar adsorption sites on the Si(111)7x7 surface. As can be seen from the Figure, probe site $f_1$ is predominantly populated. Both electric field gradients show a slight asymmetry about the z-principal axis, which points perpendicular to the (111) surface plane. With decreasing temperature the strength of electric field gradients slightly increases, following a linear temperature dependence. The complete set of hyperfine parameters is listed in Table 2.

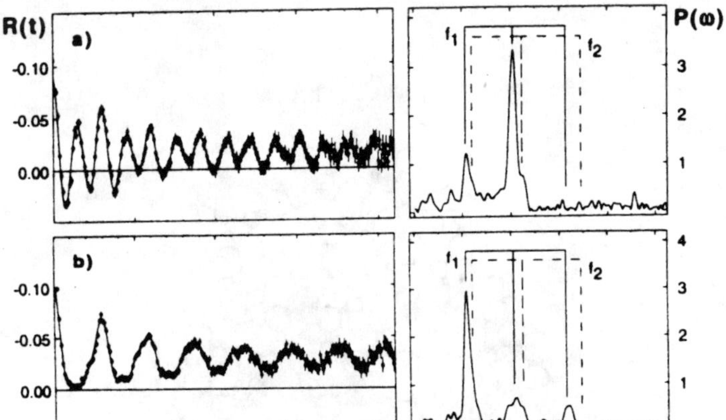


Fig. 6: PAC spectra and corresponding Fourier-transforms for $^{111}In/^{111}Cd$ on Si(111)7x7 after thermal treatment at 550 K for 10 min for two different sample-detector geometries: (a) surface normal within the detector plane, pointing under 45° between two adjacent detectors; (b) surface normal pointing under 80° to the detector plane. The solid lines represent least-squares fits to the data.

For a microscopic interpretation it is worth considering the results of a recent scanning tunneling microscopy study [15]. The authors investigated a Si(111) 7x7 surface, which had been covered by about 0.05 monolayers indium at a substrate temperature of about 700 K. After cooling to room temperature, the indium atoms were found to occupy two different adsorption sites, substituting part of the 12 silicon adatoms defining the unit cell of the (7x7) reconstruction (see Fig. 7). As a matter of fact, the adatomic positions at the edges of the unit cell were preferentially occupied on cost of the positions near the corner holes, indicating a small difference in binding energy between the respective sites. This finding is in perfect agreement with the PAC results shown above, if we identify $f_1$ and $f_2$ with $^{111}$In probes on edge positions and corner positions, respectively. Independent confirmation of this assignment comes from a theoretical work by Mohapatra *et al.* who determined the hyperfine fields for $^{111}$In/$^{111}$Cd on different adsorption sites on Si(111) in the frame of molecular cluster calculations [16]. Two different cluster geometries were taken into consideration, i.e. an atop geometry with the indium atom residing above a single silicon atom, and an eclipsed binding geometry with the indium atom being three-fold coordinated above three underlying silicon atoms. The results of this calculation are included in Table 2. Since no lattice vibrations are taken into account, the theoretical results should be compared with the experimental values extrapolated to T = 0 K. As can be seen from Table 2, the experimental results agree quite well with the data obtained for the eclipsed geometry. Neglecting possible changes in the bond length due to both reconstruction and relaxation, this geometry exactly resembles the nearest neighbourhood of the adatomic adsorption sites in the (7x7) unit cell. Deviation from axial symmetry as well as the slight difference in the strength of the

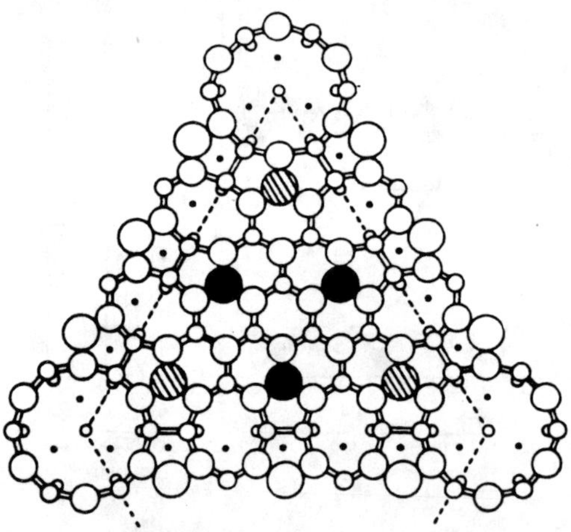

Fig. 7: Model of the (7x7) unit cell of the Si(111) surface. The first four Si layers are shown as open circles of successively decreasing size. Possible In positions are indicated by solid circles ($f_1$, edge site) and hatched circles ($f_2$, corner site).

respective field gradients are most likely due to the next-nearest neighbourhood of the $^{111}$In probes and therefore are not expected to be reproduced by this cluster calculation. The theoretical work therefore strongly corroborates the site assignment made above.

### *3.2. Si(100)2x1*

On Si(100), a different kind of experiment shall be discussed. Rather than investigating isolated $^{111}$In probes, sub-monolayer amounts of natural indium were deposited onto the clean, (2x1)-reconstructed surface and doped by the radioactive probes [14]. Indium is known to form an ordered (2x2) superstructure on Si(100)2x1 at room temperature, where the In atoms line up in dimer rows perpendicular to the silicon dimer rows of the (2x1) substrate reconstruction. This superstructure saturates at an indium coverage of 0.5 monolayers; however, already at much lower coverage the indium atoms agglomerate into long, isolated dimer rows [17]. Fig. 8 shows PAC spectra and corresponding Fourier-transforms for a Si(100)2x1 surface

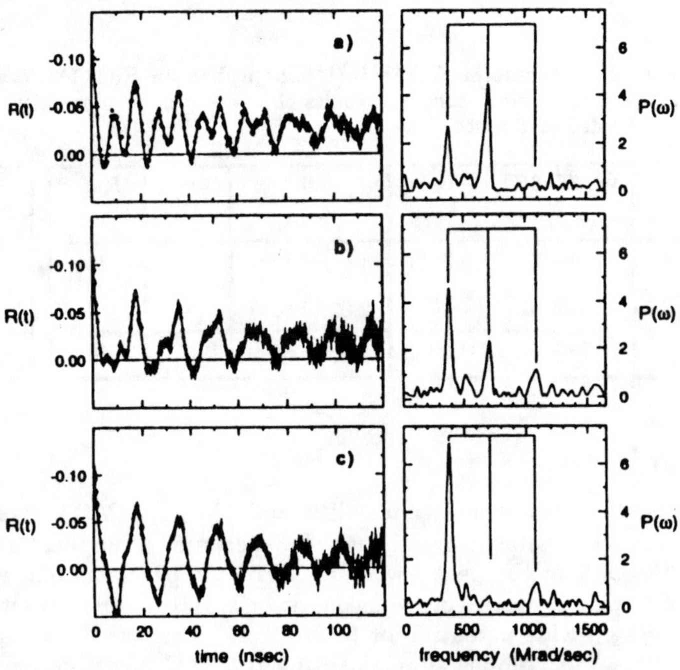


Fig. 8: PAC spectra and corresponding Fourier-transforms for $^{111}$In/$^{111}$Cd on Si(100)2x1 for different sample-detector geometries: (a) surface normal within the detector plane, pointing under 45° between two adjacent detectors; (b) surface normal pointing under 45° to the detector plane; (c) surface normal pointing perpendicular to the detector plane. The solid lines represent least-squares fits to the data.

covered with about 0.05 monolayers In. A unique, strong electric-field-gradient situation is observed, which may be attributed to $^{111}$In atoms being part of an indium dimer row. The principal axis of the electric-field-gradient tensor points perpendicular to the (100) surface plane and a small asymmetry about this axis is observed The strength of the field gradient is decreasing with increasing temperature; however, in contrast to the linear behaviour observed on Si(111)7x7, the data follow a $T^{3/2}$-dependence. The hyperfine parameters are as well included in Table 2.

This experiment has been performed for varying indium coverage; yet no coverage dependence of the hyperfine field is observed. We therefore take the observed electric field gradient as a microscopic fingerprint of the In(2x2) superstructure on Si(100)2x1. Due to the short range character of the electric field gradient, low coverage precursors of this superstructure can be identified in a coverage regime not accessible to standard diffraction techniques which sense long range order and hence require some minimum adsorbate coverage. This advantage has recently been used to study the low coverage precursor of the (2x2)→(4x3) superstructure phase transition for In on Si(100)2x1 [14].

Table 2: Hyperfine parameter for $^{111}$In/$^{111}$Cd probes on Si(111)7x7 and Si(100)2x1 surfaces. For comparison, the results of a recent molecular cluster calculation [16] are included (see text).

| | | $\nu_Q$ (T=0K) (MHz) | $\eta$ | B ($K^{-n}$) | n | Ref. | $\nu_{Q,theo.}$ (MHz) |
|---|---|---|---|---|---|---|---|
| Si(111)7x7 | $f_1$ | 240(3) | 0.20(3) | $2.81(4)\ 10^{-4}$ | 1 | 13 | 424 (atop) |
| | $f_2$ | 275(4) | 0.25(4) | $2.82(5)\ 10^{-4}$ | 1 | 13 | 282 (eclipsed) |
| Si(100)2x1 | | 393(5) | 0.23(2) | $6.80(4)\ 10^{-6}$ | 1.5 | 14 | – |

## 4. Conclusion

The present status of surface mobility studied by the PAC technique has been reviewed. It has been demonstrated that the spectrum of applications has grown considerably throughout the past few years, including problems like thermal stability of stepped surfaces. Moreover, semiconductor surfaces were investigated for the first time showing a wide potential for future work. The complementary use of different experimental techniques supplementing the PAC investigations has become increasingly important as the systems under consideration became more sophisticated. It is worth mentioning that the number of groups equipped with the necessary experimental set-ups is steadily increasing; this comprises in particular the successful installation of an in-beam UHV system at the on-line mass separator ISOLDE at CERN (Geneva) [19], where a rich variety of PAC probes will be disposable in near future.

## 5. Acknowledgment

The author is indebted to the Deutsche Forschungsgemeinschaft (Bonn), which has generously supported this work through the Sonderforschungsbereich 306.

## 6. References

1. G. Schatz, X.L. Ding, R. Fink, G. Krausch, B. Luckscheiter, R. Platzer, U. Wöhrmann and R. Wesche; Hyperfine Interactions **60** (1990) 975
2. B. Horn, E. Koch and D. Fick; Phys. Rev. Lett. **53**, 364 (1984); K. Wassmuth and D. Fick; Phys. Rev. Lett. **59**, 3007 (1987)
3. J. Korecki and U. Gradmann; Phys. Rev. Lett. **55** (1985) 2491 and Europhys. Lett. **2** (1986) 651 ; M. Przybylski and U. Gradmann; Phys. Rev. Lett. **59** (1987) 1152; J. Landes, Ch. Sauer, R.A. Brand, W. Zinn, and Z. Kajcsos; Hyperfine Interactions **57** (1990) 1941
4. T. Klas, R. Fink, G. Krausch, R. Platzer, J. Voigt, R. Wesche and G. Schatz; Europhys. Lett. **7** (1988) 151
5. R. Fink, R. Wesche, G. Krausch, T. Klas, R. Platzer, J. Voigt, U. Wöhrmann and G. Schatz; Surf. Sci. **225** (1990) 331
6. R. Fink, G. Krausch, B. Luckscheiter, R. Platzer, U. Wöhrmann, and G. Schatz; to be published
7. T. Engel: in Chemistry and Physics of Solid Surfaces VII (ed. R. Vanselow and R. Howe) Springer Series in Surface Science 10 (Berlin - Heidelberg) 1988
8. T. Klas, J. Voigt, W. Keppner, R. Platzer, R. Wesche and G. Schatz; Hyperfine Interactions **34** (1987) 577
9. R. Fink, B.-U. Runge, K. Jacobs, G. Krausch, J. Lohmüller, B. Luckscheiter, U. Wöhrmann, G. Schatz; to be published
10. R. Blender, W. Dieterich, H.L. Frisch and M. Yussouff; Z. Phys. **B71** (1988) 103
11. A. Natori, M. Baba, and N. Maruyama; Surf. Sci. **233** (1990) 392
12. K. Froböse; J. Phys. Cond. Matter **2** (1990) 5933
13. G. Krausch, T. Detzel, R. Fink, B. Luckscheiter, R. Platzer, U. Wöhrmann and G. Schatz; Phys. Rev. Lett. **68** 1992) 377
14. G. Krausch, K. Jacobs, R. Fink, J. Lohmüller, B. Luckscheiter, B.-U. Runge, U. Wöhrmann and G. Schatz; Europhys. Lett. **19** (1992) 611
15. J. Nogami, Sang-il Park and C.F. Quate; J.Vac.Sci.Tech. **B4** (1988) 1479
16. S.M. Mohapatra, B.N. Dev, L. Lou, T. Thundat, W.M. Gibson, K.C. Mishra, N. Sahoo, T.P. Das; Rev. Sol. State Sci. **4** (1990) 873
17. A.A. Baski, J. Nogami, C.F. Quate; Phys. Rev. **B43** (1991) 9316
18. R. Fink, B.-U. Runge, K. Jacobs, G. Krausch, J. Lohmüller, B. Luckscheiter, U. Wöhrmann, G. Schatz; Hyperfine Interactions 1993, in print
19. E. Hunger and H. Haas; Surf. Sci. **234** (1990) 273

Local Order
in Condensed
Matter Physics


ISBN 1-56072-220-7

# THEORY OF ELECTRIC FIELD GRADIENTS AND HYPERFINE FIELDS ON SURFACES

Bengt Lindgren

*Department of Physics, University of Uppsala*
*Box 530, S-751 21 Uppsala, Sweden*

## ABSTRACT

The hyperfine field and the electric field gradient (EFG) on Cd probe atoms at different Cu, Ag, Ni and Co surfaces are calculated self-consistently within the local density molecular/cluster approach. The EFG:s at substitutional terrace sites are of the order of $10^{18}$ V/cm$^2$ while at adatom sites they are an order of magnitude smaller . The hyperfine fields at the substitutional sites are comparable with the bulk values and the adatom values are about a factor of two smaller.

## 1. INTRODUCTION

Investigation on surfaces utilizing nuclear methods have become subject of growing interest [1]. The electric field gradients (EFG) may here give detailed information about the site of the nuclear probe, chemical bonding and the local structure, in particular when experimental results [2,3] are compared with first-principle calculations [4]. The hyperfine fields, on the other hand, gives a local information about the magnetization of surfaces, interfaces and in multilayers. Such systems with perpendicular magnetization are of large technical importance for e.g. magnetic recording. However, the hyperfine field is usually not directly proportional to the magnetization. Impurity and surface associated "resonance" states in the density of states may give different contributions to the hyperfine field and once again first principle calculations are required to relate the local magnetic field with the host magnetization.

In this paper, the EFG of Cd probe atoms at different sites at Cu, Ag, Co and Ni surfaces, are calculated and compared with experimental results obtained by perturbed angular correlation (PAC) in Konstanz [1-3]. These calculations are also extended to the magnetic Ni and Co surfaces where new results of the Cd-hyperfine field at Ni are available from the Konstanz group [5-6].

## 2. METHOD

### 2.1. Computational procedure

The theoretical basis for the electronic-structure calculations is the self-consistent one-electron local density formalism in the spin unrestricted Hartree-Fock-Slater model [7]. The essential point of this theory is the replacement of the nonlocal Hartree-Fock exchange operator by a potential depending only on the local electron density.The Hedin-Lundqvist [8] and the Barth-Hedin potential [9] was adopted for the exchange-correlation potential.

As in the usual linear-combination-of-atomic-orbital-molecular-orbital (LCAO-MO) method, the molecular-orbital eigenfunctions are expanded in terms of symmetry orbitals chosen here as linear combinations of atomic orbitals located on the different atoms in the molecule. The atomic radial wave functions are obtained numerically by solving a self-consistent free atom or ion problem. In this work a double basis composed of neutral and ionized atoms was used for the Cd atom and its nearest neighbor host atoms. This kind of basis have enough variational freedom to reproduce all major rearrangements in the valence electron density [10]. Core orbitals were "frozen", *i.e.* the cores were only present in the overlap matrix but not in the Hamiltonian matrix.

The expansion coefficients (variational coefficients) $C_{ji}$ are obtained using standard procedures by solving the matrix secular equation $(H - \varepsilon S)C = 0$, where $\varepsilon$ is the eigenvalue matrix. The Hamiltonian matrix H and the symmetry orbital overlap matrix S are obtained in the discrete variational method (DVM) [11] as a weighted sum over a set of sample points. A pseudo random Diophantine distribution [12] is mapped onto an infinite domain representing the region exterior to atomic spheres of some arbitrary radius, using a Fermi distribution and in the core regions near the nucleus, an optimized Gaussian surface mesh is used in conjunction with a radial Simpson's rule scheme in spheres around the nuclei.

The Fermi energy and occupation numbers $f_i(\varepsilon)$ for each molecular orbital (MO) are obtained by using Fermi-Dirac statistics on these MO eigenstates. The cluster charge density was then constructed by summing over all MO's. In order to calculate the potential by one-dimensional integration, this charge density was cast in a multicenter-overlapping multipolar form [10], determined by least-squares fitting to the eigenvector density.

In the self-consistent multipolar (SCM) procedure the potential calculated from this representation of the cluster density is used to determine new wave functions and hence a new cluster density, until self-consistency is obtained.

### 2.2. Cluster embedding

When representing a solid by a finite cluster of atoms, it is necessary to consider the embedding problem. For treating some properties of metals which may depend more or less sensitively on the band structure, it is desirable to broaden the discrete energy levels of the isolated cluster into bands. This can be accomplished by empirical level-smearing schemes - a technique often used to draw smooth density of state curves from the discrete energy levels, or by modifying the cluster boundary conditions. The asymptotic form of the wave function can be controlled by mixing in a selected continuum state into the cluster wave functions and solving the inhomogeneous Schrödinger equation [13]. The net result is an energy-dependent broadening of cluster levels and a smooth spectral representation of properties like the charge and spin densities.

In this work the empirical level smearing was used. Each energy level was broadened by

$$L_i(E) = (\beta/4)\cosh^{-2}[\beta(E-\varepsilon_i)/2] \qquad (1)$$

The actual population for each single-particle state is obtained by integrating up to the Fermi energy $\varepsilon_F$

$$f_i = \int_{-\infty}^{\varepsilon_F} L_i(E)\,dE = \{1 + \exp[\beta(\varepsilon_i - \varepsilon_F)]\}^{-1} \qquad (2)$$

The choice of smearing function was arbitrary. Fine details in the DOS are not significant since it is only the total occupation of each state that is controlled self-consistently through the eigenvector charge density. However, the broadening parameter $\beta$ depended on the amount of 3d-character of each state since 3d-bands are narrower than sp-valence bands. For 3d contributions to the energy states, $\beta$=300 (a.u.) was chosen while for sp contributions, $\beta$=50. Hence pure Ni and Co 3d-energy levels were 6 times sharper than the sp-states ensuring a good determination of the magnetic moments while keeping the accuracy of the hyperfine field and electric field gradient (EFG). These mainly originate from Cd s- and p-states respectively. Fig. 1 shows the DOS from a $CdCo_{54}$ cluster corresponding to bulk Co-metal. In a few cases, the more time consuming but parameter free plane-wave matching of the cluster wave functions [13], were also used but without any significant change of the results.

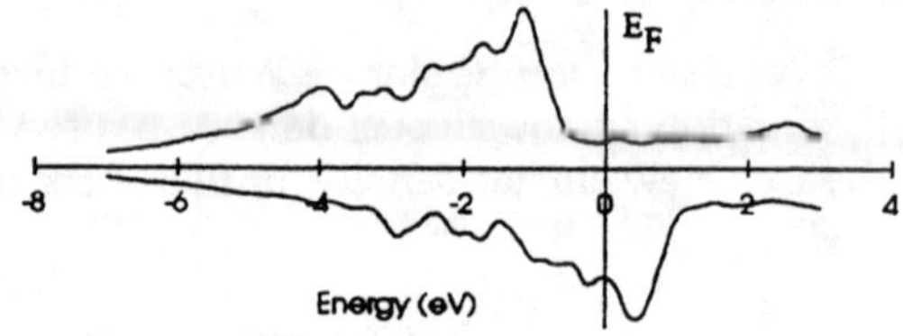


**Figure 1.** Density of states as obtained from a $CdCo_{54}$ "bulk" cluster.

### 2.3. Electric field gradient and hyperfine field

From the obtained wave functions the charge density $\rho(\mathbf{r})$ was calculated and the principle component of the electric field gradient (EFG) by:

$$V_{zz} = - \int \rho(\mathbf{r}) \, (3 \cos^2\Theta - 1) \, r^{-3} \, d\mathbf{r} + V_{zz}(\text{nuclear}) \qquad \text{(atomic units)} \qquad (3)$$

where the second term is a summation over the nuclear or ionic point-like charges with the same geometric weight factor as in the first term. The conversion factor between atomic and SI units is $9.72 \times 10^{17}$ V/cm$^2$au. The dominant contribution to the total EFG is found to originate from the p-character of the valence orbitals [14] and core states are mainly contributing through the orthogonalization requirement to the valence states.

When the point group has a three- or fourfold symmetry axis (z), the $p_x$ and $p_y$ orbitals are degenerate and the EFG is axially symmetric along this axis. For twofold symmetry, e.g. a (110) surface, an asymmetry factor is defined by $\eta = (V_{xx} - V_{yy})/V_{zz}$, with the principle axis chosen such that $|V_{zz}| \leq |V_{yy}| \leq |V_{xx}|$. For still lower point group symmetries, e.g. step sites, the coordinate system of the cluster does not necessary correspond to the EFG principle-axis system and a diagonalization of the full EFG tensor must be performed.

For the non-magnetic Cd atom without any own moment, the main contribution to the local magnetic field is coming from the Fermi contact hyperfine field ($B_{hf}$). This is calculated (non-relativistically) from the spin density at the Cd nucleus by:

$$B_{hf} = [\, \rho_{majority}(0) - \rho_{minority}(0) \,] \; 52.42 \; (\text{Tesla}/a_0^3) \qquad (4)$$

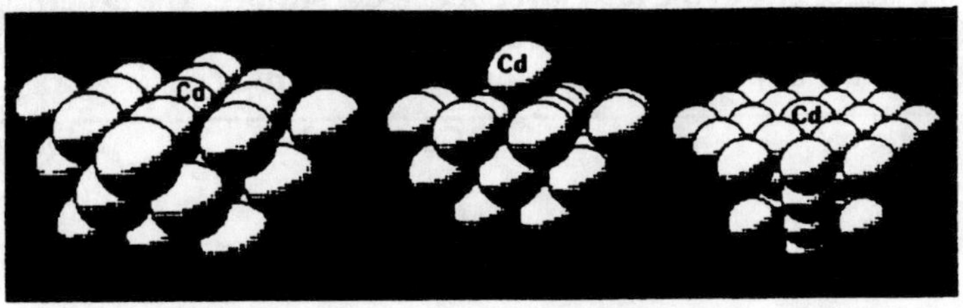


**Figure 2.** Some of the cluster geometries used to represent Cd on different fcc metal surfaces. a) $CdNi_{33}$: substitutional site on Ni(100). b) $CdNi_{25}$: adatom site on Ni(100). c) $CdNi_{31}$: substitutional site on Ni(111).

## 3. EFG - RESULTS

Different cluster-geometry representing different surfaces are shown in Fig.2. All host interatomic distances were representative for the bulk metal. An optimization of the geometry with minimization of the total energy is not possible for these large clusters with the present numerical integration accuracy.

The calculated field gradients [4,15] are summarized and compared with experimental values [1-3,5,6] in table 1 and the EFG-dependence of the Cd-probe distance to the host layer is shown in Fig. 3 for the Cu(100) and Co(100) surfaces.

The experimental results for a Cu(100) surface show an axially symmetric EFG with the symmetry axis pointing perpendicular to the surface. There are two such possible axis, one which starts from a substitutional site in the surface layer and one which starts from the interstitial bridge position and passes through a regular adatom site in the next empty layer. For all reasonable distances to the surface the EFG along the second line is much smaller than the values obtained in the first experiments while the agreement with the substitutional terrace site is quite satisfactory. Moreover, recent low-energy ion scattering (LEIS) studies on In atoms on a stepped Cu(100) surface [16] show that the In atom in the terrace have a position slightly above (about 0.4Å) the Cu surface monolayer. This explains the remaining difference in calculated and experimental EFG:s.

**Table 1.** Calculated EFG ($10^{17}$ V/cm$^2$) for different surface geometry. "Adatom" referes to a lattice site in the next layer above the surface. The experimental values (at 77 K) are from the Konstanz group [1-3,5,6] except for the Co-value (at 300K) which is from the author. The experimental sign is unknown.

| Cluster geometry | Calculated | | | | Experimental | |
|---|---|---|---|---|---|---|
| | $V_{zz}$ | $V_{xx}$ | $V_{yy}$ | $\eta$ | $V_{zz}$ | $\eta$ |
| $CdCu_{21}$ (100) substitutional | +10.3 | | | | 9.0 | 0.00 |
| $CdCu_{21}$ (100) adatom | -0.8 | | | | 0.8 | 0.00 |
| $CdCu_{38}$ (111) substitutional | +11.3 | | | | 10.1 | 0.00 |
| $CdCu_{25}$ (111) adatom | -6.2 | | | | | |
| $CdCu_{29}$ (110) substitutional | +10.6 | -12.1 | +1.5 | 0.75 | 7.9 | 0.74 |
| Surface relaxed -10% | +11.2 | -11.2 | 0.0 | 1.00 | | |
| $CdCu_{17}$ (110) substitutional | +8.4 | -7.7 | -0.7 | 0.83 | | |
| $CdAg_{21}$ (100) substitutional | +7.2 | | | | 7.5 | 0.00 |
| $CdAg_{38}$ (111) substitutional | +8.7 | | | | 8.6 | 0.00 |
| $CdNi_{33}$ (100) substitutional | +11.6 | | | | 8.2 | 0.00 |
| $CdNi_{25}$ (100) adatom | +1.3 | | | | 2.8 | 0.00 |
| $CdNi_{31}$ (111) substitutional | +11.6 | | | | 11.5 | 0.00 |
| $CdCo_{33}$ (100) substitutional | +11.5 | | | | 9.5 | 0.00 |
| $CdCo_{25}$ (100) adatom | +0.8 | | | | | |

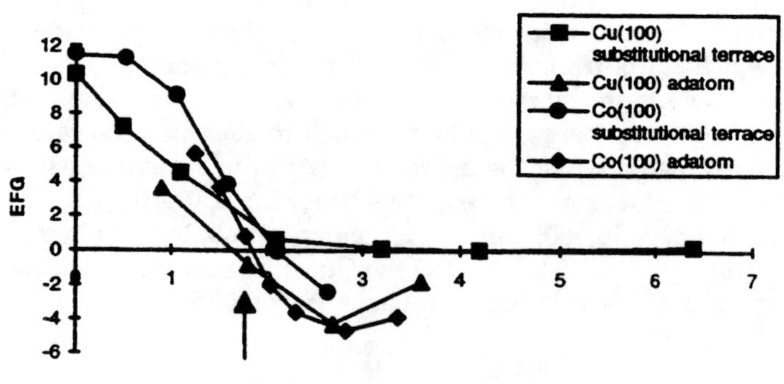


**Figure 3.** The EFG ($10^{17}$V/cm$^2$) calculated on Cd at different distances to the host surface layer. The arrow indicates the adatom site in the next empty layer.

Later experiments with In (the parent nuclei to Cd) deposited at low temperatures have also seen the adatom EFG value. The EFG at this site is an order of magnitude smaller than at the substitutional terrace site. If only equivalent charges and potentials external to the Cd atom are considered, the two sites are complementary (Fig. 4) and only a change in the sign of the EFG would be expected for the two sites. Clearly the difference must be found in the electron charge distribution local to the Cd probe atom and in particular in the 5p-character of the valence charge which contributes to the main part of the EFG. The large positive EFG at the substitutional site may either be due to a population difference, with more Cd $5p_x$ and $5p_y$ orbitals in the plane than $5p_z$ perpendicular to the plane, or to a polarization of these orbitals, i.e. they have a different radial extensions. With the used variational basis set the radial shape of the 5p charge distribution is primarily governed by the use of the neutral $Cd^0$ and the $Cd^{2+}$ atomic 5p orbitals. In table 2 the Mulliken populations [17] of these orbitals are given together with the EFG expected from only these charges. The population difference is small and would actually give the wrong sign of the EFG since e.g. the $p_z$ electrons are in majority for the substitutional site and they contributes negatively to the EFG. The polarization effect accounts for most of the final EFG. Both the $5p_z$ and the $5p_x$, $5p_y$ charge distributions are contracted compared to a free neutral Cd atom as expected since 5p is an excited state in the free atom. For the adatom site the $5p_z$ is slightly more contracted than the $5p_x$, $5p_y$ due to the interaction with the charge from the host atoms in the plane below. However, this difference is small since the nearest neighboring host atoms are not lying in the direction of the p-lobe which is the case for the $5p_x$, $5p_y$ lobes at the terrace site. Hence the 5p charge distribution in the plane is much more shifted towards the Cd nucleus with a large positive EFG as a result. The charge density maps in Fig. 5 also indicate this. At a distance of 4Å the Cd atom is essentially a free atom with no significant overlap to the surface charge density and hence, almost no EFG. At 3Å a small lobe down to the surface develops and creates a small negative EFG. Still closer to the surface layer the charge in the plane dominates and the EFG becomes positive.

The 4p-orbitals also give a considerable contribution to the total EFG. The same result is obtained with and without the Cd-4p orbitals frozen so the population difference of the $4p_z$ and $4p_{x,y}$ is entirely governed by the overlap with the host atom orbitals. However, if the EFG is calculated from the same eigenvectors but without the core othogonalization, the EFG of the substitutional sites are reduced by more than 50%. The main contribution comes from the $\langle 4p|r^{-3}|5p\rangle$ integrals with somewhat less from the $\langle 4p|r^{-3}|4p\rangle$ integrals. Hence there is no unambiguous way of separating the total EFG into a core and a valence part since the $\langle 4p|r^{-3}|5p\rangle$ contribution depends much on the 5p occupation.

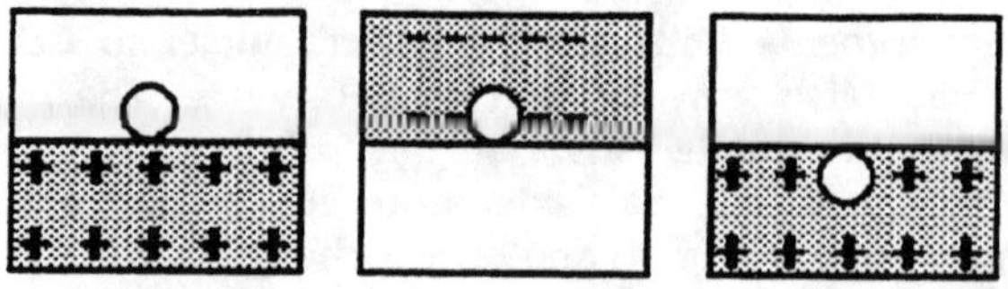

**Figure 4.** Schematic picture of the adatom site (left) and the complementary substitutional site in the layer (right). The adatom site corresponds to the bulk structure with zero EFG, but with the upper half of the host charges removed. Hence the system illustrated in the middle figure has the same EFG as the adatom and would only differ by the sign compared to the substitutional site if only external charges to the probe atom were considered.

**Table 2.** Mulliken populations of the Cd 5p-orbitals from the $Cd^0$ and $Cd^{2+}$ basis, respectively, and the EFG ($V_{zz}$) deduced from atomic $<r^{-3}>$ integrals: $V_{zz}(5p_z{}^0) = -28.4$, $V_{zz}(5p_z{}^{2+}) = -74.9\ 10^{17} V/cm^2$ and half the positive values for the $5p_x$ and $5p_y$ orbitals. From just the total population difference, with more $5p_z$ electrons than $5p_x$ and $5p_y$ respectively, a negative EFG is expected for the substitutional terrace site. Similarly a positive EFG is expected for the adatom site, opposite to what is found if also the polarization is taken into account (last column).

| Site | $5p_z$ | | | $5p_x$, $5p_y$ respectively | | | $V_{zz}$ |
|---|---|---|---|---|---|---|---|
| | $Cd^0$ | $Cd^{2+}$ | Total | $Cd^0$ | $Cd^{2+}$ | Total | $10^{17} V/cm^2$ |
| Cu(100) substitutional | -0.114 | 0.245 | 0.131 | -0.415 | 0.487 | 0.072 | +9.6 |
| Cu(100) adatom | -0.084 | 0.176 | 0.092 | 0.020 | 0.128 | 0.148 | -0.6 |

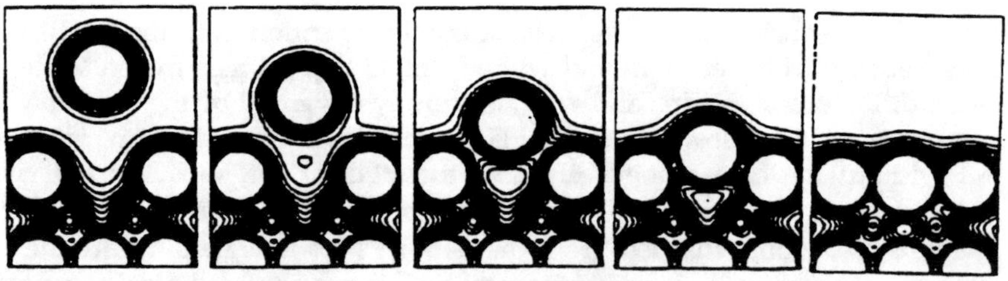

**Figure 5.** Charge densities for different Cd distances (4.2, 3.2, 2.1, 1.1 and 0 Å) to the Cu(100) surface layer. Contour level intervals are 0.002 au.

The Cu(111) results resemble those at the Cu(100) surface. The larger value is explained by the denser packing of atoms in the (111) surface compared to the (100) surface. Calculations with a much smaller cluster ($CdCu_9$) gives an EFG of $7.8 \cdot 10^{17}$ V/cm$^2$. This value is only somewhat smaller than the one obtained by using a larger cluster and shows that most of the EFG is actually determined by the interaction with the nearest neighboring atoms. This has also been observed by Li et al. [18]

Ag, Co and Ni behaves much the same with a difference to Cu that relates to the lattice constant. The Co and Ni EFG:s may also dependent slightly on the unclosed host 3d-shell but calculations with nonfrozen Cu 3d-shell do not show any significant difference. The distance dependence of the EFG (Fig. 3) is also slightly different for the Cu and Co hosts.

The Cu(110) compares less good with experiments. An unsatisfactory result in itself is also that a smaller cluster $CdCu_{17}$ gives better values than the bigger $CdCu_{29}$ cluster. However, an anisotropic local relaxation of the Cu positions is quite plausible for this kind of surface with low surface energy. Since the two largest EFG components are quite similar in magnitude, they are sensitive to any relaxation. For instance, a 10% relaxation of the surface monolayer towards the bulk, which is close to the experimental value [19] 8.5%, turns the major component perpendicular to the surface.

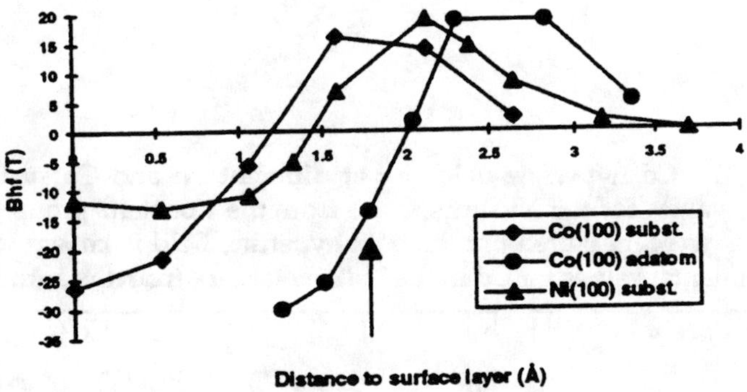


**Figure 6.** The variation of the Cd hyperfine field (Bhf) with the distance to Ni and Co surfaces. The arrow indicates the adatom site in the next empty layer.

## 4. HYPERFINE FIELDS

Table 3 summarizes the results. The calculated magnetic moment of the center Ni atom in a $Ni_{55}$-cluster is 0.52 $\mu_B$ and about the same for the corresponding Ni atom at a (100) surface. No $Co_{55}$ cluster was calculated but the average Co moment in the $CdCo_{54}$ cluster was 1.72 $\mu_B$ with almost no change for the surface clusters.

The hyperfine field on Cd in bulk Ni (from a $CdNi_{54}$ cluster) is larger in magnitude than the experimental value while in Co, with a larger band splitting, the agreement is much better. Calculations with the KKR Green's function method [20] also gives a too large value (-8.9T) in Ni. Local relaxation of the Ni sites close to the Cd impurity atom might be one possible explanation.

The calculated hyperfine field of CdNi(111) is slightly larger than the experimental value. If the Cd atom is pushed up from the surface, as is known to be the case on Cu(100) [16], the value is becoming less negative and at about 2 Å above the surface it is positive (+13 T). The same behavior for Cd on Ni(100) is shown in fig. 6. However, the difference between theory and experiments for the substitutional and adatom sites on Ni(100) are significant. There is no explanation to this at the moment except that the measurements might have been affected by carbon contamination as reported by the Konstanz group [6].

The hyperfine fields for the substitutional terrace and adatom sites scale approximately with the number of nearest host atoms but this simple relation is not valid when comparing the terrace surface site and the substitutional site in the bulk metals (table 4).

**Table 4.** The Cd hyperfine field $B_{hf}$ at different Ni and Co surfaces. The experimental values for the Ni surfaces are from the Konstanz group [5,6]. No sign of the fields were measured but the bulk hyperfine field is known to be negative. The experimental values for Cd in the bulk metals are from the author.

| Surface and position | Ni | | Co (fcc) | |
|---|---|---|---|---|
| | $B_{hf}$ (T) | Expt. (T) | $B_{hf}$ (T) | Expt. (T) |
| (100) substitutional | -11.7 | 3.9 | -26.4 | |
| (100) adatom site | -5.8 | 7.2 | -13.9 | |
| (111) substitutional | -7.3 | 6.7 | | |
| Bulk | -11.2 | -7.2 | -27.0 | -26.7 |

## 5. SUMMARY

All essential features of the EFG and the hyperfine field of Cd probe atom at various Cu, Ag and Ni surfaces are well reproduced by first-principle molecular/cluster calculations, using reasonable large clusters. In particular, the calculations can help in determining the probe site which not always is known directly from the experiments. Improvements of the calculations should involve a matching of the molecular wave functions to a bandstructure solution of the bulk, similar to what is done in the KKR Green's function method [20], but also determinations of possible local relaxation around the probe atom. For small clusters this has be done by minimizing the total energy [18] but at the moment the numerical integration accuracy is not good enough for the large clusters used in this work. Gaussian basis sets give in general more accurate total energies but the numerical basis functions are crucial for an accurate determination of local properties as the EFG and the hyperfine field.

## REFERENCES


1. G.Schatz, T.Klas, R.Platzer, J.Voigt and R.Wesche, Hyp.Int. **34**(1987)555.
2. T.Klas, R.Fink, G.Krausch, R.Platzer, J.Voigt, R.Wesche and G.Schatz, Surf.Sci. **216**(1989)270
3. R.Fink, R.Wesche, T.Klas, G.Krausch, R.Platzer, J.Voigt, U.Wöhrmann and G.Schatz, Surf.Sci. **225**(1990)331.
4. B.Lindgren, Europhys.Lett. **11**(1990)555.
5. J.Voigt, R.Fink, G.Krausch, B.Luckscheiter, R.Platzer, U.Wöhrmann, X.L.Ding and G.Schatz, Phys.Rev.Lett. **64**(1990)2202.
6. J.Voigt, thesis, Konstanz 1990.
7. J.C.Slater, *The Self-Consistent Field for Molecules and Solids* (McGraw-Hill, 1974).
8. L.Hedin and B.I.Lundqvist, J.Phys. C **4** (1971) 2064.
9- U. von Barth and L.Hedin, J.Phys.C **5**(1972)1629.
10. B.Delley and D.E.Ellis, J.Chem.Phys. **76**(1982)1949.
11. D.E.Ellis and G.S.Painter, Phys.Rev. **B2**(1970)2887.
12. D.E.Ellis, Int.J.Quantum.Chem. **2S**(1968)35.
13. B.Lindgren and D.E.Ellis, Phys.Rev. **B26**(1982)636.
14. B.Lindgren, Phys.Rev. **B34**(1986)648.
15. B.Lindgren and A.Ghandour, Hyperf. Interact., accepted 1992.
16. M.Breeman, G.Dorenbos and D.O.Boerma, Nucl.Instr. and Meth. **B64**(1992)64.
17. R.S.Mulliken, J.Chem.Phys. **23**(1955)1833, 1841.
18. Yi Li, M.R.Press, S.N.Kanna, P.Jena and M.Yussouff, Phys. Rev. **B41**(1990)4930.
19. D.L.Adams, H.B.Nielsen and J.N.Andersen, Surf.Sci. **128**(1983)294.
20. P.H. Dederichs, R.Zeller, H.Akai, S.Blügel , A.Oswald, Phil.Mag. 51(1985)137.

Local Order
in Condensed
Matter Physics


ISBN 1-56072-220-7

# DYNAMICS OF A 2D FLUID - EFFECTS OF SUBSTRATE AND LOCAL ORDER

S. D. Mahanti and Hyangsuk Seong*

*Department of Physics and Astronomy and
the Center for Fundamental Materials Research
Michigan State University, East Lansing, MI 48824, USA*

**Abstract.** Dynamics of a classical screened Coulomb (Yukawa) fluid confined to move in two dimensions (2D) in the presence of a periodic substrate potential has been studied using molecular dynamics (MD) simulation. We find that the self dynamic structure factor (SDSF), $S_S(q,\nu)$, changes dramatically as we tune the strength of the substrate potential starting from a smooth substrate (homogeneous fluid) to a strongly corrugated substrate (lattice-fluid). In the latter case, for large values of the wave vector q, SDSF can be described approximately by a superposition of two types of central peaks, one extremely narrow (solid-like) and the other relatively broad (liquid-like). This observation is in agreement with neutron scattering results of Zabel et al and earlier MD simulation results of Fan et al. In addition, we see that the SDSF has a broad finite frequency peak whose spectral weight increases with increasing q. We present a simple oscillator model to explain certain spectral features of the SDSF associated with 'solid-like' dynamics caused by local order. For small values of q, the central peak is dominated by the liquid-like diffusive dynamics even in the presence of strong substrate potential.

## 1. INTRODUCTION

Dynamics of homogeneous fluids in three dimensions (bulk) is reasonably well understood primarily due to recent advances in large scale simulation and development of mode-coupling theories [1-3] although many fundamental issues near the liquid-glass transition still remain to be sorted out.[4] During the past several years, there have been considerable experimental and theoretical interest to understand the dynamic properties of fluids subjected to different types of external constraints.[5-15] Typical examples of such systems are: two-dimensional fluids in the presence of an external periodic potential as in overlayer systems [7], intercalation compounds [8-12] and fluids confined in porous media such as vycors, zeolites etc.[13-15] In these systems, due to the inhomogeneous nature of the external constraints, fluid particles exhibit different types of dynamic behaviour in

*Département de physique, Université de Montréal, Montreal, Québec H3C 3J7 Canada.

different length and time scales. For example, when a fluid flows through a mesoporous medium, due to the strong attraction between the fluid particles and the pore wall, a solid-like layer forms near the pore walls. In this case one expects to see a co-existence of both solid like and liquid-like dynamics.

In this paper, we address the dynamics of a specific type of a constrained fluid where the fluid particles move between two atomic planes separated by a few A$^{o}$. In addition to geometrical confinement these planes exert a periodic substrate potential on the fluid particles. Intercalation compounds are excellent examples of such a fluid where the intercalant atoms are confined to move in two dimensions between two host planes. Zabel and coworkers [9,10] have carried out extensive neutron scattering measurements in graphite intercalation compounds (GIC's) to probe the dynamic properties of the intercalants in the fluid phase. In addition, Moss and coworkers [11] have carried out extensive molecular dynamics simulation for one particular system, $RbC_{24}$, a stage-2 GIC to understand the results of these neutron scattering measurements. We will discuss their major findings when we present our own results. Our main interest here is not only to see whether we can understand the experimental neutron scattering results, but also to investigate how different characteristic features of the fluid dynamics evolve as we change the strength of the periodic potential. In the limit of very large substrate corrugation, one expects the particles to be trapped at the minima of the substrate periodic potential (at a set of lattice sites determined by the substrate) and hop between these lattice sites, thus forming a lattice-liquid. One should therefore expect to see dramatic changes not only in say the single particle diffusion characteristics but in collective dynamics as well. [16]

The arrangement of the paper is as follows: In sec.II, we discuss some of the important results of neutron scattering experiments. In sec.III, we describe the model system which we have simulated and a very brief description of the simulation method. In sec. IV, we discuss our results for (1) a particular value of the strength of the substrate potential which describes $RbC_{24}$ very well [17] and compare them with earlier MD results of Fan et.al [11] and with experiment [9,10], and (2) different values of the corrugation strength focussing on the evolution of the spectral features of the SDSF $S_S(q,\nu)$ at different length ( $2\pi/q$) and frequency ($\nu$) scales. Finally in sec.V, we give a brief summary of our major findings.

## 2. DYNAMICS PROBED THROUGH NEUTRON SCATTERING

The dynamic response of a fluid can be probed through neutron scattering (NS) measurements. The physical quantity related to the NS cross section is the intermediate scattering function F(**q**,t) given by

$$F(\mathbf{q},t) = \left< 1/N \sum_{<lm>} e^{i\mathbf{q}\cdot(\mathbf{r}_l(t) - \mathbf{r}_m(0))} \right>, \qquad (1)$$

where $\mathbf{r}_l(t)$ is the position of the l-th particle at time t and the angular brackets mean average over a set of initial configurations consistent with the external constraints such as fixed temperature, pressure etc. S(**q**,$\nu$), the Fourier transform of F(**q**,t), is the dynamic structure factor which is directly related to the neutron scattering cross section. When the intercalants are predominantly incoherent scatterers or when q is large, the incoherent part $S_s(\mathbf{q},\nu)$, the Fourier transform of $F_s(\mathbf{q},t)$ (which corresponds to the $l = m$ term, the so called *self term* in Eq.(1)) dominates the NS spectra. $S_s(\mathbf{q},\nu)$ directly gives the dynamics of a tracer particle and from its small q and ν behaviour, one can extract the diffusive nature of the tracer paricle dynamics in a fluid.

In their NS experiments in stage-2 $RbC_{24}$, Zabel et al. [9,10] found that in the fluid phase, for $0.75 < q < 2.2$ (in units of $A^{o-1}$), $S(q,\nu)$ consisted of a central peak and a broad finite-frequency peak. They observed that the central peak could not be fitted by a single Lorentzian corresponding to the normal diffusive dynamics of a simple fluid. They could fit their data by a superposition of two Lorentzians, one rather broad and another very sharp and resolution limited. They identified the broad peak with liquid-like diffusion and the narrow quasi-elastic peak with a solid-like dynamics which they thought was a characteristics of a domain-wall solid (supposed to be the ground state of $RbC_{24}$) above its melting temperature. As we will discuss in sec.III, Fan et al. [11] gave a rather nice physical picture of this narrow central peak based on their MD simulation studies. In contrast to the above overall behaviour, a homogeneous fluid (either in 3d or in 2d) shows a diffusive Lorentzian peak for small q and a Gaussian for large q [1]. For intermediate values of q, the spectrum deviates from a Lorentzian and cannot be described by a superposition of two Lorentzians, one narrow and the other broad.

The self diffusion constant D can be directly obtained from the Lorentzian width ( $\Gamma_s$) of $S_s(\mathbf{q},\nu)$ for small values of q when $\Gamma_s \approx Dq^2$. On the other hand for coherent scatterers such as Rb, the small q NS cross section is dominated by the coherent scattering cross section $S(\mathbf{q},\nu)$. Zabel et al [10,11] used an approximation proposed by Sköld [18] to extract D from the width (Γ) of the coherent scattering cross section; $\Gamma \approx Dq^2 / \sqrt{S(q)}$, where S(q) is the static structure factor. Another way to access the tracer dynamics is to go to larger values of q for which the coherent and the self dynamic structure factors are nearly equal. However in this limit the SDSF is not a simple Lorentzian. Thus an accurate determination of D from NS measurements in physical systems with dominant coherent scatterers is difficult. In such systems MD simulation provides an excellent means to obtain D because one can directly calculate $S_s(\mathbf{q},\nu)$ for small values of q (consistent with finite size limitations). In addition, one can also explore how different spectral features of the tracer dynamics change at different length scales as one varies the strength of the substrate potential and approaches the homogeneous fluid limit.

## 3. MODEL SYSTEM AND MOLECULAR DYNAMICS SIMULATION

The model physical system we have studied consists of charged particles interacting via a short-range repulsive screened Coulomb potential (appropriate for intercalants in GIC's)

$$V_{int} = e^2 \sum_{<ij>} \exp(-\Gamma r_{ij}) / r_{ij}, \tag{2}$$

where e is the charge of the intercalants and Γ is an appropriately chosen screening parameter.[12] The substrate potential has the periodicity of the graphite lattice and was obtained for $RbC_{24}$ by Moss et al. [17] from experiments (through X-ray structure factor in the liquid state) and is given by

$$V_{subs}(\mathbf{r}) = -2K\,[\,2\cos(2\pi x/a)\cos(2\pi y/\sqrt{3}\,a) + \cos(4\pi y/\sqrt{3}a)]\,. \qquad (3)$$

The parameters appropriate for $RbC_{24}$ are $\Gamma = 0.49$ $A^{o-1}$, $a = 2.46$ $A^o$, $2K = 0.9$ (in units of equivalent temperature 300K). The planar density of intercalants is taken to be 0.0318 $A^{o-1}$. The barrier height ($\Delta$) for the potential $V_{subs}$ is $8K$ and for $RbC_{24}$ it is equal to 1080K. To study the evolution of the dynamics from a homogeneous to lattice-fluid limit, we increase $2K$ from 0.0 to 0.9 through intermediate values, 0.15, 0.36, 0.55, and 0.75. In our present study, we have chosen a 864 particle system with periodic boundary condition. To extract the low-frequrncy behaviour of SDSF accurately, our maximum MD time runs are of the order of 100-120 ps. Details of our simulation studies can be found in ref.[12].

## 4. RESULTS

### 4.1 Homogeneous Fluid Limit

Let us first discuss the dynamics in the homogeneous limit. In Fig.1, we give the frequency dependence of $S_s(q,\nu)$ for different values of q starting from q = 0.4 to 2.0 in increments of 0.1 and for q = 4.0. The temperature is chosen to be 254K. In our earlier

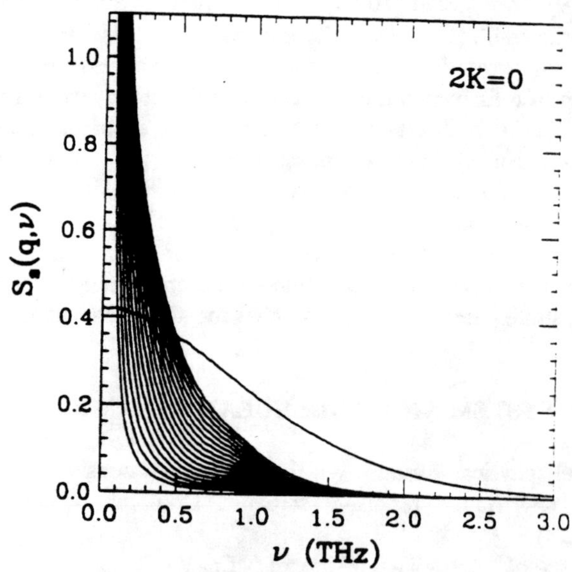


**Figure 1.** Self dynamic structure factor (SDSF) $S_s(q,\nu)$ vs. frequency $\nu$ for a system with a smooth substrate potential ($2K = 0$) at 254K (in the liquid phase). Various curves are for different q (in $A^{o-1}$), q increases from 0.4 (the narrowest curve) to 2.0 in steps of 0.1 and the curve starting from 0.42 is for q = 4.0 showing a Gaussian free-particle like response.

simulation studies of the thermodynamic properties of 2D screened Coulomb systems [19], we had found that for this system the liquid to solid transition temperature was about 200K; thus we are not too far above the transition temperature. Fig.1 gives a general picture of the dynamics of a 2D homogeneous fluid. The diffusive Lorentzian central peak for small q evolves to a Gaussian for large q, the latter characterizing the dynamics of a free particle.[1] Note that one may expect to see small deviation from a Lorentzian behaviour as $\nu \to 0$ for small q, an effect of the hydrodynamic long time tail in velocity autocorrelation function [20]. However our maximum simulation time ($2t_r$) is not large enough to pick up the effects of this anomalous dynamics accurately. On the other hand, we must make finite time corrections to a Lorentzian line shape to extract the diffusion constant correctly as discussed below, a point missed in earlier simulation studies. [10]

For *small* q (= 0.4), $F_s(q,t) = e^{-Dq^2 t}$ and its Fourier transform over MD total run time ($2t_r$) is given by

$$S_s(q,\upsilon) = \frac{2Dq^2}{\omega^2 + \left(Dq^2\right)^2}\left[1 - e^{-Dq^2 t_r}\cos\omega t_r\right] + \frac{2\omega}{\omega^2 + \left(Dq^2\right)^2} e^{-Dq^2 t_r}\sin\omega t_r \qquad (4)$$

where $\omega = 2\pi\nu$ and D is the diffusion constant. When $t_r \to \infty$ , Eq.(4) recovers the Lorentzian central peak. For systems with small $Dq^2t$ compared to 1, it is important to take into account the effects of finite $t_r$ while extracting D from $S_s(q,\nu)$. For the present system we obtain $D = 0.62 \times 10^{-4}$ cm$^2$/sec, which agrees very well with a direct calculation of D using the relation $< (\Delta r)^2 > = 4$ Dt, for large t. For *intermediate* values of q, a weak broad peak grows at a finite frequency, centered about 0.9 THz , and its origin can be traced to "cage dynamics", i.e the dynamics of a particle trapped in the potential of its neighbours for a short duration. For *large* q ( = 4.0), the central peak takes a Gaussian shape.

## 4.2 Lattice-Fluid Limit

Next we discuss the dynamics in the presence of a rather strong substrate potential ($2K$ = 0.9) which corresponds to $k_BT/\Delta \approx 1/4$. The length scale corresponding to the substrate periodicity, $\Lambda$, is 2.46 A$^O$ compared to the average inter-particle spacing of about 6 A$^O$. Since the particles spend a considerable amount of time in the substrate potential well, their dynamics is more appropriately described by a lattice-liquid. As we can see from Fig.2, the dynamic structure factor is drastically different from the homogeneous fluid case. First the diffusive Lorentzian central peak for *small* q ( < 0.4 i.e $q\Lambda < 1$ ) narrows considerably indicating a dramatic reduction in D. By fitting the spectrum to a Lorentzian with finite MD run time correction we obtain $D = 0.3 \times 10^{-5}$ cm$^2$/sec, nearly a factor of 20 smaller than the homogeneous fluid. Note that this value also agrees very well with D obtained directly from the mean square displacement. As we increase q, the central peak starts to deviate from a simple Lorentzian and in addition, a broad finite frequency peak grows near $\nu \approx 1$ THz. Finally for *large* q ($q\Lambda >> 1$), the spectral function deviates drastically from a Gaussian spectrum indicating a deviataion from free particle behaviour.

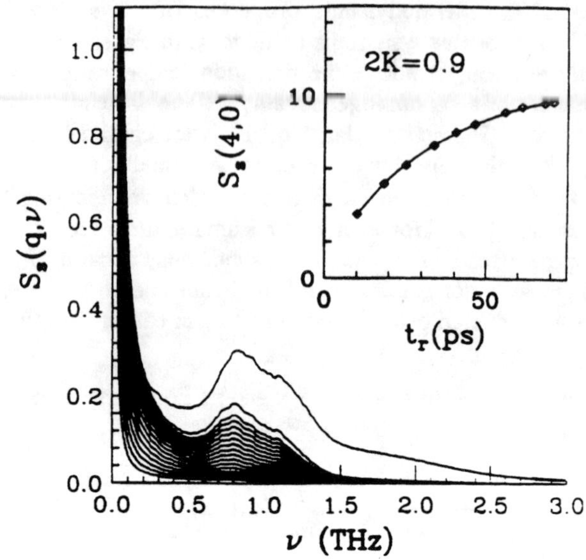


**Figure 2.** Self dynamic structure factor (SDSF) $S_S(q,\nu)$ vs. frequency $\nu$ in the presence of a strong substrate potential ($2K = 0.9$) at 248K (in the liquid phase). Various curves are for different q (in $A^{o-1}$) which are the same as in Fig.1. The upper curve ($q = 4.0$) is no longer a simple Gaussian. Inset gives $S_S(4,0)$ vs $t_r$ where $2t_r$ is the MD run time.

As discussed before, for $0.75 < q < 2.2$ Zabel et al. [10] could not fit their central peak to a single Lorentzian and suggested that the central peak consisted of two Lorentzians, one representing the usual liquid diffusion and the other (extremely narrow and resolution limited) to a solid-like peak. Even for a large q (= 4.0), the narrow central peak still existed on the top of a broad liquid-like central peak. [10] Fan et al.[11] carried out a MD simulation study to understand the origin of this narrow solid-like peak. They came up with an interesting explanation for this peak by calculating, as a function of time, the fraction of fluid particles which remained in the neighbourhod (within a small distance) of their starting position. This quantity decayed very slowly with a characteristic time $\tau$ which they estimated to be about 26 ps in reasonable agreement with the time scale extracted from the narrow central peak observed in the experiment. Since in the liquid phase, this fraction does not decay to zero during the time of the simulation, one can identify the narrow peak to be associated with the dynamics of "solid-like" particles albeit with a finite but long life time.

To see the sensitivity of this long $\tau$ to MD runtime we have fitted $S_S(q = 4, \nu = 0)$ to $C_1[1 - e^{-t_r/\tau}] + C_2$, where the contribution from the second term comes from the usual liquid-like dynamics. Finite time correction to this second term is negligible because the associated correlation function decays much faster in the time scale of $\tau$. From the $t_r$ dependence of the above quantity we conclude that the narrow central peak is not a delta-function. We find $\tau \approx 33$ ps which is slightly larger than 26 ps found by Fan et al.; this difference could be due to the difference in temperature of two simulations. Thus we conclude that fluid particles on the average get trapped in the potential well for about 33 ps at T = 254K.

### 4.3 From Homogeneous Fluid to Lattice-Fluid

To see how $S_s(q,\nu)$ changes from $2K = 0.9$ to $2K = 0$, we give in Fig.3 its $\nu$ dependence for two different length scales, q = 4.0 and q = 0.4 (inset). Again the temperature is about 250K. In the short length scale, the complex spectrum corresponding to the lattice-fluid changes to a Gaussian of a homogeneous fluid by modifying all the spectral characteristics. The finite frequency peak softens and broadens. The central peak also broadens. For $2K = 0.15$ ($\Delta = 180$K), corresponding to $k_BT > \Delta$, the spectrum is closer to a Gaussian whereas for $2K = 0.36$ ($\Delta = 432$K), corresponding to $k_BT < \Delta$, a narrow central peak appears whose intensity grows with increasing strength of the barrier height $\Delta$, and concurrently a finite frequency peak grows (also its width narrows a little). This is understandable because as the single particle barrier height decreases, the fluid particles spend less amount of time inside the well and consequently the strength of the solid-like component decreases.

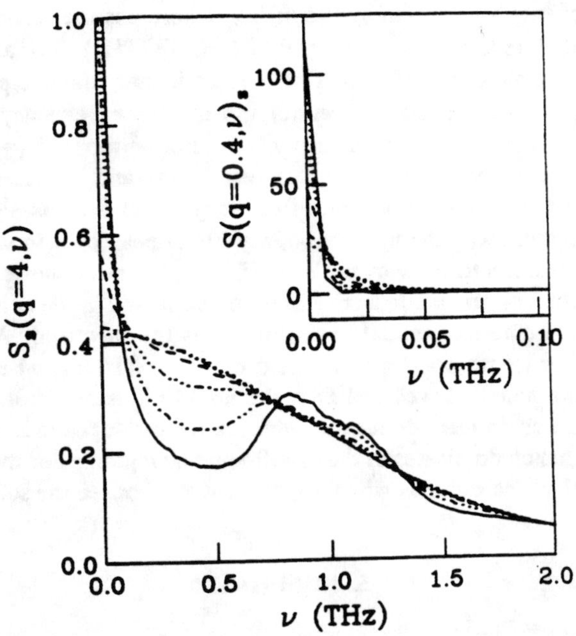


**Figure 3.** Self dynamic structure factor (SDSF) $S_s(q,\nu)$ vs. frequency $\nu$ for different strengths of the substrate corrugation ($2K$ = 0, 0.15, 0.36, 0.55, 0.75, and 0.9, the solid line corresponds to 0.9) at about 250K for q = 4.0 $A^{o-1}$ showing the growth of a sharp central peak when the barrier height $\Delta$ becomes larger than $k_BT$. Inset is $S_s(q,\nu)$ for q = 0.4 $A^{o-1}$, showing a dramatic decrease in the Lorentzian peak width as $2K$ increases from 0 to 0.9 indicating slowing down of diffusion.

To probe the nature of the central peak further, we go to a smaller q i.e. to a longer length scale. This is a distinct advantage of MD over NS experiment where the coherent Rb scatterers donot allow one to study the dynamic self correlation in long length scales because of the dominating coherent contribution to the NS spectra. For q=0.4, we find that the entire spectrum is confined to the central peak with almost vanishing weight in the finite frequency peak. The central peak can be fitted very well to a single Lorentzian with finite MD run time correction; and its width narrows with increasing $2K$. We donot see any strong evidence of the narrow solid-like component in this long lengh scale. Thus a proper physical understanding of the origin of the solid-like component must address this strong q-dependence. It is also possible that the q-dependence of the intensity of the finite frequency peak is correlated with this observation.

A simple way to understand the origin of the narrow quasi-static peak and the q-dependent intensity of the entire central peak and the finite-frequency peak is to look at the dynamic response of a harmonic oscillator. In this case, one has

$$S_H(q,\nu) = e^{-y} \Sigma\, \delta(\nu-n\nu_o)\, I_n(y) = \Sigma\, A_n\, \delta(\nu-n\nu_o) \;, \tag{5}$$

where $y = k_B T q^2/m(2\pi\nu_o)^2$ and $I_n(y)$ is the modified Bessel function of the first kind. The characteristic frequency $\nu_o$ is given by $\surd(c/m)/2\pi$, where c is the second derivative of the substrate potential at its minimum i.e. $\nu_o = 0.982\ \surd(2K)$ (THz). For a given y (i.e. T and q), $I_n(y)$ decreases as n (i.e. $\nu$) increases. Thus we can identify the two peaks of $S_H(q,\nu)$ at $\nu = 0$ and $\nu = \nu_o$ with the 'solid-like' central and the finite frequency peak respectively. Peaks at $n\nu_o$ for $n \geq 2$ have very little intensity [12]. Furthermore, $A_o = 1$ when $y = 0$ and decreases with increasing y whereas $A_1 = 0$ when $y = 0$, and increases with y. Thus the spectral weight of the central peak (finite frequency peak) decreases (increases) with q. We can now understand why the finite frequency broad peak of $S_s(q,\nu)$ is not seen at q = 0.4 but at q = 4.0. Finally to see why the 'solid-like' central peak shows up only for large q inspite of the fact that its amplitude increases with decreasing q, we note that at low q, the contribution of the liquid-like central peak dominates the spectrum. We conjecture that there is a cross-over length scale characterized by $q^* = \surd(1/D\tau)$ where $\tau$ is the average resident time in the potential well and D is the liquid diffusion constant such that for $q \gg q^*$, the solid-like component dominates whereas for $q \ll q^*$, the Lorentzian associated with the liquid diffusion dominates in the low-frequency region. For the $2K = 0.9$ system, we estimate $q^* = 1$ which explains why for q = 0.4 we donot see the solid-like component.

## 5. SUMMARY

In summary we have discussed how the dynamics of a fluid geometrically constrained to move in two-dimension in the presence of a substrate potential evolves from a homogeneous to a lattice-liquid limit. In the latter case, the fluid particles are trapped for sufficiently long times at the bottom of the single particle potential well and undergo activated diffusive motion (in long length and time scales). These trapped particles give rise to the solid-like narrow central peak which can be clearly observed only at large q values (short length scales) because the liquid-like diffuson component dominates the central peak at small q. Concurrent with the narrow central peak one sees a broad finite freqency peak whose weight increases with q. These two features can be understood within a simple harmonic oscillator model and can be associated with the existence of long lived local order in the lattice-fluid. In the other extreme limit i.e. for a homogeneous fluid, we donot see any evidence (in our MD simulation) of such a narrow peak even when we approach the liquid to solid transition temperature. This we believe is due to the dynamic nature (same time scale as the diffusive motion) of the cage potential which traps the fluid particles.

## ACKNOWLEDGMENT

We thank Dr. Surajit Sen for useful discussions and Dr. Tahir Cagin for making his MD code available. The work was partially supported by NSF DMR Grant 90-24955 and NSF CHE-9224102.

## REFERENCES

1. J. P. Hansen and I. R. McDonald, *Theory of Simple Liquids*, Academic Press, London, 2nd edition, 1986.
2. M. P. Allen and D. J. Tildesley, *Compute Simulation of Liquids*, Oxford University Press, Oxford, 1989.
3. Jean Pierre Boon and Sidney Yip, *Molecular Hydrodynamics*, McGraw Hill Inc., 1980.
4. F. Yonezawa, *Solid State Physics*, Ed. H. Ehrenreich and D. Turnbull, (Academic Press, NY) vol 45 page 179.
5. J. Koplik, J. R. Banavar, and J. F. Willemsen, Phys. Rev. Lett. **60**, 1282 (1988); M. Sun and C. Ebner, Phys. Rev. Lett. **69**, 3491 (1992).
6. P. A. Thompson and M. O. Robbins, Phys. Rev. Lett. **63**, 766 (1989) and references therein.
7. W. A. Curtin, Phys. Rev. Lett. **59**, 128 (1987); D. Zhu and J. G. Dash, Phys. Rev. Lett. **60**, 433 (1988).
8. *Graphite Intercalation Compounds, Vol 1: Structure and Dynamics*, H. Zabel and S. A. Solin, eds, Springer Series in Topics in Current Physics, Springer, 1990.
9. H. Zabel, A. Magerl, J. J. Rush, and M. E. Meisenheimer, Phys. Rev. B **40**, 7616 (1989).
10. H. Zabel, A. Magerl, A. J. Dianoux, and J. J. Rush, Phys. Rev. Lett. **50**, 2094 (1983); H. Zabel et al., Phys. Rev. Lett. **57**, 2041 (1986); W. A. Kamitakahara and H. Zabel, Phys. Rev. B **32**, 7817 (1985)
11. J. D. Fan, G. Reiter, and S. C. Moss, Phys. Rev. Lett. **64**, 188 (1990); J. D. Fan, O. A. Karim, G. Reiter, and S. C. Moss Phys. Rev. B **39**, 6111 (1989).
12. Hyangsuk Seong and S.D. Mahanti, Phys. Rev. B **49**, 5042 (1994); Hyangsuk Seong and S. D. Mahanti, Proceedings of the MRS Fall meeting (Nov 30 - Dec 4, 1992), Symposium on Materials Theory and Modelling.
13. D. L. Jhonson, J. Koplick, and L. M. Schwartz, Phys. Rev. Lett. 57, 2564 (1986).
14. P. P. Mitra, P. N. Sen, and L. M. Schwartz, Phys. Rev. B **47**, 8565 (1993)

15. J. Karger and D. M. Ruthven, Diffusion in Zeolites and Other Microporous Solids, Wiley, New York, 1993.
16. H. Seong and S. D. Mahanti, Proceedings of the MRS Fall meeting (Nov 30 - Dec 4, 1992), Symposium on Materials Theory and Modelling; H. Seong, Ph.D Thesis, Michigan State University (1993).
17. S. C. Moss, G. Reiter, J. L. Robertson, C. Thompson, J. D. Fan, and K. Oshima, Phys. Rev. Letters 57, 3191 (1986).
18. K. Sköld, J. M. Rowe, G. Ostrowski, and P. D. Randolph, Phys. Rev. A 6, 1107 (1972)
19. H. Seong, S. D. Mahanti, S. Sen, and T. Cagin, Phys. Rev. B 46, 8748 (1994).
20. F. den Hollander, J. Naudts, and F. Redig, J. of Stat. Phys. 69, 731 (1992); Comm. Math. Phys. (to be published).

Local Order
in Condensed
Matter Physics


ISBN 1-56072-220-7

# MAGNETISM IN REDUCED SIZES AND DIMENSIONS

S. N. Khanna, B. V. Reddy and P. Jena

*Physics Department, Virginia Commonwealth University*
*Richmond, VA. 23284-2000*

## ABSTRACT

The magnetic properties of low dimensional or finite systems are shown to be strongly affected by size, symmetry, dimension and composition. The magnetic moments of ferromagnetic transition metals are shown to increase as one lowers the dimension from the bulk to the atom, e.g. planes are more magnetic than bulk, chains are more magnetic than planes, isolated atoms are more magnetic than those in chains. Magnetic moments per atom of small clusters of ferromagnetic elements are influenced by large surface to volume ratio and are larger than bulk moments. However, the cluster magnetic moments are free to rotate and the clusters exhibit super-paramagnetic behavior. Whereas Rh is non-magnetic in the bulk, small clusters of Rh are ferromagnetic because of the reduced size and symmetry. The electronic mechanisms responsible for this unusual magnetism in reduced size and dimension are reviewed.

## I. INTRODUCTION

The electronic and magnetic properties of individual atoms are now well understood using quantum mechanics. Atoms with closed electronic shells have paired electrons and zero multiplicities while atoms with unfilled shells display finite spin multiplicities. The Hund's rules link these multiplicities to the degeneracies of the electronic shells and hence provide a framework for understanding magnetism in non heavy atoms. It is interesting to note that the spin multiplicities are governed by electronic degeneracies and for atoms where the effective potential has the radial symmetry, these are derived from the angular momentum quantum number. The atomic magnetism is therefore governed by the angular momentum state of the valence electrons. The other extreme to the atomic picture occurs in solids. Here the effective electronic potential has the symmetry of the lattice. The atomic electronic states are broadened into bands with a width and total number of states governed by the degeneracy and the mixing between the starting atomic orbital. Some of the localized electronic states in atoms become partially or completely delocalized in solids. The magnetism now arises due to moments[1] localized at the atoms and coupled together ferromagnetically. What is remarkable is that the new features are so detrimental to magnetism that out of a vast majority of atoms with non zero spin multiplicities, only seven elements are left ferromagnetic in the solid state. It is therefore important to ask how the magnetism of atoms is modified as they are brought together to form agglomerates. Such a study is timely from another point of view also. Developments in experimental techniques such as supersonic beams[2], molecular beam epitaxy[3], tunneling electron microscopy[4], are enabling scientists to generate materials with controlled size, dimension and thickness. These include atomic clusters containing two to several thousand atoms, chains, mono-layers, and multi-layers of different elements. The new materials have features which can be very different from conventional solids. For example, the bond lengths and geometrical arrangement of atoms in clusters[5] can be very different from those in the bulk. Atoms in mono- and multi- layers experience the quantum effects due to reduced dimensions. These features can give rise to novel electronic, magnetic, and chemical properties.

The purpose of this review is to focus on the effect of size, dimensionality, symmetry, and composition of materials on their magnetic properties. We start by considering the case of solids first and discuss how the local moment on a given atom changes as one goes from bulk to slabs to free surfaces to planes to finally chains of atoms. These studies show[6] that the moment is primarily governed by the local atomic coordination and as the coordination number is reduced, the moment increases. For small clusters marked by the preponderance of surface sites, this would predict larger moments than bulk. We then discuss the actual Stern-Gerlach experiments on free transition metal clusters [7-8] which showed that they have less magnetization per atom than bulk in contradiction to the above expectation. This apparent controversy between theory and experiment is resolved[9] by noting that small clusters behave like giant atoms and exhibit paramagnetic behavior. Finally, we shall discuss the magnetism in 4d and 5d metals. While atoms belonging to these series have non-zero spin multiplicities, none of these solids are magnetic. The only way to magnetize these materials is to induce magnetization by implanting ferromagnetic impurities[10]. For example, Fe impurities in Pd induce host magnetization exceeding 10-12 Bohr magneton. These giant induced moments discovered several years ago[10] arise due to spin polarization of the host atoms surrounding the magnetic impurity. The effect is, however, specific to Pd. We[11], on the other hand, recently proposed that the 4d elements like $Rh_n$ can be spontaneously magnetized by reducing their size. Indeed, we proposed that a $Rh_{13}$ cluster would be highly magnetic. These predictions have been borne out by recent experiments.

In section II, we discuss the effect of dimension on magnetism in bulk systems. Section III is devoted to experimental results on the clusters of ferromagnetic transition metals and in section IV we provide a simple model for the experimental observations. Section V is devoted to a discussion of ferromagnetism in 4d elements. Finally, section VI contains the main conclusions.

## II. FERROMAGNETIC TRANSITION METALS IN REDUCED DIMENSIONS

Ferromagnetism in Fe, Co and Ni arises due to 3d electrons[1]. As the atoms are brought together, the atomic d-states become partially delocalized and a positive exchange interaction between the localized atomic moments develops, favoring the ferromagnetic coupling. This persists until the distance is so small (solids under pressure) that the exchange integral changes sign. For these metals, the band theory[12] provides a simple description of magnetism. The ferromagnetism is favored when the solid has a high density of electronic states at the Fermi energy. Under these conditions, the bands corresponding to the spin up and down electrons split due to exchange interactions and the net moment is determined by the amount of splitting as well as the relative occupancy of the spin up and down bands.

To examine the effect of dimension we studied the moment of Fe, Co and Ni in various local environments. These included sites around defects in solids, slabs of various thicknesses, planes, and chains of atoms. The calculations were carried out using a modified ab-initio recursion scheme[6] within a tight binding framework. In this scheme, the tight binding parameters are obtained from first principles local density calculations on small clusters. The electronic structure calculations were carried out by assuming that the interatomic distance in various geometries is the same as in bulk.

The resulting moments were found to be primarily determined by the local coordination. This is shown in Fig. 1 where we have plotted the change in the local moment for Fe, Co, and Ni as a function of the number of near neighbors. One notices that as the coordination number is reduced, the moment increases towards the atomic limit. This progression can be understood from the fact that within a tight-binding description, the band-widths are governed by the local coordination. As the coordination number is reduced, the bands become narrower. If the band

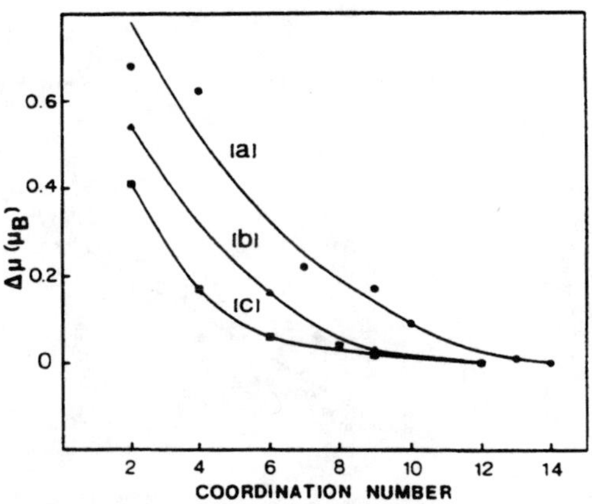


Fig. 1. Deviation from the bulk magnetic moment $\Delta\mu$ in Fe, Co, and Ni as a function of nearest neighbor coordination number (in various structures). (a), (b) and (c) correspond to Fe, Co and Ni respectively. The smooth lines are drawn simply to guide the eye.

splitting is maintained, this narrowing results in minority spins being spilled over to the majority band, causing the moment to rise. These results suggest that small clusters where most atoms reside at the surface, should be more magnetic than those in the bulk. Note that in the above analysis we have ignored any change in inter-particle distance. A decrease in interatomic distance increases the band-width and therefore lowers the moment. The interatomic distances in clusters are indeed a few percent shorter than that in bulk. However, this effect is expected to be much smaller than the enhancement due to lower coordination at the surface. It is with this expectation that experiments were carried out on small Fe, Co and Ni clusters. The results were startling.

## III. EXPERIMENTS ON FREE Fe, Co, AND Ni CLUSTERS

Several groups have undertaken an experimental determination of the moments in small clusters. The experimental procedure was similar to the one used by Stern and Gerlach to demonstrate space quantization. The clusters were generated in molecular beams by a laser vaporization of the metal and passed through the gradient field. The earliest results came from the Exxon group[13] who tried to use depletion in cluster beams to determine moments. The actual deflection profiles were, however, first observed by deHeer et al.[7] on $Fe_n$ clusters containing 56-256 atoms. Their results showed that unlike the case of atoms where the total angular momentum quantization splits the beam symmetrically, all clusters deflected in the same direction. Further the observed magnetization per atom increased with magnetic field but even at the highest fields, the magnetization was much below the bulk value. Bucher et al.[8] performed similar experiments on $Co_n$ clusters. Their experiments, however, permitted a better control on cluster temperatures. These experiments provided two important results. Firstly, the measured moment per atom in clusters were found to increase linearly with size (See Fig.2). Secondly, the moment was found to decrease as the cluster vibrational

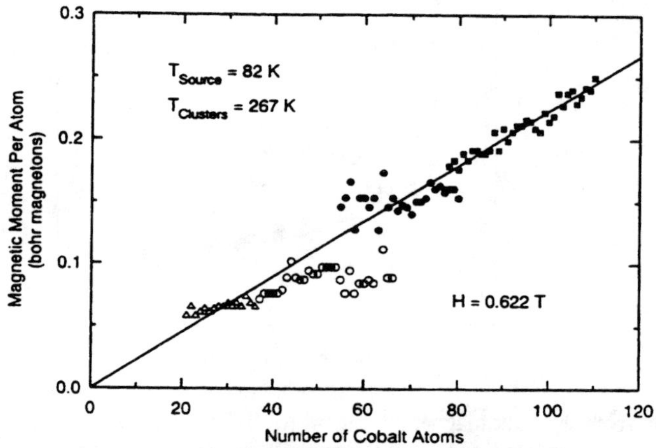


Fig.2. Measured $\mu_{eff}$[8] as a function of cluster size for $Co_n$ clusters. Solid line is based on Eq.(2).

temperature was increased. In this paper we will mainly focus on these experiments on $Co_n$. Following is a summary of salient experimental results: (1) As opposed to atoms, the clusters undergo unidirectional deflection. (2) The measured moments per atom are less than in bulk. (3) the moment per atom increases linearly with cluster size. And (4) the moment per atom decreases as the temperature is increased. Observation (2) is in clear contradiction to the theoretical expectations outlined in section II.

# IV. THEORETICAL INTERPRETATION OF EXPERIMENTS

A simple theoretical model which explains the above observations and removes the apparent discrepancy between theory and experiment was proposed by one of the present authors[9]. To illustrate the theoretical arguments entering this model, note that the clusters sizes studied in the above experiments are much smaller than the typical domain sizes in ferromagnetic solids. At the ambient cluster temperatures, the atomic moments are therefore aligned due to the exchange coupling and the cluster of N atoms behaves like a giant atom with a moment of $N\mu$ where $\mu$ is the moment per atom. As this cluster is placed in a magnetic field, it starts to precess around the field and acquires a spin rotational angular momentum (SRAM) due to precession. This is shown in Fig. 3. Note that in case of atoms, the conservation of the z-

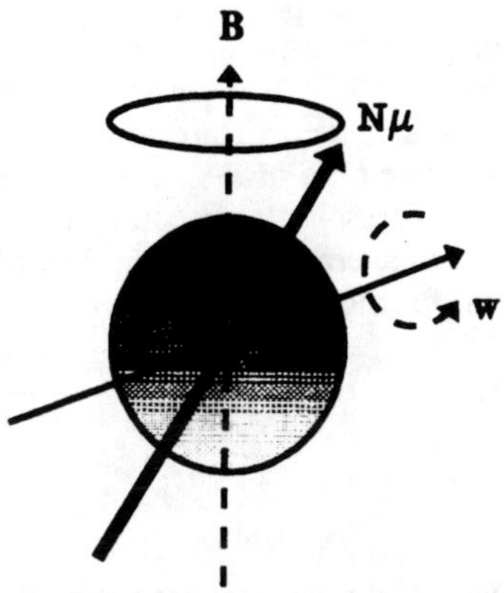


Fig. 3. Schematic diagram of a cluster rotating with an angular velocity w placed in a magnetic field B.

component of total angular momentum splits the atomic beam into various components. The question remains as to why the clusters do not undergo a similar decomposition ?

We first discuss the case of transition metal clusters. Here, the anisotropy energy per unit volume increases with decreasing size due to surface, shape and other contributions. A rough calculation shows that in clusters containing up to about few hundred atoms, the total anisotropy energy is, however, smaller than the thermal energies at ambient cluster temperatures. The thermal disorder can therefore change the direction of the cluster spin. For ferromagnetic particles in fluids, this relaxation gives rise to the well known superparamagnetic behavior[14]. For clusters in beams, however, additional constraints arise because a change in the direction of cluster moment involves a change in the corresponding angular momentum and since the clusters are in interaction free environment, the z-component of the total angular momentum needs to be conserved just as in case of atoms. The difference between atoms and clusters arises because, unlike single atoms which have no rotational classical angular momentum due to small size, a cluster has a finite rotational inertia and corresponding angular momentum (see Fig.2). Therefore, a coupling between the classical rotational angular momentum to the SRAM, would allow the cluster-spin to re-orient in various directions. The re-orientation could also occur via quantum tunneling[15]. These possibilities make even a free cluster behave like a super-atom whose total spin $N\mu$ can fluctuate in all directions as it passes through the field. This is similar to the case of super-paramagnetic relaxation of ferromagnetic particles in fluids. The relaxation times are given by[9,16]

$$\tau = \tau_0 e^{CV/k_B T} \tag{1}$$

Here C is the magnetic anisotropy energy per unit volume, T is the temperature and V is the volume of the cluster. The value of $\tau_0$ depends on the gyromagnetic precession time and is typically[16] $10^{-10}$ - $10^{-13}$ s. For granular alloys of Fe in matrices, Xia et al.[17] measured a value of C to be 2 x $10^7$ ergs/$cm^3$ for 20-40 A° particles. Using this value one finds that $Fe_n$ clusters having 120-140 atoms and at a temperature of 100 K would have $\tau$ around 1-1000ps. This is much smaller than the passage time of the cluster through the magnet which is around 100 $\mu$s. Each cluster therefore explores its large configuration space in passing through the magnet. Hence, all clusters undergo average deflection. The magnetization per atom ( $\mu_{eff}$ ) is then given by the usual Langevin function, namely

$$\mu_{eff} = \mu[\coth[\frac{N\mu B}{k_B T}] - \frac{k_B T}{N\mu B}] \tag{2}$$

where B is the magnetic field and $\mu$ is the moment per atom. Note that the experiments measure $\mu_{eff}$ and not $\mu$. In Fig. 2 we also show our calculated $\mu_{eff}$ for $Co_n$ clusters as a function of size (solid line). The agreement is remarkable. Similar agreement[8] is obtained when one compares measured $\mu_{eff}$ with calculated values as a function of magnetic field. One can also use the experimental measured $\mu_{eff}$ and Eq.(2) to calculate $\mu$ . In Table 1 we compare the calculated $\mu$ for clusters with the corresponding values in the bulk. It is gratifying to note that the moment per atom in clusters are about 30 % higher than in bulk. This is in complete agreement with the predictions in section III.

We would like to add that what we have proposed above is a simple model. We have ignored any effects due to anisotropy energy. These have been recently studied by Jensen and Bennemann[18]. They show that the measured $\mu_{eff}$ can be further reduced from its value in Eq.(2) when the anisotropy is included. Recently Billias et al.[19] have measured magnetization in $Fe_n$ clusters containing 25-700 atoms and have reported effects which could be assigned to a resonance between the spin-rotation and the rotational temperature. These results needs detailed investigations.

Table 1. Magnetic moments per atom in Fe, Co, and Ni obtained by using the experimental $\mu_{eff}$ and equation (2).

| Element | Atom | Cluster | Bulk |
|---|---|---|---|
| Iron | 4.0 | 2.85 | 2.22 |
| Cobalt | 3.0 | 2.24 | 1.72 |
| Nickel | 2.0 | 0.88 | 0.62 |

# V. POSSIBILITY OF MAGNETISM IN 4d-ELEMENTS

The 3d, 4d, and 5d transition metals, all have unfilled d-states. However only 3d-solids Fe, Co and Ni exhibit ferromagnetic order and none of the 4d or 5d solids are magnetic. The only way to magnetize groups of the 4d elements is to induce moments[10,20] by implanting ferromagnetic impurities ( eg. Fe) as was demonstrated by the famous giant induced moments in Pd[1].

We have recently proposed[11] that an alternate method of making 4d elements magnetic is to reduce their size. We carried out ab-initio electronic structure calculations on 13 atom clusters of Rh, Pd and Ru. The cuboctahedric and icosahedric geometries were examined and the energy was minimized by optimizing the radial bond length. In all cases, the icosahedral clusters were found to be more stable than the FCC counterparts. Interestingly, all clusters had large ground states multiplicities. These are given in Table 2. The most surprising was the case of $Rh_{13}$ which had a moment of 21 $\mu_B$ in our calculations. An analysis of the electronic states showed that the moment was due to an abnormally high density of electronic states due to icosahedral geometry and the reduced size.

Following our predictions Cox et al.[21] experimented with $Rh_n$ clusters containing 12-32 atoms using Stern-Gerlach arrangement. They indeed found that the clusters are magnetic confirming the theoretical prediction. However, the measured moment for $Rh_{13}$ was 0.5 $\mu_B$ per atom as opposed to a predicted value of 1.5 $\mu_B$ per atom. The reasons for this will be outlined in a forthcoming paper[22]. The moment was also found to vary strongly with size. In particular, clusters containing 15, 19, 22, and 26 atoms were far more magnetic than others. Finally, the moments decreased to zero for clusters containing more than 100 atoms.

Our preliminary studies[22] on the magic magnetic cluster $Rh_{15}$ show that the moment is highly sensitive to the geometrical arrangement. For example, the moment disappears for a BCC like 15-atom arrangement but acquires a high value for a 13-atom icosahedron decorated with a dimer. These investigations are under study and will be reported later.

Table 2. Magnetic moment ($\mu_B$) at the central and outer sites of cuboctahedral and icosahedral 13 atom clusters of Pd, Rh and Ru.

| Cluster | Geometry | Radial dist. (a.u.) | B.E (eV) | Moments Central | Moments Outer |
|---|---|---|---|---|---|
| $Pd_{13}$ | Cubo | 5.20 | 19.8 | 0.0 | 0.0 |
| | Icos | 5.12 | 20.3 | 0.43 | 0.12 |
| $Rh_{13}$ | Cubo | 4.90 | 41.3 | 1.54 | 1.45 |
| | Icos | 4.84 | 42.6 | 1.58 | 1.62 |
| $Ru_{13}$ | Cubo | 4.86 | 55.5 | -0.28 | 1.19 |
| | Icos | 4.81 | 56.0 | -0.29 | 1.02 |

## V. CONCLUSIONS

To conclude we have shown that the dimension, size and symmetry can have a profound effect on the magnetic properties of materials. Due to space limitations, we have not discussed here, the behavior of clusters of antiferromagnetic materials and rare earth materials. In fact, the effect of geometry is most transparent in systems with anti-ferromagnetic coupling. We had earlier shown[23] that icosahedral clusters of a anti-ferromagnetic material would behave like frustrated paramagnets because of the icosahedral geometry. This may be the reason why the measured moments in $Cr_n$ clusters[24] are extremely small. The other case refers to the clusters of rare earth metals like $Gd_n$. These have been studied by Douglass et al.[25] and show a different behavior than Fe, Co, or Ni clusters. At certain cluster sizes, the anisotropy energy is higher than the thermal energy and the cluster moment does not fluctuate in direction as in case of superparamagnetic clusters. If one combines these findings with those discussed in this paper, one can summarize the magnetic behavior in bulk and in systems with reduced sizes in the following table.

Table 3. Behavior of metals in bulk and at reduced sizes.

| Metal | Bulk | Cluster |
|---|---|---|
| Na, K | Paramagnetic | Ferromagnetic |
| Fe, Co, Ni | Ferromagnetic | Super-paramagnetic |
| Gd, Tb | Ferromagnetic | Rotors/Super-paramagnetic |
| Cr | Anti-ferromagnetic | Frustrated paramagnetic |
| Rh | Paramagnetic | Ferromagnetic |

## REFERENCES


1. M. B. Stearns, Physics Today, April 1978, p. 34.
2. O. F. Hagena, Surface Sci. 106, 101(1981).
3. E. Kaldis, Ed., Crystal Growth of Electronic Materials (North-Holland, Amsterdam, 1985).
4. G. Binning and H. Rohrer, Rev. Mod. Phys. 59, 615 (1987).
5. P. Jena, S. N. Khanna, and B. K. Rao, Int. Jour. Mod. Phys. 6, 3657 (1992).
6. F. Liu, M. R. Press, S. N. Khanna, and P. Jena, Phys. Rev. B39, 6914 (1989).
7. W. A. de Heer, P. Milani, and A. Chatelain, Phys. Rev. Lett. 65, 488 (1990).
8. J. P. Bucher, D. C. Douglass, and L. A. Bloomfield, Phys. Rev. Lett. 66, 3052 (1991).
9. S. N. Khanna, and S. Linderoth, Phys. Rev. Lett. 67, 742 (1991).
10. R. M. Bozarth, P. A. Wolff, D. D. Davis, V. B. Compton and J. H. Wernick, Phys. Rev. 122,1157 (1961); G. J. Nieuwenhuys, Adv. Phys. 24, 515 (1975).
11. B. V. Reddy, S. N. Khanna, and B. I. Dunlap, Phys. Rev. Lett. 70, 3323 (1993).
12. W. M. Lomer in Theory of Magnetism in Transition Metals, ed. W. Marshall, (Acad. Press, 1967), p. 1.
13. D. M. Cox, D. J. Trevor, R. L. Whetten, E. A. Rohlfing, and A. Kaldor, Phys. Rev. B32, 7290 (1985).
14. L. Neel, C. R. Acad. Sci. 224, 1488 (1947).
15. E. Chudnovsky and L. Gunther, Phys. Rev. Lett. 60, 661 (1988); D. D. Awshalom, M. A. McCord, and G. Grinstein, Phys. Rev. Lett. 65, 783 (1990).

16. G. Xia, S. H. Liou, A. Levy, J. N. Taylor, and C. L. Chien, Phys. Rev. B34, 7573 (1986).
17. L. Bloomfield (Private Communication).
18. P. J. Jensen and K. H. Bennemann (Private Communication).
19. I. M. L. Billas, J. A. Becker, A. Chatelain, and W. A. de Heer, Phys. Rev. Lett. 71, 4067 (1993).
20. J. W. Cable and L. David, Phys. Rev. B16, 297 (1977); B. H. Verbeek, G.J. Nieuwenhuys, J. A. Mydosh, C. Van Dijk and B. D. Rainford, Phys. Rev. B22, 5426 (1980).
21. A. J. Cox, J. G. Louderback, and L. A. Bloomfield, Phys. Rev. Lett. 71, 923 (1993).
22. B. V. Reddy and S. N. Khanna (to be Published).
23. B. V. Reddy and S. N. Khanna, Phys. Rev. B45, 10103(1992).
24. D. C. Douglass, J. P. Bucher, D. B. Haynes, and L. A. Bloomfield in Physics and Chemistry of Finite Systems: From Clusters to Crystals, Ed., P. Jena, S. N. Khanna, and B. K. Rao (Kluwer Acad. Pub., 1992) p. 759.
25. D. C. Douglass, J. P. Bucher, and L. A. Bloomfield, Phys. Rev. Lett. 68, 1774 (1992).

Local Order
in Condensed
Matter Physics


ISBN 1-56072-220-7

# FIRST-PRINCIPLES INVESTIGATION OF LUMINESCENCE PROCESSES IN SOLIDS: COPPER IN ZINC OXIDE AND ZINC SULFIDE

K. C. Mishra, K. H. Johnson* and P. C. Schmidt**

*Central Research, OSRAM SYLVANIA INC., 100 Endicott Street, Danvers, MA 019*
**Department of Materials Science, Massachusetts Institute of Technology, Cambridge, MA 02139*
*** Institut fur Physikalische Chemie, Technische Hochschule Darmstadt, Germany*

## 1. INTRODUCTION

The electronic structures of solids, particularly those involving f- and d-electrons have long presented a challenge to solid state theorists interested in the luminescence properties of materials. The main difficulty to study the excited state properties using the first principles theories is to calculate the eigenvalues and eigenfrequencies associated with the excited states accurately enough to predict transition frequencies and moments in a reliable manner. One obvious way is to calculate the total electronic energies and eigenstates for the ground state and excited state configuration rigorously, and then to compute the luminescence properties using these calculated parameters. Except for very few simple molecular systems, such calculations are computationally prohibitive and are often not reliable.

Traditionally, the spectroscopists have approached the radiative problems using the crystal field theory in conjunction with the experimental data. in a semi-empirical manner. This alternative approach to a rigorous first-principles method cannot describe many solid state processes which involves state due to the host system. The covalency effect which plays a very important role for the transition metal ions is usually studied in a parametric manner.

During the last few years, we have been studying the electronic excitation processes in solids using the self-consistent field scattered wave molecular orbital cluster approach (SW-X$\alpha$) and the augmented scattered wave (ASW) band structure supercell approach[1-3]. Using the results from the ASW approach and the SW-X$\alpha$ cluster method, it is possible to generate a reliable energy level diagram which provides a satisfactory explanation for the excitations localized at the impurity ions. and also for transitions involving host materials and the impurity ions or defects. In this paper, we will apply these two techniques in a complementary manner to explain the luminescence mechanism of copper in zinc oxide and zinc sulfide.

The luminescence of copper in zinc oxide and zinc sulfide provides an interesting case for a theoretical reasons for a number of reasons. First, there exist several models of luminescence in the literature to explain the electronic transitions in copper-activated phosphors, yet there does not exist a clear theoretical description of the luminescent mechanism for this system. Although these models have been proposed mostly on the basis of experimental data, they have not been tested by theoretical calculations. Second, copper in both its valence state, i.e., +1 or +2 is a relatively simple transition metal ion. The complication arising from spin orbit interaction and relativistic effects requiring to compute many electron states within L-S or j-j coupling schemes is expected to be rather minimal here. It is therefore an ideal system to investigate within an one-electron framework.

## 2. REVIEW OF EXPERIMENTAL RESULTS

### 2.1 Copper in Zinc Oxide

Zinc oxide crystallizes in the hexagonal wurtzite structure[4] and is transparent to visible radiation, the energy gap being 3.4 eV[5]. It is believed that copper substitutes for zinc in divalent state. The subsequent lowering of symmetry due to the crystal field removes the orbital degeneracy of the d levels. Dietz et al.[6] have measured the crystal field splitting of the d -like levels to be 0.71 eV which compares reasonably well with that of copper in other oxides.

The luminescence properties of copper in zinc oxide were extensively studied by Dingle[5]. The excitation spectrum leading to the green luminescence has a major peak at 2.86 eV. The green luminescence has been identified to be due to divalent copper based on a comparison of the ratio of the intensity of the two zero phonon lines to the ration of the natural abundance of $^{63}$Cu to $^{65}$Cu . On the basis of the measured components of g tensor, the green fluorescence has been assigned to transitions involving the d-hole in the divalent copper ion. The binding energy of the d-hole was estimated to be greater than 3.3 eV and the excited state of the d-hole is expected to be approximately 0.45 eV above the valence band based on the observed band gap and excitation peak. The measured values of g tensors in the ground state and excited state suggest that the while the hole is fairly localized at the copper center, the excited state is very delocalized indicating a charge transfer transition. These experimental data describe adequately the relative orientation of various energy levels participating in the luminescence process in zinc oxide.

### 2.2 Copper in Zinc Sulfide

Zinc sulfide crystallizes both in the cubic zinc blend and the hexagonal wurtzite structure[7]. However, in this study we will consider only the wurtzite structure. Copper is assumed to enter substitutionally, although other sites have also been proposed in the literature[8]. One can have both monovalent and divalent copper ions by adding appropriate codopants for charge compensation.

A number of emission sites have been associated with copper in zinc sulfide. We will focus on three of them which are unambiguously assigned to copper[9-11]. They are the green center (Cu-G), the blue center (Cu-B) and the infrared center (Cu-ir). The copper green center is associated with the green emitting phosphor whose emission peaks at 2.38 eV, the corresponding excitation peak being at 2.95 eV[10]. Emission from the blue center peaks at 2.67 eV with the excitation peak at 3.35 eV[10]. The copper ir center peaks at 0.95 eV which correlates strongly with absorption peaks at 1.7 eV and 2.3 eV[11]. The band gap transition occurs at 3.8 eV[11]. No electron paramagnetic resonance signal has been found to be associated with green and blue centers indicating that these centers are due to the monovalent copper ions. The copper ir transition has been identified as being due to a crystal field transition in a divalent copper.

A number of luminescence models have been proposed in the literature to explain cooper luminescence in zinc sulfide[12-15], a brief description of which will be found in References 1 and 2. In this paper, it will be shown that the transitions associated with the divalent copper can in explained by the Lambe-Klick model[13] and that of the monovalent copper can only be explained by the Riehl, Schoen and Klasens model[12] only in a qualitative sense since the relative ordering of the energy levels is far different from what has been assumed by these two models.

## III. THEORY

The electronic structure calculations have been performed using the scattered wave Xα (SW-Xα) molecular orbital procedure[16] and the augmented spherical wave band structure method (ASW)[17]. There are a number of similarities between these two methods: the sphorically symmetric potential, local density approximation of the exchange correlation potential etc. The main differences are in the choice of basis set and numerical procedure used to solve the appropriate Schrodinger equation. First, the ASW method is a linear combination of atomic orbital type of approach with fixed energy basis set while the scattered wave method constructs the molecular orbitals from the energy dependent solutions of the radial Schrodinger equation within each muffin tin sphere. Thus the latter approach is not limited by the finite basis set. Second, the one electron energy eigenvalues are determined by the Rayleigh-Ritz procedure in the ASW approach, whereas the scattered-wave method uses a multiple scattering formalism, similar to the Korringa-Kohn-Rostoker method [18], to calculate the molecular orbital eigenvalues.

Neither of the two procedures is specifically desingedd to study defective solids. The cluster approximation is used to simulate the defective solid as a cluster with a proper embedding scheme to account for the electrostatic field due to rest of the crystal. On the other hand, we will use a supercell approach for the band structure calculation which describes the solid with impurity ions as a periodic system with large translation vectors. For a large cluster or very dilute systems, results from both the calculation are expected to have a nice correspondence. The complementarity between the two methods, the cluster approach describing rigorously the local interaction and the band structure approach providing a good description of the long-range interactions of the impurity states with respect to rest of the lattice, will be utilized in this work to develop a reliable model for luminescence.

## 3. RESULTS

### 3.1 Zinc oxide

Zinc oxide was simulated by a tetrahedral cluster of one zinc atom and four oxygen atom surrounded by a Watson sphere carrying six units of positive charge. The atomic separation between zinc and oxygen atoms was taken to be 1.95 A ( 3.68 a.u ). The actual site symmetry of the zinc atom is $C_{3V}$. However, a $C_{3V}$ distortion of the $T_d$ symmetry does not affect relative separation of the energy eigenvalues by more than 0.2 eV.

The occupied states due to oxygen 2s- and 2p-like atomic levels and zinc 3d- like levels lead to two a1, two e , one $t_1$ and four $t_2$ molecular orbitals. The highest molecular orbital consists (HOMO) belongs to the $t_2$ irreducible representation. These molecular orbitals appear in four distinct groups, the relative binding energy of which with respect to the outermost group on the energy scale are found to be 2.06 eV, 5.03 eV and 15.5 eV. These values compare reasonably well with 2.8 eV, 5.8 eV and 18.0 eV derived from the observed UPS peaks at 3.1 eV, 5.9 eV, 8.9 eV and 21.1 eV[19]. These results suggest that the calculated energy eigenvalues for the tetrahedral cluster of $ZnO_4$ describe reasonably well the electronic structure of the states associated with the electrons in the valence band.

The lowest unoccupied molecular orbital (LUMO) defining the band edge belongs to the $a_1$ irreducible representation. This molecular orbital has zinc 4s-like character and is highly delocalized. The gap between the LUMO and HOMO energy levels is 3.9 eV, which is in reasonably good agreement with experiment. The lowering of the $T_d$ to $C_{3V}$ site symmetry reduces this gap by 0.2 eV which brings the calculated band gap closer to the observed energy gap.

Next we discuss the results from the band structure calculation for pure zinc oxide. The lattice parameters for zinc oxide are a=3.25 eV and c/a =1.6021[4]. The radii of the oxygen and zinc atoms were taken to be 1.588 A and 1.189 A respectively. For pure zinc oxide, zinc 3d-like bands are found to be lying between oxygen 2p and 2s like bands. The band gap at the Γ-point is found to be 3.4 eV. The purpose of the calculation involving the pure system is to optimize the sphere radii for zinc and oxide atoms which provide a reasonable description of the pure system. It is observed that the relative separation of different bands is strongly correlated to the sphere radii. However, the shapes of the energy bands do not change appreciably with minor changes in the sphere radii. It is also observed that the sphere radii when chosen closer to their ionic radius provide a better description of the optical properties of the material. These optimized sphere radii are used to calculate electronic structure of defects by the supercell approach.

### 3.2 Copper in Zinc Oxide

The molecular orbital calculations for divalent copper is performed using a copper centered cluster in Td symmetry using the spin polarized version of the scattered wave program. Except for the spin splitting due to exchange correlation, the composition and symmetry of the valence orbitals are very similar to that for the zinc centered cluster. The copper d-like level splits into one $t_2$ and one e like orbital, the separation between the two being 0.86 eV which compares well with the observed splitting of 0.71 eV. These two orbitals are further split by 0.47 eV due to exchange interaction. The up-spin states associated with $t_2$ and e like levels have 54% and 67% copper d character respectively; the down-spin states have 48% and 58% copper d character associated with the same states. This reflects the expected covalency of copper-oxygen bonds anticipated on the binding energies for oxygen p and copper d electrons, namely 9.53 eV for copper and 8.935 eV for oxygen from atomic calculations.

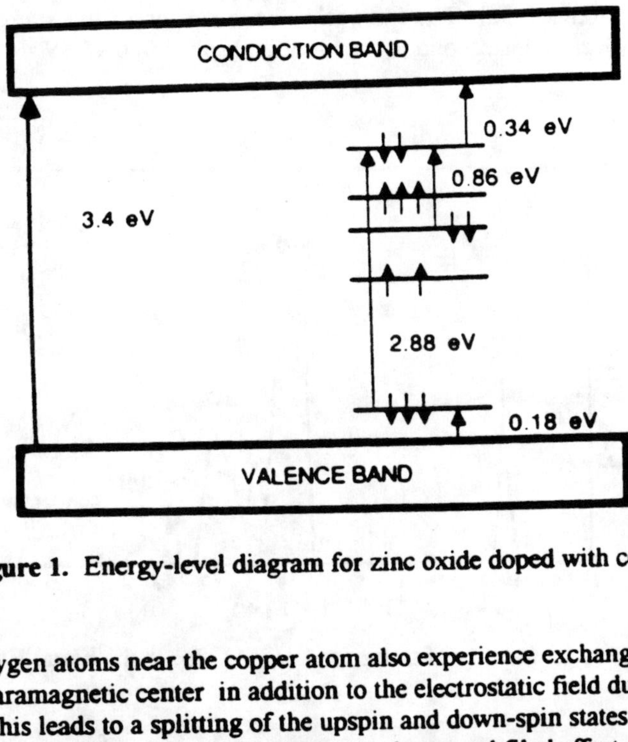


**Figure 1. Energy-level diagram for zinc oxide doped with copper.**

The oxygen atoms near the copper atom also experience exchange field due to the copper paramagnetic center in addition to the electrostatic field due to the charge in copper. This leads to a splitting of the upspin and down-spin states associated with the oxygen p-like orbitals. Since the change in the crystal filed effect due to substitution of zinc by copper is not expected to be significant, it is assumed that the unperturbed levels of the host material will lie near the centroid of the split levels of the oxygen atoms adjacent to the copper atom. These considerations lead to a $t_2$ spin down state 0.18 eV above the valence band which lies 2.88 eV below the $t_2$ state associated with the copper d hole. This energy difference agrees well with the observed excitation peak at 2.86 eV suggesting that this transition involves a charge transfer transition from oxygen to copper, a conclusion arrived at by Dingle from experiment. Thus assuming the band gap to be 3.4 eV, one can develop an energy level diagram as shown in Figure 1 which is very similar to one that can be developed using data from Dangle's measurements, the only important difference being the fact that the Fermi surface lies 0.34 eV below the conduction band instead of 0.1 eV inferred by Dingle[5].

In developing the energy level diagram from the metal centered clusters a number of assumptions have been made regarding the energy levels regarding the host system. For example its assumed that the top of the valence band remains essentially unperturbed and therefore, the copper d-like level is inferred to be in the band gap. In order to test the validity of this assumption, a supercell band structure calculation for Cu/ZnO was performed using unit cells containing 32 atoms, the atom at the origin being the copper atom with lattice parameters doubled compared to the host material. The 3d-like bands due to zinc atom range from -7.5 eV to -5.5 eV and the band due to oxygen p-like states from -5 to -1.5 eV. The copper 3d-like bands are found to be 1.1 eV above the valence band. The two peaks due to copper 3d-like states in the density of states plot (Fig. 2) correspond to a crystal field splitting 0.45 eV.

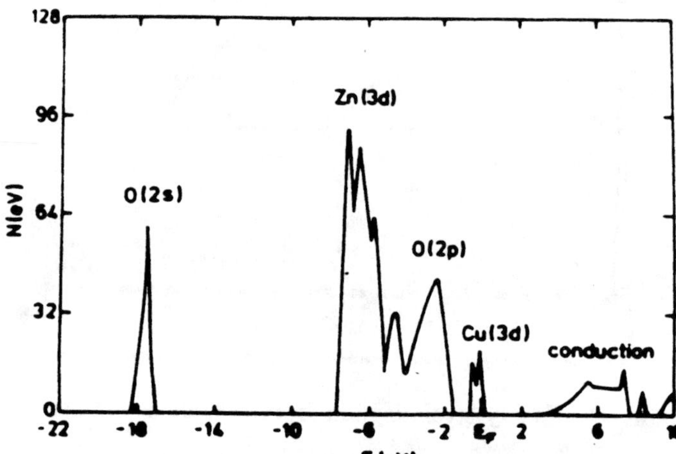


**Figure 2. Density of states for Cu/ZnO. $E_F$ is the Fermi Energy.**

In Figure 3, we have shown the joint density of states corresponding to transitions from copper d-like bands and the valence band to the d-hole in copper. The isolated peak at 0.45 eV corresponds to the infrared transition. The peak at 2.5 eV correspond to the charge transfer transition at 2.86 eV[5]. The large maxima between 5 eV and 8 eV is due to excitation from zinc 3d-like bands to copper d-hole. In general, the band calculations provide a similar picture as the cluster calculation but the transition frequencies are estimated to be lower than the observed values and also the cluster results.

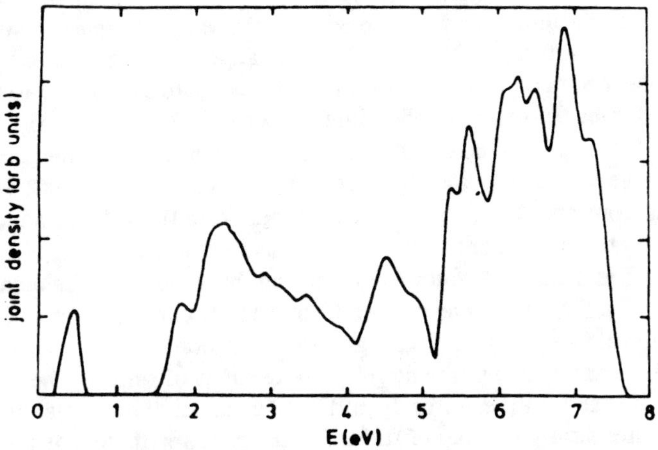


**Figure 3.** Joint density for excitation into the Cu hole state as determined from a spin restricted band-structure calculation.

### 3.3 Zinc Sulfide

The electronic structure of zinc sulfide was investigated using a metal centered cluster with four sulfur atoms in tetrahedral coordination. The nature of the molecular orbitals in the valence band is very similar to that in zinc oxide. The HOMO is found to have 69% sulfur 3p character and approximately 6% contribution from zinc 4p and 3d like states. The zinc 3d like levels are found to be 4.0 eV below the bottom of the valence band. The splitting of the 3d levels is found to be 0.6 eV. The lowest unoccupied level, $a_1$ is due to zinc 4s like state and is found to be 4.15 eV above the HOMO level. The charge transfer transition associated the energy gap is measured to be 3.8 eV[11] which is in reasonable agreement with the calculated value of 4.15 eV. Similar to zinc oxide, the cluster method overestimates the excitation energy in general.

The band structure calculation for zinc sulfide was performed using the crystal structure data for the hexagonal structure from the literature[7]. The sphere radii for zinc atoms and the empty spheres are taken to 2.22 a.u. and that for sulfur to be 3.47 a.u.. These choices of the sphere radii yields the observed band gap of 3.8 eV at the $\Gamma$ point. In the supercell calculations we will be using these sphere radii for zinc, empty spheres and the oxygen atoms. For the copper atom, the sphere radius will be taken to be the same as that for zinc.

### 3.4 Divalent Copper In Zinc Sulfide

The splitting of the copper d-like level in zinc sulfide is similar to that in zinc oxide. The crystal field splitting characterized by the separation between the $t_2$ and e molecular orbitals are found to be 0.93 eV which compares reasonably well with the observed splitting of 0.85 eV. The exchange splitting is found to be 0.23 eV. The highest occupied level due to sulfur 3p like states belonged to $t_2$ representation and is found to be below the nonbonding copper $t_2$ state by 1.98 eV. Another e like level is found to be 2.3 eV below the same copper state carrying the d-hole. These separations agree very well with the observed transitions at 1.7 eV and 2.3 eV[11]. These transitions are charge transfer excitations from the valence band to the copper d-hole similar to the divalent copper transitions in zinc oxide.

The density of states from supercell band structure calculation for $Zn_{15}CuS_{16}$ is shown in Figure 4. The copper 3d-like bands contribute to a minor peak near the top of the valence band. The energy bands between -6.0 eV and -7.0 eV are predominantly due to the atomic 3s-like states of sulfur. The 3d-like bands due to zinc lie between the sulfur 3p and 3s like bands. The partial density of states analysis indicates strong overlap of sulfur 3d-like bands with zinc 4s and 4p like bands.

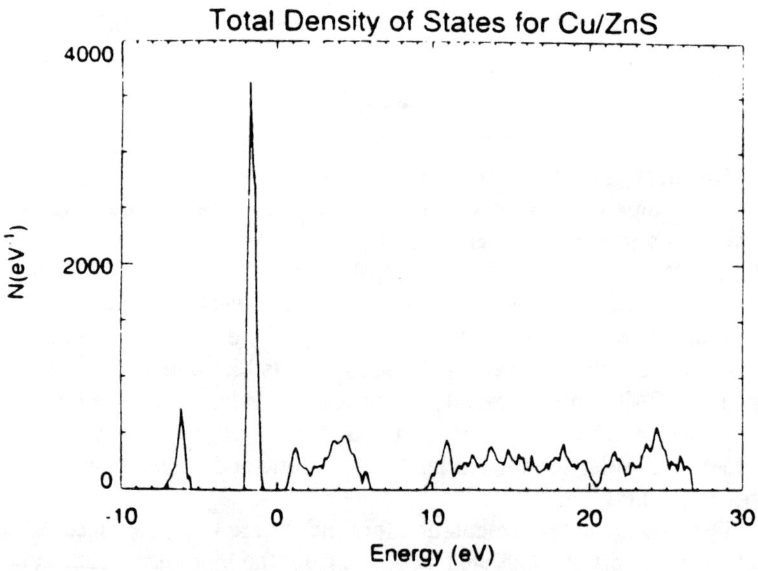


**Figure 4. Density of status due to Cu/Zns. The peak near 6.0 eV is due to the 3d bands of divalent copper.**

The joint density plot for excitation to the copper d-hole is shown in Figure 5. The three major peaks are at 0.75 eV , 1.53 eV and 2.07 eV. The first peak can be identified as the crystal field splitting in good agreement with 0.93 eV calculated by cluster approach and 0.85 eV measured experimentally. The other two peaks compare rather well with the charge transfer transitions described earlier.

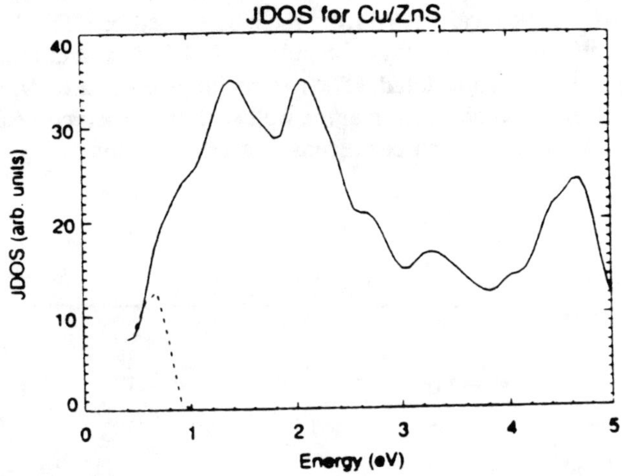


**Figure 5.** Joint density of state for excitation into the Cu-hole state in Cu/ZnS. The dashed curve gives the peak due to crystal field transition.

### 3.5 Monovalent Copper

The supercell used for this calculation consisted of one copper, one chlorine, fifteen zinc and fifteen sulfur atoms. The energy bands are shown in Figure 6. The

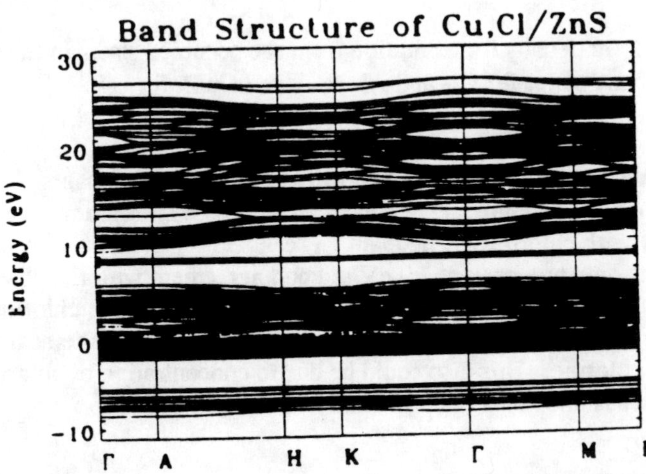


**Figure 6.** Energy bands for Cu, Cl/ZnS based on a calculation of hexagonal $Zn_{15}$ Cu cl $S_{15}$. The Cu 3/d band appears between 8.0 eV and 9.0 eV.

copper d like bands now move into the energy gap. The 3d-like states due to copper lead to rather flat bands indicating the localized nature of the copper 3d like states. The d bands are approximately 1.5 eV above the top the valence band which compares reasonably well with 1.7 eV calculated from the cluster method. The JDOS for this system is shown in Figure 7. Since the d-bands are full, only excitations to the conduction band states are considered. The two major peaks at 2.6 eV and 3.1 eV before the onset of the band absorption agree well with the observed peaks at 2.95 eV and 3.35 eV. The lower excitation peak leads to green emission at 2.38 eV.

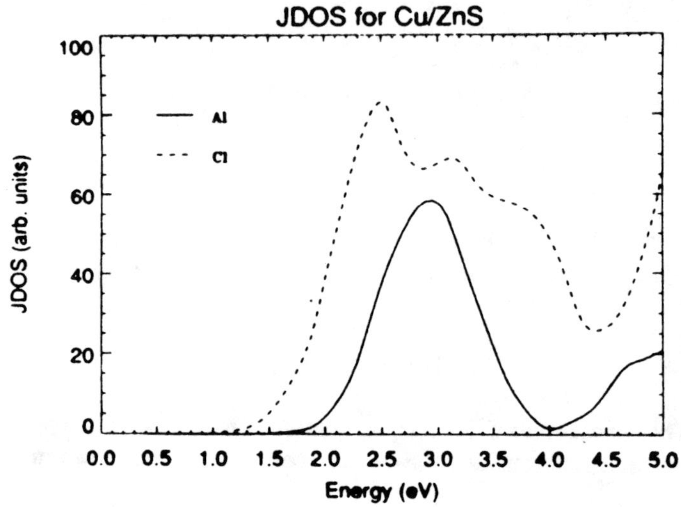


**Figure 7.** Joint density for excitation from the occupied states to the conduction band states for Cl- and Al- co-doped Cu/ZnS.

One can have monovalent copper with aluminum as codopants. The electronic structure of monovalent copper codoped with aluminum is very similar to that with chlorine as codopant. The JDOS for this system shown in Figure 7 has only one peak at 2.9 eV in good agreement with the observed peak at 2.95 eV. This peak position is different from that due to chlorine which may be due to the fact that the aluminum ions cannot be the nearest neighbor like chlorine. This also could be due to concentration of chlorine and aluminum assumed in this calculation.

## 4. CONCLUSION

Our main objective was to identify the luminescence mechanisms associated with divalent and monovalent copper in zinc sulfide and zinc oxide. We found that excitation peaks associated with transitions of the d hole to the occupied levels lead to better agreement with experiment. This is conceptually similar to the Lambe and Klick[13] model although the actual position of the copper d-like level is far different from what was implied in the model. In ZnO this level is very close to the conduction band while in zinc sulfide this level lies next to the valence band. In the case of monovalent ions, all the d levels of copper are filled; therefore there cannot be any excitation to the d-like levels of copper. Our calculations, both from the band structure and cluster methods indicate that the excitation peaks associated with monovalent copper can be explained by considering transitions from the occupied states to the conduction band. These transitions are in keeping with the Riehl, Schoen and Klasens model[12] discussed earlier.

Besides explaining the excitation process of copper in zinc oxides and zinc sulfides, the present approach indicates that in studying processes involving impurity ions, it is essential to use both the cluster and band structure approaches. We found that the cluster method in general provides a better agreement with the optical transitions while the band structure approach is essential for processes involving interactions between the host lattice and the impurity ion. Unless one uses fairly large clusters, it is not possible to describe the latter processes reasonably using the cluster approach alone.

This calculation also provides an explanation for the success of crystal field approach in solids. The results from band structure suggests that the impurity states are fairly localized. Thus by including the charge transfer effects in an empirical manner one can easily explicate all the information regarding the impurity ion by assuming an isolated ion whose energy levels have been split by the crystal field. But for transitions involving the host system and the impurity ions, the crystal field approach will not succeed since this approach does not include the host system in a rigorous manner.

Finally, we are hopeful that the first principles calculations using both the band structure and the cluster method can be used to design new luminescent materials. By limiting the analyses to one approach, one will obtain only a partial and incomplete picture of the system. It should be borne in mind that for such many electron systems, the appropriate Schrodinger equation cannot be solved exactly. The important question for a theorist is to find a way in which these approximate approaches can render a complete scenario. Our approach is a possible way to achieve this goal. The results presented in this work suggests that the first principles method can in deed be used not only to explain the observed data but also predict new materials and novel properties of the existing materials.

## 5. ACKNOWLEDGMENT

Authors express grateful acknowledgments to Mr. E. Dale, Dr. B. DeBoer, Dr. J. K. Berkowitz and Mr. J. Olsen for many helpful discussions.

## 6. REFERENCES

1. K. C. Mishra, P. C. Schmidt, K. H. Johnson, B. G. DeBoer, J. K. Berkowitz and E. A. Dale, Phys. Rev. B. 42, 1423 (1990).
2. K. C. Mishra, K. H. Johnson, P. C. Schmidt, B. G. DeBoer, J. Olsen and E. A. Dale, Phys. Rev. B 43, 14188 (1991).
3. K. C. Mishra, J. K. Berkowitz, K. H. Johnson and P. C. Schmidt, Phys. Rev. B 45, 10902 (1992).
4. S. C. Abrahams and J. L. Bernstein, Acta Crystallo. B 25, 1233 (1969).
5. R. Dingle, Phys. Rev. Lett. 23, 579 (1969).
6. R. E. Dietz, H. Kamimura, M. D. Sturge and A. Yariv, Phys. Rev. 132, 1559 (1963).
7. R. W. G. Wyckoff, Crystal Structures (Interscience, New York, 1951).
8. For example, an interstitial site for Cu-B center was proposed by H. Blicks, N.Riehl and R. Sizmann, Z. Phys. 163, 594 (1961).
9. J. L. Birman in Proceedings of the International Conference in Luminescence, edited by G. Szigeti (Hungarian Academy of Sciences, Budapest, 1966), p-919.
10. S. P. Keller and G. D. Petit, Phys. Rev. 115, 526 (1959).
11. I. Broser, H. Maier and H. J. Schultz, Phys. Rev. 140, A2136 (1965); I. Brose, K. H. Franke and H. J. Schultz in II-VI Semiconducting Compounds, edited by D. G. Thomas (Benjamin, New York, 1967)
p. 81.
12. M. Schoen, Z. Phys. 119, 463 (1942); H. A. Klassens, J. Electrochem. Soc. 100, 72 (1953); N. Riehl and H. Ortmann, J. Phys. Radiat. 17, 620 (1956).
13. J. Lambe and C. C. Klick, Phys. Rev. 98, 909 (1955).
14. J. S. Prener and F. E. Williams, J. Phys. Radiat. 17, 667 (1956).
15. J. L. Birman, Phys. Rev. 121, 141 (1961).
16. J. C. Slater and K. H. Johnson, Phys. Rev. B 5, 844 (1972); K. H. Johnson and F. C. smith, Jr., ibid. 5, 831 (1972).
17. A. R. Williams, J. Kubler and C. D. Gellat, Jr., Phys. Rev. B 19, 1990 (1979).
18. J. Korringa, Physica 13, 392 (1947); W. Kohn and N. Rostoker, Phys. Rev. 94, 1111 (1954).
19. L. Ley, R. A. Pollak, F. R. McFeeley, S. P. Kowalczyk and D. A. Shirley, Phys. Rev. B 9, 600 (1974).

Local Order
in Condensed
Matter Physics


ISBN 1-56072-220-7

# ROLE OF HYPERFINE INTERACTIONS IN UNDERSTANDING LOCAL ORDER IN CONDENSED MATTER SYSTEMS

Tara Prasad Das

*Department of Physics*
*State University of New York at Albany*
*Albany, New York 12222, USA*

As examples of the role of hyperfine interactions in providing useful information about condensed matter systems at atomic and electronic levels, we have discussed two classes of systems of current interest, namely high-temperature superconducting materials and fullerenes. In the first class of systems we have used Hartree-Fock cluster methods to study the electronic structures of $La_2CuO_4$, $YBa_2Cu_3O_6$ and $YBa_2Cu_3O_7$ system as well as muon in $La_2CuO_4$. The calculated electronic wave-functions are used to determine the charge and spin-distributions in these materials which in turn are used to derive the nuclear magnetic and quadrupole interaction parameters in these systems. Comparison is made with hyperfine interaction parameters determined from experimental data from nuclear magnetic resonance, nuclear quadrupole resonance and muon spin rotation measurements. There is overall satisfactory agreement between theory and experiment and possible sources for bridging the quantitative differences are discussed. For the muon in $La_2CuO_4$ system, the location of the muon determined through the energy minimization procedure is shown to compare favorably with the experimentally determined position from relaxation data from zero field muon spin rotation measurements. Possible improvements in both theory and experiment to obtain better quantitative agreement between the two have been proposed. The position of the muon inside $La_2CuO_4$ is important to know accurately because the muon can be used to obtain information about the magnetic moment distribution inside the crystal in the antiferromagnetic state and to study the nature of possible coexistence between magnetism and superconductivity near the transition temperature. The knowledge of the electron distribution in these one- and two-layered compounds is expected to be valuable for quantitative analysis of the mechanisms for origin of superconductivity in the superconducting systems associated with these compounds.

In the second class of systems, we have studied the location and associated hyperfine properties of muonium in fullerene again using the Hartree-Fock cluster procedure. Our results for the hyperfine interaction of muon including the influence of relaxation in the positions of all the carbon atoms and muonium, when compared with experimental results from muon spin rotation measurements, indicate that while the center with larger hyperfine interaction is associated with the muon at the center of the fullerene, the situation regarding the center with smaller hyperfine interaction is not completely settled. Both exohedral and endohedral sites where a surface carbon atom bonds to a muonium respectively outside and inside the fullerene can explain the observed hyperfine constant. The endohedral site however, seems to obviate the problem concerning the influence of the muonium on the dynamics of the fullerene in the solid state where muon spin rotation and nuclear magnetic resonance data suggest that the influence of the muonium is minimal. These results are expected to be useful in providing insights into the locations of other impurity atoms like metallic atoms when introduced into fullerene, as well as the nature of their bonding with the fullerene host.

## 1. INTRODUCTION

It has been a very valuable and emotional experience for me to spend these past four days in the idyllic surroundings of Jekyll Island and in particular the Jekyll Island Club, the venue of this International Symposium. I have been greatly touched by the kindness and thoughtfulness of my past students Professor Purusottam Jena and Professor Subhendra Deb Mahanti, who organized this symposium on the occasion of my sixtieth birthday and by the warmth of the tributes by my other past students who have been present at this Conference, Professor Bijan K. Rao, Professor Lambros Tterlikkis, who is presiding over this session and Professor Shukri B. Sulaiman and my past and present post-doctoral associates Dr. Kailash C. Mishra and Dr. Frank Hagelberg. I am also very grateful to all of them and all my past friends who have come from different parts of this country and the world and the new friends I have made, for taking precious times out of their busy schedules to attend this Conference, and present very important and interesting talks which have educated me immensely in their areas of great expertise. I have also been very appreciative and humbled by the kind remarks made about my work in the field of condensed matter physics and in particular hyperfine interactions.

The topic of hyperfine interactions that I will be talking about has been one of the common denominators of the research program that the members of our research group and I have been involved over the past three decades. It has been a collective effort involving both our group (including forty-one graduate students who have received their Ph.D. degrees already and nine currently working on their thesis projects and fifteen post-doctoral associates) and a number of international collaborations with active groups at Delft, Netherlands, Leuven, Belgium, Darmstadt, Munich and Erlangen in Germany, Aarhus, Denmark, Uppsala, Sweden, Ferrara, Italy, Tokyo, Japan and Bombay, Madras and Bhubaneswar in India. So I will be representing the collective efforts of a large number of fellow scientists that I have had the fortune to be associated with over the years.

Hyperfine interactions[1] are a valuable tool to investigate local order in condensed matter, both because they are sensitive to, and provide valuable information about, the structures and symmetries of solid state systems at both atomic and electronic levels. Thus, hyperfine properties are sensitive both to the spacing between atoms and ions and the symmetry of their arrangement. The strength of the isotropic contact contribution[2] to the magnetic hyperfine field at a nucleus as well as that of the anisotropic dipolar component[2] and the nuclear quadrupole interaction energy[3], are dependent on the spacing between the atom containing the nucleus under study and the neighboring atoms. Additionally the latter two properties are sensitively dependent[2,3] on the symmetry of the environment around the nucleus, vanishing because of their anistropic nature when the environment has high symmetry such as spherical in atomic systems and cubic, octahedral or tetrahedral in solid state systems. Additionally when there is departure from axial symmetry, as one would expect when the symmetry is less than three-fold, one obtains a departure from axial symmetry for both the anisotropic component of the hyperfine field and the electric field-gradient tensor responsible for the nuclear quadruple interaction, the departure from axial symmetry being represented in the latter case by the asymmetry parameter, $\eta$. More significantly, the hyperfine properties, which include the isomer shift[4] in Mössbauer studies, nuclear quadrupole interaction[3] and isotropic and anisotropic components[2] of the magnetic hyperfine field, provide detailed and quantitative information[5] about the electron density in the vicinity of the nucleus. The isomer shift provides information about the net charge density dependent on the total electron density at the nucleus, the nuclear quadruple interaction about the anisotropy of the charge density distribution around the nucleus while the isotropic and anisotropic components of the magnetic hyperfine field provide information regarding the strength and symmetry of the spin distribution which involves the difference between electronic densities at the nucleus corresponding to electrons in different spin states. Thus, these different hyperfine properties, taken collectively, provide a rather complete picture of the electron distribution in the system under study. Further, the analysis of hyperfine properties provides a direct link between electronic structures calculated theoretically and a large number of experimental techniques including nuclear magnetic resonance (NMR)[6], nuclear quadruple resonance (NQR)[7], Mössbauer effect[8], perturbed angular correlation and distribution (PAC and PAD)[9] and the muon spin rotation ($\mu$SR)[10] methods. The procedure involved in implementing this relationship between theory and experiment is to carry out theoretical investigations of the electron distributions in the system under study and see if the hyperfine properties that these calculated distributions predict agree with the results from experimental measurements by the techniques listed before, and use the nature of the agreement found to suggest either improvements in theory or in the experimental measurements or both.

The systems in which we have studied hyperfine interactions over the years are:

1. Atomic Systems with Emphasis on Many-Body Effects Studied by Perturbation Theoretic Methods (both Non-Relativistic and Relativistic)

2. Atomic Systems in Atmospheres of Rare Gases

3. Molecular Systems and Molecular Solids

4. Biomolecules-Hemoglobin and Derivatives and Reaction Centers in Photosynthesis

5. Metallic Systems - Non-magnetic and Ferromagnetic-Theory of Megagauss Transient Fields Experienced by Highly Charged Ions in Ferromagnetic Materials

6. Impurities in Semiconductors including Muonium

7. Semiconductor Surfaces

8. Ionic Crystals - Perfect and with Point Impurities-Solid Solutions and Color Centers

9. High-$T_c$ Superconductors

10. Fullerene Systems with Impurity-Atoms including Muonium

11. Antimony-Based Cluster Systems

I had originally planned to speak briefly on each of these eleven topics illustrating how the studies on these systems provide useful information about local order. The first two topics do not involve condensed matter systems but the study of atomic systems, where one can obtain very accurate results often in agreement within 1 to 2 percent with experiment. The study of atomic systems allows one to understand quite thoroughly the mechanisms contributing to hyperfine properties and thus obtain valuable insights regarding the expected importance of these effects in condensed matter systems, where due to the complexities of these systems, such effects cannot be studied as accurately. Among my earlier students and associates present here, Professor Bijan K. Rao had been associated with me on topics #1 and 2 dealing with free and interacting atomic systems. Professor Subhendra D. Mahanti had worked in both atomic systems and metallic systems as also had Professor Lambros Tterlikkis. Professor Purusottam Jena had concentrated primarily on metallic systems, including both on their magnetic and nuclear quadruple hyperfine properties. Professor Shukri B. Sulaiman had worked on both muonium and other impurities in semiconductors and systems in the field of high temperature superconductivity. Dr. K.C. Mishra had worked extensively on atomic and molecular systems, as well as in biological systems, semiconductor systems including surfaces and in ionic crystals. All of them are now making very valuable contributions in a number of other current fields and I feel fortunate that I can learn about these fields both from their talks here and through their kind efforts in keeping me updated about their work. My present collaborator, Dr. Frank Hagelberg has been working on both the origin of the megagauss and higher transient fields in ferromagnetic systems which he has already talked about at this symposium and also on the interesting antimony-based cluster systems.

I soon discovered, while preparing my talk, that covering all the eleven topics listed even very briefly would be too impracticable to do in a limited time. I have therefore decided to concentrate mainly on our work on high-$T_c$ systems and talk briefly on fullerene-muon systems, using both these classes of systems to illustrate the usefulness of hyperfine interactions for obtaining information on condensed matter systems, especially with respect to their electronic structures.

## 2. HIGH TEMPERATURE SUPERCONDUCTING MATERIALS-STUDIES ON ELECTRONIC STRUCTURES AND HYPERFINE INTERACTIONS

In discussing our investigations[12-19] on the electronic structures and hyperfine properties of high-$T_c$ systems which we have been involved in over the past three years, I would first want to thank my collaborators on this project without whose dedicated efforts these investigations would not have been possible. They include my past students, Professor Narayan Sahoo, who is currently on the faculty at Albany Medical College in Albany, New York, Professor Shukri B. Sulaiman, currently on the faculty in Universiti Sains Malaysia in Penang, Ms. Sudha Srinivas, presently working on her Ph.D. thesis in our group, Professor Kanetada Nagamine, Director, Meson Science Laboratory, University of Tokyo, Japan, Professor Eiko Torikai, Department of Computer Science and Electrical Engineering at Yamanashi University, Japan and Professor Onofrio Donzelli, Department of Physics, University of Ferrara, Italy.

The major motivation for our investigations on high-$T_c$ systems has been the desire to understand from first-principles the electronic structures of these systems, since we believe that the successful theory or theories that emerge from among the number of them that have been proposed in the literature for the origin of superconductivity in these systems will have to utilize the actual electron energy level structures and electron distributions in these materials. As we have mentioned earlier, the hyperfine data provide quantitative information about the isotropic and anisotropic charge and spin distributions in solid state systems and this is also true of the high-temperature superconductivity systems. We have been working on the normal state of these systems and have confined ourselves to the single and double layer systems $La_2CuO_4$, $YBa_2Cu_3O_6$ and $YBa_2Cu_3O_7$. I will present the results of our investigations[12-19] on the hyperfine properties of $^{63}Cu$, $^{17}O$, $^{135}Ba$, $^{139}La$ nuclei in these systems and the hyperfine properties associated with muon in $La_2CuO_4$, making extensive comparison with available experimental data[20-28]. After briefly presenting the Hartree-Fock-Roothaan procedure[29] utilized in our investigations and the methods for using the electronic structure results from our investigation to obtain the various associated hyperfine properties, we shall present our results and compare them with experiment. The effective charges on the various atoms in these materials will also be presented and discussed because they have important bearing on the nature of the covalent bonding in these systems.

### 2.1 Procedure

The procedure we have used for our investigations of the electronic structures of the high-$T_c$ systems is the Hartree-Fock Cluster procedure[30]. This procedure has been shown to be very successful in investigating the electronic energy levels and wave-functions for a variety of systems including semiconductor systems involving impurities in the bulk[31] and adsorbed at surfaces[32] and in perfect[33] and imperfect ionic crystals[34]. For the latter systems Professor John M. Vail has presented a very comprehensive talk at this Conference about his extensive investigations in the field. The Hartree-Fock procedure has also been shown[35] to provide satisfactory results for large biologically important molecules like hemoglobin derivatives.

With the Hartree-Fock procedure[29], one solves the Hartree-Fock equation.

$$(H_{\mu} - E_{\mu})\psi_{\mu} = 0 \tag{1}$$

where $H_{\mu}$ is the one-electron self-consistent Hamiltonian involving the kinetic energy, nuclear attraction energy from the nuclei in the molecular cluster and the direct and exchange interaction energies between an electron in the state $\psi_{\mu}$ and electrons in the other occupied states. In the high-$T_c$ systems we have studied[12-19], as in ionic crystal systems[33,34], we choose a sizable cluster around the ion containing the nucleus under study so that its immediate environment up to the third and fourth neighbors is well represented. Additionally the influence of the ions outside the cluster is represented by including in the Hartree-Fock potential, the potential due to the point charges on these external ions. In effect therefore, one includes in the simulation of the solid state system, the whole crystal. To extend the size of the cluster treated in detail in the Hartree-Fock procedure[30,33,34], we sometimes use pseudopotentials for some of the ions more distant from the nucleus under study, handling the effects of the ions beyond these distant ions in the point charge approximation.

Since, because of the multicenter nature of the cluster system, one cannot solve for the eigen-values and eigen-functions of the Hartree-Fock differential equations (1) by a direct integration procedure, one applies a variational procedure as in the Hartree-Fock-Roothaan formalism[29-32], expressing the molecular orbital eigen-function, $\psi_{\mu}$, in terms of orbitals $\chi_i$ based on the atoms in the cluster in the form

$$\psi_{\mu} = \sum_{i} C_{\mu_i}\chi_i \tag{2}$$

where the parameters $C_{\mu_i}$ are determined variationally. The natures and numbers of basis functions can be chosen with both accuracy and practicability in mind and in the past, both numerical atomic Hartree-Fock functions as well as exponential Slater-type functions have been used. In our recent investigations by the cluster procedure, as is the common practice currently, because of the enhanced speed of computations using them, we have been employing Gaussian orbitals[36] and making use of the available Gaussian software[37]. The results obtained by the variational procedure have of course to be tested with respect to the sizes of the basis sets in Equation (2) and this has been done in all our investigations[12-19][31-33] so far, including the high-$T_c$ systems. In our high-$T_c$ calculations[12-19], we have used extensive basis sets for the ions of copper, oxygen, lanthanum and barium involved, such as for instance, 4s (6,2,1,1), 4s (4,1,1,1), and 2d (4,1) Gaussian basis functions for copper and 4s (4,4,3,2) and 2p (4,3) for oxygen. The conventional notations used imply, for instance 4 basis s functions for copper involving respectively contracted (linear combinations) of functions involving 6 Gaussians, 2 Gaussians and two separate Gaussians. These functions had been found to give satisfactory results for nuclear quadrupole interactions in other ionic crystals[38,39] like cuprous and zinc oxides. The convergence of our results has been studied with respect to cluster sizes by using different sized clusters and studying the variation of hyperfine properties with cluster size. In handling clusters with finite spin as we have to do when dealing with magnetic hyperfine properties of antiferromagnetic systems, such as[17] $YBa_2Cu_3O_6$ and[13] $La_2CuO_4$, we have made use of the Unrestricted Hartree-Fock procedure[40] in which the wave-functions for states of opposite spin are allowed to be different, permitting the inclusion of exchange polarization contributions to the hyperfine fields at the nuclei from paired spin states.

Before describing our results for the hyperfine properties and charge distributions in the systems we have studied, we list the expressions for these properties in terms of the calculated one-electron wave-functions for the clusters used. Thus, the effective charge on an atom A in the cluster is given in the Mulliken approximation[41] by:

$$\zeta_{eff,A} = \zeta_A - n_A \tag{3}$$

where $\zeta_A$ is the charge on the nucleus A. The term $n_A$ is the total electronic population on atom A given by:

$$n_A = \sum_{\mu} \sum_{i \in A} [\sum_{j \notin A} (C^2_{\mu_i} + S_{ij} C_{\mu_i} C_{\mu_j})] \tag{4}$$

the summation over $\mu$ referring to all the occupied molecular orbitals and $S_{ij} = \langle \chi_i | \chi_j \rangle$ is the overlap integral between the basis functions $\chi_i$ and $\chi_j$, the summation over i and j in Eq. (4) being carried out over basis functions both centered on A and not on A respectively.

For nuclear quadrupole interaction, one needs the principal components of the electronic field-gradient tensor at the nucleus under study. To obtain these, one first obtains the components of the field-gradient tensor in any chosen axis system where the tensor will in general be non-diagonal and the component $V_{ij}$ is given by[42]:

$$V_{ij} = \sum_n \frac{\zeta_n e(3x_{in} - r_n^2 \delta_{ij})}{r_n^5} - \sum_{\mu} \int e |\psi_\mu|^2 \frac{3x_i x_j - r^2 \delta_{ij}}{r^5} d\tau + \sum_{N(ext)} \frac{\zeta_N e(3x_{iN} x_{jN} - r_N^2 \delta_{ij})}{r_N^5} \tag{5}$$

where the values 1,2,3 for i and j refer to the x,y and z components, the first term referring to the contributions from the nuclear charges $\zeta_n$ in the cluster, the second to the contributions from the electrons in the cluster and the third from the effective point charges on the ions outside the cluster. Since the influence of the potential due to these point charges on the electron distribution in the cluster is incorporated, the influence of antishielding effects[43] is already included in the second term obviating the need for any antishielding parameters[43] in the third term. Once the components $V_{ij}$ in any chosen system are calculated, one can obtain the principal axes (X',Y',Z') and principal components $V_{i'i'}$ by diagonalizing Vij. In the accepted convention in the literature[44], since only two of the principal components are independent in view of Laplace's equation, only two parameters are used to characterize the principal components, namely,

$$q = V_{z'z'} \tag{6}$$

related to the quadrupole coupling constant $e^2qQ$ and the asymmetry parameter

$$\eta = \frac{V_{x'x'} - V_{y'y'}}{V_{z'z'}} \tag{7}$$

where $|V_{z'z'}| \geq |V_{y'y'}| \geq |V_{x'x'}|$ (8)

Of course, the process of obtaining the principal components can be substantially abbreviated if one has some knowledge of the principal axes symmetry considerations[45].

For the magnetic hyperfine interaction there are two main contributions[46], isotropic and anisotropic. The isotropic term which leads to a term $A\vec{I} \bullet \vec{S}$ in the spin-Hamiltonian used to analyze the experimental signal in magnetic resonance, Mossbauer effect or other radioactive techniques, has the isotropic hyperfine constant A given by

$$A = \frac{8\pi}{3S}\gamma_e \gamma_I h^2 \sum_u |\psi_{u\uparrow}(0)|^2 + \sum_p \left\{ |\psi_{p\uparrow}(0)|^2 - |\psi_{p\downarrow}(0)|^2 \right\} \tag{9}$$

where $\psi_{u\uparrow}$ refers to the electronic wave-functions for the unpaired spin states and $\psi_{p\uparrow}$ and $\psi_{p\downarrow}$ for the paired spin states with spins parallel and antiparallel to the unpaired. The quantities $\gamma_e$ and $\gamma_I$ refer to the gyromagnetic ratios for the electronic and nuclear spins and S to the total electronic spin of the system. The anisotropic component, which occurs when there is departure from cubic or tetrahedral symmetry about the nucleus, leads[46] to a term of the form $B(2I_zS_z - I_xS_x - I_yS_y)$ for axial symmetry and is given by

$$B = \frac{2}{3S}\gamma_e \gamma_I h^2 \{ \sum_u \left\langle \psi_{u\uparrow} \left| \frac{3\cos^2\theta - 1}{r^3} \right| \psi_{u\uparrow} \right\rangle + \sum_p [ \langle \psi_{p\uparrow} \left| \frac{3\cos^2\theta - 1}{r^3} \right| \psi_{p\uparrow} \rangle - \left\langle \psi_{p\downarrow} \left| \frac{3\cos^2\theta - 1}{r^3} \right| \psi_{p\downarrow} \right\rangle ] \} \tag{10}$$

In Eq. (10), $\theta$ refers to the angle between the radius vector of the electron and the direction of axial symmetry. Usually, the constants A and B are expressed in frequency units, so one has to divide A and B by $2\pi h$. The hyperfine field $H_{hyp}$ is then obtained by applying the factor $\frac{2\pi S}{\gamma}$ to the components (A+2B) and (A-B) parallel and perpendicular to the symmetry axes respectively. In the work reported here, we shall be concerned primarily with axial symmetry. For the general case one needs three parameters $B_{xx}$, $B_{yy}$ and $B_{zz}$ in a principal axis system which can be treated as in the case of the field-gradient tensor discussed earlier in the section.

## 2.2 Results and Discussions

We shall present and discuss the electronic structure and hyperfine interaction results first for the compounds $YBa_2Cu_3O_6$ and $YBa_2Cu_3O_7$, followed by that for $La_2CuO_4$ and subsequently for the $La_2CuO_4$-muon systems. The interrelationships of the results in the three sets of systems and the insights they provide regarding the electron distributions in them shall be carefully discussed.

2.2.1 $YBa_2Cu_3O_7$ and $YBa_2Cu_3O_6$ Systems. We have studied[12,14,17,18] the $^{63}Cu$, $^{17}O$ and $^{135}Ba$ nuclear quadrupole interactions in both the systems $YBa_2Cu_3O_7$ and $YBa_2Cu_3O_6$, and the magnetic hyperfine interaction for the $^{63}Cu$ nucleus in $YBa_2Cu_3O_6$ in the antiferromagnetic state. The experimental results for these properties have been obtained by nuclear magnetic resonance measurements[20-23] in both powdered and single crystal systems.

The nuclear quadrupole interactions will be discussed first. The structure[47] of $YBa_2Cu_3O_7$ is presented in Fig. 1. For the investigation of $^{63}Cu$ nuclear quadrupole interactions, we have chosen[12] a cluster $Cu_3O_{12}$ involving two planar Cu(2) ions directly above and below each other and their four oxygen neighbors (O(2), O(3) and their equivalent counterparts) bonded to the same Cu(2) in each case, as well as the chain Cu(1) ion on the c-axis in between the two Cu(2) ions, the four neighboring oxygen ions of Cu (1), two, O(4), along the c-axis between Cu(1) and each of the Cu(2) in the cluster and the two O(1) ions on the chain containing the Cu(1) in the cluster. The same cluster is used

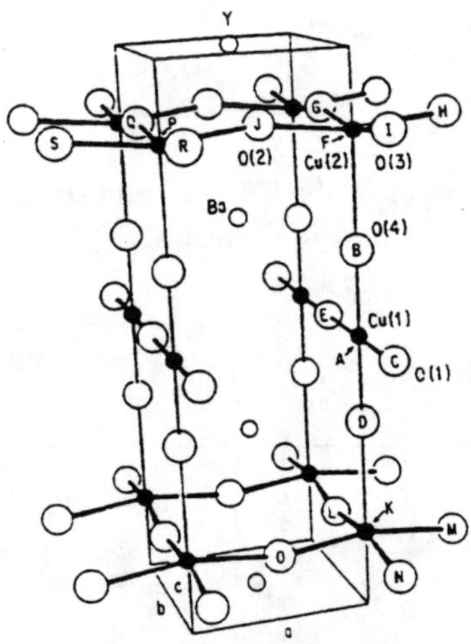


Figure 1. The orthorhombic unit cell of $YBa_2Cu_3O_7$. The various copper and oxygen atoms are designated in the way they are referred to in the text.

for the $^{17}O$ nuclear quadrupole interaction[12] associated with O(4), since it has within the same cluster, the two copper ions Cu(2) and Cu(1) it is directly bonded to as well as their oxygen neighbors. A similar cluster $Cu_3O_{10}$ was considered adequate[17] for $^{63}Cu(1)$, $^{63}Cu(2)$ and $^{17}O(4)$ in $YBa_2Cu_3O_6$ where the chain oxygens O(1) are not present in the lattice and therefore do not occur in the cluster used. For the $^{17}O(2)$ and $^{17}O(3)$ nuclei, clusters $Cu_2O_9$ were used[12], which involve the oxygen atom containing the $^{17}O$ nucleus under study and its two nearest neighbors Cu(2) on the plane and for each of the latter, its three other oxygen neighbors on the plane and the O(4) neighbor on the c-axis.

In Table 1, we have listed the charges on the Cu(1), Cu(2) and O(1) through O(4) ions, the O(1) of course being absent in $YBa_2Cu_3O_6$. The charges depart significantly from the formal charges +1, +2 and -2 for Cu(1), Cu(2) and O(2) (and O(3)) in both cases, -1 and -2 for O(4) and O(1) in the (1,2,3,7) system and -2 for O(4) in the (1,2,3,6). These departures suggest the presence of significant covalent bonding between copper and oxygen ions in these systems, the covalent bonding between Cu(1) and O(4) in the (1,2,3,7) compound being stronger than all other Cu-O bondings because of the shorter[47] Cu(1)-O(4) distance of 1.843 A as compared to all the other Cu-O distances in this system. For the (1,2,3,6) systems, the O(4) ion carries a charge close to the formal charge -2 than the -1 in the (1,2,3,7) systems due to the different numbers of electrons involved in the two cases because of the different stoichiometry for oxygen.

In Table 2, the $^{63}Cu$ nuclear quadrupole coupling constants ($e^2qQ$) and asymmetry parameters $\eta$ obtained by our theoretical investigations and experiment are listed. For the theoretical values of the $e^2qQ$ in Table 2 , we have used the most recent value[38] of $Q = -0.18 \times 10^{-24} cm^2$ available in the literature.

**Table 1** Charges on the Copper and Oxygen Ions in $YBa_2Cu_3O_7$ and $YBa_2Cu_3O_6$ Systems

| System / Ion | $Ba_2Cu_3O_7$ | $YBa_2Cu_3O_6$ |
|---|---|---|
| Cu(1) | 1.21 | 0.92 |
| Cu(2) | 1.67 | 1.77 |
| O(1) | -1.84 | ---- |
| O(2) | -1.95 | -1.94 |
| O(3) | -1.96 | -1.94 |
| O(4) | -1.11 | -1.97 |

The theoretical results in Table 2 show many features in agreement with experiment[20,22] and some differences. We shall discuss both these successes and the departures from experiment and possible avenues of improvement. Thus, considering the quadrupole coupling constants in $YBa_2Cu_3O_7$, the experimental trend of substantial increase in going from $^{63}$Cu(1) to $^{63}$Cu(2) is well explained by theory. Also from $YBa_2Cu_3O_7$ to $YBa_2Cu_3O_6$ there is a substantial increase in $e^2qQ$ for $^{63}$Cu(1) and an opposite trend of substantial decrease in the case of $^{63}$Cu(2), both features being in agreement with experiment. However, considering the $e^2qQ$ in $YBa_2Cu_3O_6$ alone, there is a significant decrease observed experimentally. On the other hand, the theoretical values are nearly equal. The asymmetry parameters for both Cu(1) and Cu(2) vanish both theoretically and experimentally. This is a consequence of the tetragaonal symmetry. For the $YBa_2Cu_3O_7$ system, the asymmetry parameter $\eta$ for $^{63}$Cu(2) is found to be small but finite and the same is found from theory. For $^{63}$Cu(1), on the other hand, $\eta$ is found to be very large experimentally, close to the maximum value of unity. The theoretical value of $\eta$ increases in going from $^{63}$Cu(2) to $^{63}$Cu(1) but only by a small amount. So there is a very significant difference between theory and experiment for $\eta$ of $^{63}$Cu(1) in the (1,2,3,7) system. Additionally, the theoretical values of $e^2qQ$ for both the (1,2,3,7) and (1,2,3,6) systems are significantly larger than experiment, the ratio being about 1.5 for $^{63}$Cu(1) in both systems, and significantly larger for $^{63}$Cu(2), about 1.7 and 1.9 in the two systems.

We have examined[17] the case of the $e^2qQ$ for $^{63}$Cu(2) in $YBa_2Cu_3O_6$ in some detail and found that a small but significant admixture of $d_{z^2}$-like hole character to the dominant $d_{x^2-y^2}$-like hole character could lead to agreement with experiment. Similar comments could be made for the other Cu systems. Thus, in particular in the case of Cu(1) for $YBa_2Cu_3O_7$, differences in the admixtures of $d_{xz}$ and $d_{yz}$ characters to the $d_{z^2}$ and $d_{x^2-y^2}$-like orbitals or $4p_x$ and $4p_y$ characters to the $4p_z$-like orbitals, since the 4p-like orbitals

**Table 2** $^{63}$Cu Nuclear Quadrupole Interaction Parameters in $YBa_2Cu_3O_7$ and $YBa_2Cu_3O_6$

| System | $YBa_2Cu_3O_7$ | | | | $YBa_2Cu_3O_6$ | | | |
|---|---|---|---|---|---|---|---|---|
| Nucleus | Theory[a] | | Experiment[a,b] | | Theory[a] | | Experiment[a,c] | |
| | $e^2qQ$ | $\eta$ | $e^2qQ$ | $\eta$ | $e^2qQ$ | $\eta$ | $e^2qQ$ | $\eta$ |
| $^{63}$Cu(1) | 56.6 | 0.11 | 38.4 | 1.00 | 86.5 | 0.0 | 59.1 | 0.0 |
| $^{63}$Cu(2) | 105.1 | 0.07 | 62.9 | 0.02 | 89.5 | 0.0 | 47.6 | 0.0 |

a. $e^2qQ$ in MHz; b. Ref. 20, C. H. Pennington et. al. and T. Shimizu et. al.
c. Ref. 22.

make major contributions to the electric field gradient tensor in this ion (with close to $Cu^+$ like character and a nearly filled 3d shell), could explain the large observed $\eta$. The question is what types of effects can lead to these types of orbital admixtures. One type of effect that can be suggested is the influence of spin-orbit interaction which has been found to be very effective[48] in influencing the nuclear quadrupole interactions for the Mossbauer isotope $^{57m}Fe$ in heme compounds among biological systems. The other possibility is the influence of many-body effects, for instance correlation between 3d-like orbitals and between 3d-like and 4s-like and 4p-like orbitals, effects of the latter type having been found to be rather important in copper atom[49]. These effects[50] need to be explored quantitatively in the future. However the fact that the difference with experiment is stronger for $e^2qQ$ for planar $^{63}Cu(2)$, as compared to chain $^{63}Cu(1)$ in $YBa_2Cu_3O_6$ suggests that part of the correlation effects may be associated with the mechanisms connected with the origin of superconductivity.

Table 3 lists the nuclear quadrupole interaction parameters $e^2qQ$ and $\eta$ for the $^{17}O$ nuclei in $YBa_2Cu_3O_7$ from experimental nuclear magnetic resonance measurements[21] and results from our cluster calculations[12]. No calculations were carried out on the $^{17}O$ nuclei in $YBa_2Cu_3O_6$, since at the time of our work[17], no experimental data on the nuclear quadrupole interactions in these nuclei were available. Since $^{17}O(2)$ and $^{17}O(3)$ in $YBa_2Cu_3O_7$ corresponding to the oxygen neighbors of planar Cu(2) are nearly equivalent, both from their geometrical parameters[47] and the experimental results in Table 3, we have carried out[12] calculations for only $^{17}O(2)$. The cluster $Cu_2O_9$ chosen for this work has already been described earlier in this section. Of the other two $^{17}O$ nuclei, only $^{17}O(4)$ has been investigated in our work[12]. For this nucleus, the immediate environment that one needs to include in a cluster investigation is covered in the cluster chosen[12] for investigations on the nuclear quadrupole interaction in the $^{63}Cu$ nuclei. No theoretical investigations were carried out on the chain $^{17}O(1)$ which are the neighbors of the chain Cu(1) ions. As in the case of $^{17}O(2)$, one would require a different cluster than that for the $^{63}Cu$ nuclei. It would be helpful to have theoretical results to compare with experiment[21] in view of the very significant differences from experiment found for the $^{63}Cu(1)$ nucleus, especially for the asymmetry parameter $\eta$.

It should be noted that while the signs of $e^2qQ$ cannot be obtained from NMR measurements and therefore only the magnitudes are given in Table 3, the theoretical values for $^{17}O(4)$ as $^{17}O(2)$ (and one expects the same sign in $^{17}O(3)$ as in $^{17}O(2)$) are both negative. The value of Q ($^{17}O$) is taken from the literature[51] as $-0.026 \times 10^{-24} cm^2$ to obtain $e^2qQ$ from our calculated values of q.

Our results for $^{17}O(2)$ in Table 3 show quite satisfactory agreement with experiment[21]. The value of $\eta$ agrees rather well with experiment. The value of $e^2qQ$ is somewhat larger than experiment[21], the overestimation, by a factor of about 1.2 being smaller than that (about 1.7) for the neighboring Cu(2) ion on the plane(Table 2). The calculations on the cluster containing O(2) were carried out with a smaller basis set than for the cluster containing the $^{63}Cu$ nuclei, for economy in computing effort, and it would be helpful in the future to test the result using the larger basis set.

**Table 3** $^{17}O$ Nuclear Quadrupole Interaction Parameters in $YBa_2Cu_3O_7$

| | Theory[b] | | Experiment[c] | |
|---|---|---|---|---|
| Nucleus[a] | $e^2qQ$ | $\eta$ | $e^2qQ$ | $\eta$ |
| $^{17}O(1)$ | -- | -- | 10.0 | 0.1 |
| $^{17}O(2)$ | -8.0 | 0.27 | 6.6 | 0.22-0.23 |
| $^{17}O(3)$ | -- | -- | 6.4 | 0.22-0.23 |
| $^{17}O(4)$ | -14.2 | 0.20 | 7.3 | 0.32 |

a. All $e^2qQ$ are expressed in MHz; b. Ref. 12; c. Ref. 21.

For $^{17}O(4)$, the theoretical result[12] in Table 3 is in less satisfactory agreement with experiment than for $^{17}O(2)$ and is more similar in this respect to the results for $^{63}Cu(1)$, although the asymmetry parameter is closer to experiment than in the case of $^{63}Cu(1)$ in Table 2. However, the magnitude of $e^2qQ$ is about 1.95 times that from experiment.

In looking for causes to bridge the differences between theory and experiment for the $^{17}O$ nuclei, one could invoke the same causes as discussed for $^{63}Cu$ nuclei. The influence of spin-orbit interactions[48] at the Cu(1) and Cu(2) sites leading to mixing of molecular orbitals with different d-like characters can of course influence the adjacent oxygen atoms O(1), O(4), O(2) and O(3). The ion O(4) is adjacent to both Cu(1) and Cu(2), but closer to Cu(1) and more covalently bonded as indicated by the effective charges in Table 1. It is therefore interesting that there is greater departure for O(4) from experimental quadrupole coupling parameters than for O(2) which is bonded to Cu(2), the same trend that is observed in this respect for Cu(1) and Cu(2). This situation is partly understandable since any causes that influence the nuclear quadrupole interactions for the copper ions are expected to also influence the oxygen ions to which they are bonded. In addition to the influence of spin-orbit effects[48] associated with the copper ions, one also expects the influence of conventional many-body effects[49,50] associated with both the copper and oxygen ions as well as special many-body effects associated with the origin of the superconducting states of the high-$T_c$ systems based on these materials.

Table 4 lists our calculated values[14,18] of $e^2qQ$ and $\eta$ in $^{135}Ba$ in $YBa_2Cu_3O_6$ and $YBa_2Cu_3O_7$ which are compared with experimental values[23,27] from nuclear magnetic resonance measurements.

**Table 4** $^{135}Ba$ Nuclear Quadrupole Interaction Parameter in $YBa_2Cu_3O_7$ and $YBa_2Cu_3O_6$

| System | Theory | | Experiment | |
|---|---|---|---|---|
| | $e^2qQ$ | $\eta$ | $e^2qQ$ | $\eta$ |
| $YBa2Cu_3O_7$ | -24.2[b] | 0.82[b] | 16.0[c] | 0.80[c] |
| $YBa_2Cu_3O_6$ | -51.6[d] | 0[d] | 54.0[e] | 0[e] |

a. $e^2qQ$ in MHz; b. Ref. 18; c. Ref. 27; d. Ref. 14; e. Ref. 23.

The cluster used in these calculations[14,18] was a $BaO_4$ cluster involving a $Ba^{2+}$ ion above the center of a rectangle containing four O(4) ions. The rectangle in the case[18] of $YBa_2Cu_3O_7$ was close to a square because of the slight orthorhombic character of the lattice and in the case[14] of $YBa_2Cu_3O_6$ which had a tetragonal structure, it was exctly a square. For $YBa_2Cu_3O_7$, we also used[18] a $BaO_6$ cluster in which, in addition to the four O(4) ions mentioned above, we also included the next nearest O(1) neighbors which were along the a axis below the plane containing the O(4) ions. The results for the latter cluster differed very little from the $BaO_4$ cluster indicating that it did not make too much difference whether one used the two O(1) ions as point charges or ions with extended electron distributions.

From Table 4, it can be seen that the result[14] for $e^2qQ$ in $YBa_2Cu_3O_6$ was in good agreement with experiment[27], the asymmetry parameter of course being zero because of tetragonal symmetry at the $Ba^+$ site. For $YBa_2Cu_3O_7$, the theoretical $e^2qQ$ had[18] a magnitude of about a factor of 1.5 times the experimental value, similar to the situation[12] for Cu(1). However, the asymmetry parameter $\eta$ which is rather large[18], namely 0.82, is in very good agreement with experiment[27]. What is also interesting is that the Z-direction in the principal axis system, corresponding to the largest principal component of the field-gradient tensor, was in the direction of the a-axis in agreement with experimental data[27] which indicate that the Z-axis is perpendicular to the c-axis in the crystal. The large value of $\eta$ indicates that the electric field-gradients along the principal Z and X axes are nearly equal and that for the Y-direction is small. This situation is expected to be a result of the interplay of the presence[47] of the two next nearest O(1) ions which removes the equivalence of the b and a directions, and the hole in the p-shell in O(4) which has a configuration close to $O^-$. The situation is totally different for the $YBa_2Cu_3O_6$ lattice[52] where there are no O(1) ions and the four O(4) ions having spherical $O^{2-}$ configurations and arrayed symmetrically on a square in the plane below $Ba^{2+}$. The charges[14,18] on Ba in the (1,2,3,7) and (1,2,3,6) systems were 1.88 and 1.94 respectively, indicating stronger ionic characters than for Cu(1) and Cu(2) as suggested by the charges in Table 1.

We turn next to the magnetic hyperfine interaction in the $YBa_2Cu_3O_6$ system in the antiferromagnetic state. The antiferromagnetism appears to be associated with the planar Cu(2) ions and so there are unpaired spin distributions in their vicinity. This would lead to magnetic hyperfine fields at the sites of the $^{63}$Cu(2) nuclei and the neighboring $^{17}$O nuclei. The fields at the $^{63}$Cu(2) nucleus have been measured in $YBa_2Cu_3O_{6+x}$ for zero[53] and small values[54] of x. These measurements suggest that the experimental value for x = 0 is about 80 kiloGauss and it is oriented on the ab plane. This orientation would be expected because neutron diffraction measurements[55] indicate that the magnetic moments in the antiferromagnetic state are aligned on the ab plane with no definitive information available about the exact orientation on the ab plane. In our work[13], we have used the orientation of the above spin as along the bisector of the a and b axes, not an unlikely orientation. From our UHF calculations[17], one can obtain, using equations (9) and (10), and the discussion following them, the hyperfine field at $^{63}$Cu(2) in the direction perpendicular to the symmetry axis. This is the appropriate procedure to use because the axial symmetry direction in the tetragonal $YBa_2Cu_3O_6$ lattice is along the c-axis and the hyperfine field would be oriented along the direction of the magnetic moment on the Cu(2) ion.

For a proper calculation of the electron distribution in the antiferromagnetic state we have to deal with a cluster including a number of Cu(2) ions with aligned antiparallel spins. Apart from the fact that such a cluster would involve a large number of atoms, it is difficult to include the charge and spin transfer mechanisms that lead to the antiferromagentic alignment. We have therefore used[17] the approximation involving a $CuO_5$ cluster with a Cu(2) ion, its two O(2) neighbors, two O(3) neighbors and the O(1) neighbor in Fig.1, with the influence of the rest of the lattice being approximated as point charges for their Coulomb potentials. The unpaired spin on the Cu(2) ion in the cluster leads[17] to the major contribution to the hyperfine field at the $^{63}$Cu nucleus and one has to make an approximate analysis for the contribution from the spin distributions outside the cluster.

By this procedure, the contributions from the isotropic contact and anisotropic dipolar mechanisms for the $^{63}$Cu nucleus from within the cluster come out as -20.9 kiloGauss and 137.8 kiloGauss respectively, leading to a net hyperfine field of 116.9 kiloGauss. The contribution to the hyperfine field from the magnetic dipole moments at the other copper ions outside the cluster, considering them as point dipoles and summing over the entire lattice, comes out as rather small, less than 2 kiloGauss and can be neglected. The experimental result[53,54] is about 25 percent lower than the calculated net hyperfine field. The mixing of $d_{z^2}$ like character with the $d_{x^2-y^2}$ -like character of the hole on the Cu(2) ion does not influence the hyperfine field as sensitively as the nuclear quadrupole interaction discussed earlier. Thus, spin-orbit interaction[48] and many-body effects[49,50] responsible for this mixing are not as effective in influencing the hyperfine field at the $^{63}$Cu nucleus.

One therefore has to examine other sources that could bridge the 25 percent gap between the theoretical and experimental hyperfine fields. Thus, the magnetic moment on the Cu(2) ion we have obtained[17] from a Mulliken population analysis[41] of the unpaired spin population on this ion is 0.85 Bohr magnetons as compared to the value of 0.6 Bohr magnetons suggested by neutron diffraction measurements[55]. Since the hyperfine field at the Cu(2) ion arises from the unpaired spin population, a simple scaling by this factor leads to a reduced theoretical value of about 83 kiloGauss in good agreement with the experimental value. It has recently been shown[56] that the Mulliken approximation[41] in antiferromagnetic systems leads to an overestimate and that careful analysis including the magnetic moment distribution including the influence of the antiparallel spin alignment on adjacent ions, and the effect of spin fluctuations[57], leads to good agreement with experiment[53,54]. A similar analysis of the contributions to the hyperfine field from the antiparallel spin distributions on the adjacent clusters centered on the corresponding Cu(2) ions would be expected to lead to a quantitative reduction in the hyperfine field. An analysis of this type would represent the quantitative evaluation of the influence of the mechanism suggested by Mila and Rice[58], and is distinct from the negligibly small point dipole contribution from these and other copper ions in the lattice pointed out in the preceding paragraph.

2.2.2 $La_2CuO_4$ System. For this system, we have studied[13,14,19] the $^{63}Cu$, $^{17}O$ and $^{139}La$ nuclear quadrupole interactions as well as the magnetic hyperfine interaction[13] at $^{63}Cu$ in

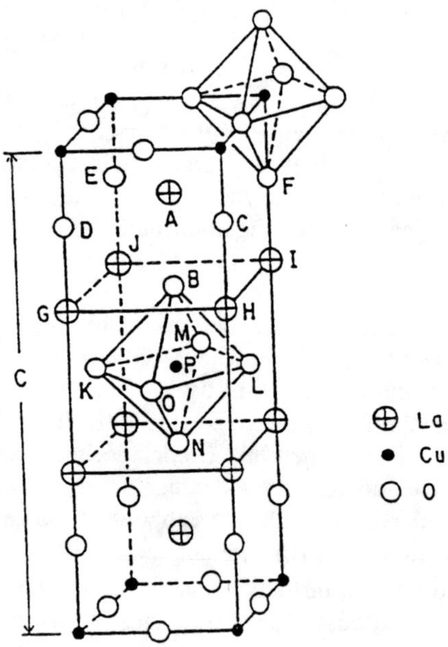


Figure 2. The crystal structure of $La_2CuO_4$ lattice in the tetragonal phase which is similar to the orthorhombic structure used in our work.

the antiferromagnetic state. Experimental data[24,25,26,59] are available for all these properties to allow comparison between theory and experiment as in the Yttrium Barium systems. Considering the nuclear quadrupole interaction results first, for the $^{63}Cu$ nucleus, we have made use of the bipyramidal cluster $CuO_6$ represented by the copper ion P, the planar oxygens K,M,L and O and apical oxygens B and N in Fig. 2 which shows the lattice structure in $La_2CuO_4$. The influence of the rest of the lattice is included, as described earlier, by treating the ions as point charges in terms of their potential as seen by the electrons in the cluster. For studying the convergence with respect to cluster size, a cluster $Cu_5O_6$ was employed, including the nearest neighbor copper ions of the planar oxygens K, M, L and O. The charges on the ions in the cluster changed by no more than 5 percent, indicating satisfactory convergence. The averages of the charges on the copper, the planar and apical oxygens using the two clusters were found to be 1.55, -1.89 and -1.95. The departures of these results from the formal charges of 2, -2 and -2 indicate significant covalent binding as in the case of the YBa copper oxide systems (Table 1). The charges on the planar and axial oxygens indicate that there is stronger covalent bonding of the former with copper than the latter. This is expected since the distance[60] between the copper and planar oxygens is 1.904A, significantly smaller than the corresponding distance of 2.397A for apical oxygen. The antiferromagnetic coupling between the neighboring copper ions is expected to take place through super-exchange involving the planar oxygen ions.

Nuclear magnetic resonance measurements[24] with both a little excess or deficiency of oxygen compared to $La_2CuO_4$ indicates that the value of the quadrupole coupling constant for $^{63}Cu$ in the latter system is about 65 MHz, the asymmetry parameter $\eta$ being very small, of the order of 0.03, but not zero, because of the slight departure of the lattice from tetragonal symmetry. Our calculation for the $CuO_6$ and $Cu_5O_6$ clusters lead respectively to $e^2qQ = 86.5$ MHz and $\eta = 0.020$ and $e^2qQ = 84.1$ MHz, $\eta = 0.027$. Like the effective charges, these results indicate that there is reasonable convergence with respect to cluster size. The ratio of the theoretical and experimental[24] quadrupole coupling constants is about 1.3, the agreement being significantly better than in the cases of the YBa copper oxide systems (Table 2). In looking for mechanisms to improve agreement with experiment we can consider the same ones as those discussed for Cu(2) ion in both $YBa_2Cu_3O_6$ and $YBa_2Cu_3O_7$, since for the copper ion in $La_2CuO_4$, the electronic configuration is close to $Cu^{2+}$ as in the other two systems The mechanisms such as spin-orbit interaction[48] and many-body effects leading to admixture of $d_{z^2}$ and $d_{x^2-y^2}$ like hole configurations are thus expected to be significant in effect for the $La_2CuO_4$ system.

For the $^{17}O$ nuclei, as in the case of the planar oxygens in the YBa Copper Oxide systems, one has to consider clusters centered around the oxygen atoms of interest. For the $^{17}O$ nucleus in the apical oxygen ion B, we have taken[19] the cluster containing all the numbered atoms A through P, in Fig. 2. It thus involves a $La_5CuO_{10}$ cluster involving the bipyramid to which B belongs and the four oxygen ions C through F and the lanthanum ions A,G,H,I and J. This allows us to include upto the sixth sets of neighbors of the oxygen ion B. For the planar oxygen nuclei, typically that in the oxygen ion L in Fig. 2, we have used[19] a cluster $La_4Cu_2O_{11}$ involving the bipyramid with the copper ion P at the centre and the corresponding one on the right to which L also belongs and the four $La^{3+}$ ions H and I and their counterparts in the BC plane below L. In Table 5, we have presented the results obtained for the quadrupole coupling constants $e^2qQ$ and asymmetry parameters $\eta$ for apical and planar $^{17}O$ nuclei and the directions of the Z' axis corresponding to the largest component and compared them with experimental results.

**Table 5** Nuclear Quadrupole Interaction Parameters for Planar and Apical $^{17}O$ Nuclei in $La_2CuO_4$

| Nucleus | Theoretical Results[a] | | | Experimental Results[a,b] | | |
|---|---|---|---|---|---|---|
| | $e^2qQ$ | $\eta$ | Z'-axis | $e^2qQ$ | $\eta$ | Z'-axis |
| $^{17}O(p)$ | -4.5 | 0.40 | a | 4.67 | 0.36 | a |
| $^{17}O(a)$ | -2.3 | 0.00 | c | 1.30 | 0.00 | c |

a. Results for $e^2qQ$ are in MHz; b. Ref. 25

The theoretical results for the $^{17}O$ quadrupole interaction properties are seen from Table 5 to be in generally good agreement with experiment. For the planar $^{17}O$, the theoretical vales of $e^2qQ$ and $\eta$ are very close to the experimental values, with the principal Z' direction being along the a-axis for both theory and experiment. For the apical $^{17}O$ nucleus the Z' principal axis is found to be along the c-axis from both theory and experiment[25]. The asymmetry parameter $\eta$ comes out as zero because of the near tetragonal symmetry of its environment and the unit cell. The magnitude of $e^2qQ$ however is significantly larger than experiment, the ratio being about 1.75. This difference could possibly be bridged by the influence of the mechanisms discussed earlier for $^{17}O(2)$ in the case of $YBa_2Cu_3O_7$, namely those involving spin-orbit effects[48] associated with the neighboring copper ion and the oxygen ion itself as well as many-body effects[49,50]. However, the fact that there is excellent agreement with experiment[25] for the planar $^{17}O$ nucleus while there is significant difference for apical $^{17}O$ suggests another possible contribution. Thus in our work, in common with other investigations on electronic structures of ionic crystals, only the influence of the potential from charges on the ions outside the cluster has been included. However, for the highly deformable ions like the doubly negative oxygen ions, significant dipole moments are expected to be present on them, especially when there is no reflection symmetry at the oxygen site as is the case for the apical oxygens. Since the influence of these dipole moments has not been included on the potential seen by the electrons in the cluster, the contribution of the dipole moments[61] to the electric field gradient has also not been included. In the $La_2Cu0_4$ system, as may be seen from Fig. 2, there are four apical oxygen ions C,D, E and F quite close to the apical oxygen ion B and therefore a significant influence on the electric field gradient from the dipole moments is expected. In general, such effects are expected to be effective for all the nuclei in the three different systems we have studied and in particular could be significant contributors for the $^{17}O(2)$ and $^{17}O(4)$ nuclei in $YBa_2Cu_3O_7$ and perhaps lead to a narrowing of part of the difference between theory and experiment in Table 3.

For the study of $^{139}La$ nuclear quadrupole interaction, we have used the $LaO_5$ cluster which involves the $La^{3+}$ ion A and its five apical oxygen ion type neighbors, C,D,E and F on a plane parallel to the ab plane and B in Fig. 2. Using the value[62] of $0.21x10^{-24}$ $cm^2$ for the nuclear quadrupole moment of $^{139}La$ and our calculated result for the largest of the principal components of the field gradient tensor, which is oriented along the c-axis, the value of the quadrupole coupling constant $e^2qQ$ comes out as -90.81MHz. The asymmetry parameter $\eta$ small, 0.02, but non-vanishing because of the slight orthorhombicity[60] of the $La_2CuO_4$ lattice. These results are in good agreement with the experimental results of $e^2qQ$ = 89.4 MHz and $\eta$ = 0.027. This observed better agreement with experiment for $^{139}La$, as compared to the $^{63}Cu$ and $^{17}O$ nuclei, indicates that if many-body effects[50,51] represent the major contributors to the difference from experiment, then they are less pronounced near the $La^{3+}$ ion, away from the active region involving the copper and oxygen ions which are considered to be responsible for the superconducting character of the $La_2CuO_4$ based systems. The same observation can also be made for the YBa copper oxide systems. The charge obtained for the lanthanum ion from our cluster investigations is 2.98 indicating that, like barium in the YBa copper oxide systems, the lanthanum ion is strongly ionic and close to a $La^{3+}$ configuration.

For the magnetic hyperfine field at the $^{63}Cu$ nucleus, we have followed the same procedure for antiferromagnetic $La_2CuO_4$, as was described for $YBa_2Cu_3O_6$ earlier in this article. The same bipyramidal cluster with a copper ion at the center was used as for the investigation of the $^{63}Cu$ nuclear quadrupole interaction. The contact hyperfine field was found[13] to be -31 kiloGauss and was dominated by the dipolar field of 136 kiloGauss oriented along the direction of the unpaired spin on the copper ion at the center of the cluster, which has been found from neutron diffraction measurements[63] to be in the direction of the b-axis in the orthorhombic axes system[64], that is at 45° to the line joining the lanthanum ion and a planar oxygen. The contribution from the magnetic dipole moments on the copper ions outside the cluster, considering them as point dipole moments and summing over the entire lattice again comes out negligible, similar to the case of antiferromagnetic $YBa_2Cu_3O_6$ discussed earlier. The total hyperfine field 105 kiloGauss is 25 kiloGauss higher than the experimental value of about 80 kiloGauss (close to that for the $YBa_2Cu_3O_6$ system) suggested from nuclear magnetic resonance measurements[65] for slightly higher and lower contents of oxygen as compared to $La_2CuO_4$. The same discussion as for $YBa_2Cu_3O_6$ considered earlier applies about the sources that could improve agreement with experiment. Thus, while many-body effects[49,50] and spin-orbit effects[48] could have some influence on the hyperfine field, although perhaps not a major one, the main reduction in the theoretical hyperfine field is expected to arise from the influence of the opposite spin distributions in adjacent clusters. The magnetic moment on the copper ion from our electronic structure calculations using the Mulliken approximation was found to be 0.88 Bohr magneton, substantially higher than the value 0.5 found from neutron diffraction measurements[63]. The explanation discussed for bridging a similar gap between our result[17] and experiment[55] in $YBa_2Cu_3O_6$ should apply here. This would involve the incorporation of spin fluctuations[57] and a more accurate treatment as for the magnetization[56] for including the influence of the opposite spin distributions on neighboring clusters as discussed for the $^{63}Cu$ hyperfine field in $YBa_2.Cu_3O_6$.

2.2.3 $La_2CuO_4$ - Muon System. I like to finish this discussion of our electronic structure investigations in high-$T_c$ materials by briefly describing our recent work[15,16] on the interaction between muon and $La_2CuO_4$. This work was stimulated by the extensive muon spin rotation measurements[66,67,68] on this system by a number of groups including the group at University of Tokyo.

There are a number of reasons for studying the muon position and associated hyperfine properties. The first is to test how well we can explain the interaction of the muon with $La_2CuO_4$ and thus obtain a check on the understanding of the electron distribution in the latter which is complimentary to that one gets by working on the pure material as just discussed. The second reason is to see how well theory can explain the experimentally observed hyperfine field at the muon from muon spin rotation measurements[67] in the antiferromagnetic state. The third reason is to make comparison with the position derived from the relaxation effect[10] in zero-field muon spin rotation measurements[68], this effect arising from the distribution in the magnetic field[10] at the muon site due to the distribution in orientations of the copper nuclear spins in the presence of the nuclear quadrupole interactions at their sites.

Earlier analysis to determine the position of the muon based on approximate theoretical analyses[69,70] of the energy of interaction of the muon with $La_2CuO_4$ and the estimation of the hyperfine field[71] assuming that it arose from localized moments on the copper ions had suggested that it is located on the ac plane, the a and b axes in this work

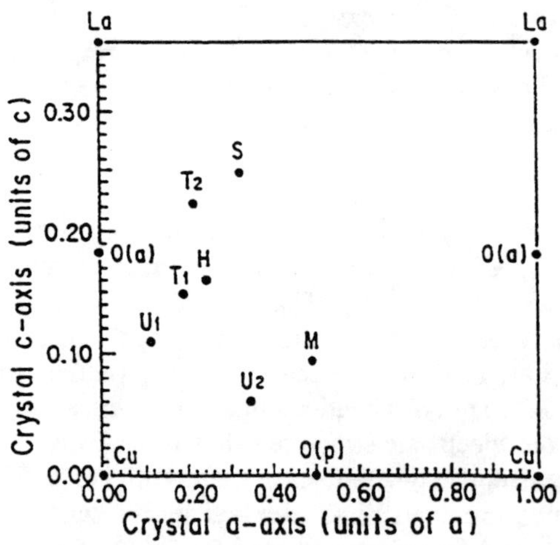


Figure 3. Position of $\mu^+$ on the ac plane of the tetragonal unit cell determined in earlier work (Refs. 68-71) and present work.

being taken as the tetragonal system of axes pointing along the Cu-O lines in Fig. 2 for the planar oxygens instead of the orthorhombic axes system used earlier for our discussion of the $^{63}$Cu hyperfine field in $La_2CuO_4$ in the antiferromagnetic phase. A similar conclusion regarding the location of the muon on the ac plane was made from the analysis of relaxation effects in zero-field muon spin rotation measurements[68] although the data also seems to fit a position on a plane involving the c-axis and the bisector of a and b axes. In view of these results we have concentrated our analysis[15,16] with the muon on the ac plane. Since the muon, like hydrogen or a proton, would tend to attach itself to an oxygen negative ion, we have explored the vicinities of both the apical and planar oxygens. For muon next to the apical oxygen B in Fig. 2, we have used a seventeen atom cluster, involving the muon together with the sixteen atom cluster used[19] for studying the $^{17}$O nuclear quadrupole interaction as described earlier. For the vicinity of the planar oxygen ion L in Fig. 2, we have used a sixteen atom cluster involving the muon in addition to the fifteen atom cluster used[19] for the study of planar $^{17}$O nuclear quadrupole interaction. The energies of each of these two clusters was studied as a function of the angle made by the line B$\mu$ with the c-axis for the vicinity of the apical oxygen B and between the line L$\mu$ and the line LP joining L and the copper ion P for the vicinity of the planar oxygen L. The results show a number of energy minima, the deepest one being found near the apical oxygen with the line B$\mu$ making an angle of 25° with the c-axis. This position is denoted by U1 in Fig. 3, the minimum position found near the planar oxygen being indicated by U2, which corresponds to a higher energy than that for U1. The total energy of the cluster was also minimized as a function of the B$\mu$ bond distance and positions perpendicular to the ac plane, the latter procedure showing that the position U1 on the ac plane is the lowest in energy and therefore the most likely position for the muon from our investigations. The positions T1 and T2 represent the positions found from zero-field $\mu$SR investigations[68] in the paramagnetic state, H from fitting[71] the hyerfine field at the muon from $\mu$SR measurements[66] in the antiferromagnetic state using a point dipole approximation for the magnetic moments on the copper ions and M and S from earlier theoretical investigations[69,70] of energy minima. We have calculated the hyperfine field at the muon at the position U1 following the steps described in the Procedure section (2.1) and find the net hyperfine field to be -559.2, -657.3 and 515.3 Gauss along the a, b and c directions in the antiferromagnetic state. Use has been made of the fact that the magnetic moments on the copper ions are aligned parallel and antiparallel to the 110 direction, from neutron diffraction measurements[63]. A weighting factor has been applied as in the case of $^{63}$Cu hyperfine field to reduce the calculated results to take account of the fact the observed magnetic moments[63] on the copper ions are 0.5 Bohr magnetons rather than the 0.85 Bohr magnetons found from our cluster calculation. A more sophisticated estimate including the contributions from neighboring clusters with antiparallel spins on the copper ions could be done later, as pointed out earlier for the hyperfine interaction in $^{63}$Cu in this system, to incorporate the Mila-Rice type of effect[58], which would obviate the use of the approximate weighting procedure.

In making comparisons with experiment, for the muon position, the experimental reference points most appropriate are expected to be the ones, T1 and T2, determined from zero-field $\mu$SR measurements[68] because the point H in Fig. 3 is based on an estimate[71] of the hyperfine field from point dipoles on the copper ion and does not take account of the local contributions from the electron spin distribution in the vicinity of the muon. This local contribution is found to be about 30 percent of the distant contribution from our work[15,16]. Our muon position U1 in Fig. 3 is found to be closer to T1 than T2 and our calculated hyperfine field is found[15,16] to have a magnitude of 1005 Gauss somewhat more than twice as large as the experimental value[66,67] of 400 to 430 Gauss.

The direction described by the polar and azimuthal angles $\theta = 59°$ and $\theta = 49.6°$ are in reasonable agreement with single crystal measurements[72] leading to $\theta = 65° \pm 5°$ and $\phi = 28° \pm 8°$. Possible sources that could lead to improved agreement with experiment are many-body effects and the lattice distortion effects associated with the presence of the muon. We are in the process of exploring the latter effect which should also influence the position T1 determined form zero-field $\mu$SR data in the paramagnetic state based on the magnetic field distribution[68] at the muon site from the copper nuclear moments. These latter contributions are dependent on the nuclear quadrupole interactions at the $^{63}$Cu nuclei. The earlier analysis[68] for T1 used the $^{63}$Cu nuclear quadrupole interaction for the pure $La_2CuO_4$. Revised analysis is under way in our group in collaboration with the Tokyo group including the changed $^{63}$Cu nuclear quadrupole interaction in the presence of the muon as determined from our cluster investigations[15,16].

In summary, the results of all our Hartree-Fock cluster investigations[12-19] described here in Section 2 have provided an overall satisfactory explanation[20-28,66,72] of the hyperfine properties of the family of high-temperature superconducting systems $La_2CuO_4$ (including that for muon in $La_2CuO_4$), $YBa_2Cu_3O_6$ and $YBa_2Cu_3O_7$. The agrement with experiment is more satisfactory at the $^{139}$La and $^{135}$Ba nuclei which are away from the active regions for superconductivity, namely, those involving the $^{63}$Cu and $^{17}$O sites. It appears that both standard many-body effects[49,50] as among the electrons in conventional atomic, molecular and solid state systems, as well as many-body effects associated with the origin of superconductivity in this family of systems may be responsible. In this respect, the agreement with experiment seems to be overall somewhat better for the $La_2CuO_4$ system[13,14,19] than the $YBa_2Cu_3O_6$ and $YBa_2Cu_3O_7$ systems[12,14,17,18], which suggests that the role of many-body effects may be different in the two cases. This should be useful for the understanding of the origin of the superconductivity in these systems. Also expected to be useful in this respect are the general conclusions that there is significant covalent bonding among the copper and oxygen ions in these systems, especially between the chain copper and bridge oxygen in $YBa_2Cu_3O_7$ systems, the charges on the various ions being significantly different from the expected formal charges.

## 3. POSITION AND ASSOCIATED ELECTRONIC STRUCTURE AND HYPERFINE INTERACTIONS FOR MUONIUM AND HYDROGEN IN FULLERENE $C_{60}$

Before ending my talk, I would like to briefly discuss our recent investigations[73,74] on muon and hydrogen in fullerene. This work has been carried out recently in collaboration with Professor Onofrio Donzelli of Department of Physics, University Ferrara, Italy and Ms. Tina Briere of our research group at State Unviersity of New York at Albany, presently working on her Ph.D. thesis. This work represents a good example of how hyperfine properties can provide valuable information about the positions of impurity atoms in a system, the associated electron distributions and the ability to distinguish, and choose, from alternate models for the system when it contains an impurity.

Fullerenes are very interesting systems which are the subject of extensive current attention not only because of their intriguing structures and as examples of the largest highly symmetric clusters available, but also for the several interesting properties that make them important for many potential applications in technology and other branches of physics besides molecular and cluster physics. Among these are their ability in the presence of impurities to have very interesting magnetic[75] and conducting properties[76], the superconducting nature of alkali fullerides[77], their relationship to the interesting carbon nanotubes[78] which promise to have valuable technological applications and the likelihood of their being present in outer space[79] both by themselves and as complexes involving abundant elements like hydrogen, oxygen and the alkali elements. There have been a number of very interesting and insightful talks in this conference on the synthesis[80,83] and interesting properties of pure fullerenes[81,82,83] and fullerene compounds[80,83], fullerenes with impurities[82] and carbon nanotubes[84], by a number of experts in the field. I shall deal with only one example of a fullerene system with impurities, namely that with muonium (hydrogen), for two main reasons. The first is because first-principles Hartree-Fock theory in this system can increase our insight into more complex impurities like for instance alkali metal and transition metal atoms. The second reason is that the study of hyperfine properties by muon spin rotation[85,86,87] and electron paramagnetic resonance techniques[88] have raised interesting questions regarding the locations of muonium (and hydrogen) in this system.

The experimental situation from muon spin rotation measurements[85-87] with muonium in $C_{60}$ fullerene and electron paramagnetic resonance[88,89] on hydrogen atom in fullerene is the following. The $\mu$SR measurements have revealed two signals, one at low frequency and one at high frequency, $\nu_{LF} = 325$ MHz and at $\nu_{HF} = 4256$ MHz. The EPR measurements on the $C_{60}H$ system have revealed only a single signal whose frequency, after applying the ratio of the gyromagnetic ratios of the proton and muon, appears to correspond to $\nu_{LF}$ for the muonium. Of the two frequencies for the muonium, the high frequency $\nu_{HF}$, because of its closeness to the hyperfine frequency 4463 MHz for muonium in vacuum[90], suggests[85,87] that it is associated with muon at the center of $C_{60}$ where it is expected to be weakly bonded to the fullerene. It is referred to in the literature as the Mu@$C_{60}$. For the second resonance, two centers have been studied earlier[91,92] by methods involving approximations applied to the Hartree-Fock procedure, refered to as the MNDO[93] and PRDDO[94] methods. The two centers studied correspond to the muonium, inside and outside the fullerene, bonded to a carbon atom, and referred to as endohedral and exohedral respectively (fig. 4). The endohedral center was found[89,90] by the MNDO and PRDDO methods to be higher in energy than the exohedral and the Mu@$C_{60}$. In our first investigation, we have tested[73] this conclusion by the Unrestricted Hartree-Fock procedure allowing for only the muonium and its nearest neighbor carbon atom to relax, for economy of computational effort. Our results also showed the same ordering of the energy levels. Using our calculated Unrestricted Hartree-Fock wave-functions and

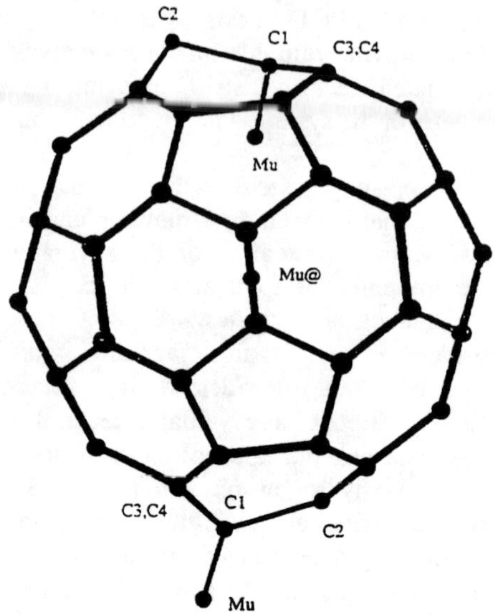


Figure 4. Various centers (Mu@, exohedral and endohedral) for muonium in fullerene $C_{60}$.

carrying out vibrational averaging[95] of the magnetic hyperfine interaction, we obtained[73] a hyperfine frequency of 591MHz for the exohedral position. The unpaired spin for the exohedral system was located not on C1, which was next to the muonium, but on the adjacent carbon C2 which had a near single bond with C1 in Mu-$C_{60}$ but shared a double bond in pure fullerene $C_{60}$. The contribution to the muonium hyperfine constant was primarily of the isotropic contact type (Equation (9)), with a small contribution, about 10 MHz, arising from the dipolar effect. The smallness of the anisotropic dipolar contribution is in keeping with the conclusion from the small line-width from low temperature μSR measurements[86] at 5°K where the motion of the fullerene system is expected to be frozen out and the line-width arises from a distribution of dipolar fields at the muon site in a powdered sample due to a distribution in the crystalline axes. Another UHF calculation[96] on the exohedral model besides ours, allowing for thirteen other neighboring carbon atoms to relax, led to a 33% reduction in the isotropic hyperfine constant, compared to our result[73] with only Mu and C1 allowed to relax, and closer agreement with experiment[85,86].

Thus, the exohedral model appeared to be the appropriate model for the trapped muonium responsible for the low frequency μSR signal from both Hartree-Fock investigations[73,96] in the literature as well as earlier work[91,92] using approximations to the Hartree-Fock procedure. However there were some problems with this assignment. One was that the calculated position of the muonium from the UHF investigations[73,96] places it at an almost equal distance from a carbon atom at an adjacent fullerene in the solid state (where the μSR measurements[85,87] are carried out), as the distance of about 1.1-1.2 A from C1 in the system involving a muonium attached to a single fullerene considered in both UHF investigations. Thus the conclusions from the total energy minimization and hyperfine interaction may not apply to the solid state system where the μSR measurements were carried out. The second important question arises from the consideration that the μSR line narrowing[86] at higher temperatures near about 242°K has been ascribed[86] to a reduction in the distribution of the hyperfine fields at the muon due to the anisotropic dipolar contribution because of the rotational motion of the $C_{60}$ system. An analysis[86] of this narrowing effect has suggested that the associated reorientational dynamics is very similar to that for the pure $C_{60}$ system as obtained from nuclear magnetic resonance studies[97]. This would indicate that the influence of the neighboring fullerenes in the pure $C_{60}$ system is virtually unaltered by the presence of the muonium between the fullerenes in the exohedral system, which is difficult to understand.

The endohedral Mu-$C_{60}$ system would not be subject to these problems because the muonium would be inside the buckyball containing it. In view of this, we have recently carried out[74] an extensive calculation on the endohedral model allowing all the carbon atoms in $C_{60}$ to relax. This required very substantial computer time on the Cray C90 at the CINECA supercomputer facility in Italy and showed that the stability of the endohedral model was considerably enhanced, making its total energy comparable to the Mu@$C_{60}$ system and much closer to the exohedral system than was found[73] when only the Mu and its nearest neighbor carbon atom was allowed to relax. Also the isotropic hyperfine constant came out as 339MHz, in good agreement with that from the low frequency μSR signal[85,86] and the dipolar contribution was small (about 19MHz), in agreement with expectations from the μSR measurement[86] at 5°K.

These recent results[74] show that one cannot consider at present time that the exohedral center is unequivocally the correct one for the Mu-$C_{60}$ system that explains the hyperfine interaction obtained from μSR measurements[85,86] and that much additional work remains to be done in the future to decide on the correct model for muonium bound to carbon atoms at the fullerene surface. These include the study of the interaction of the muonium with more than one fullerene in the exohedral system, the influence of larger and more flexible basis sets for the UHF procedure and the influence of the inclusion of many-body effects[49,50,98,99].

In summary, as mentioned at the onset of this discussion of our muon-fullerene investigations, hyperfine interaction studies are continuing to help very significantly in the attempts to understand the geometry and nature of interaction of muonium with fullerene $C_{60}$.

## ACKNOWLEDGMENT

In this talk I have pointed out with two examples of current interest, both how hyperfine properties can be quantitatively investigated through first-principle study of electronic structures and how hyperfine properties can be used to draw conclusions about structures of molecular and solid state systems at atomic and electronic levels and atomic systems at the latter level. This is the area of Physics on which I have concentrated my major efforts almost all through my scientific career. I am very grateful that I have had the opportunity to talk about this primary interest of mine before this audience consisting of my past and present students and associates that I am so proud of, my friends who have helped and supported me intellectually over the years, the other distinguished scientists whose work I have admired over the past and have had the opportunity to hear and learn from at this meeting, and my wife, who does not often attend my talks, but is very supportive and constructively critical when she does attend.

I am deeply indebted to all the people who have helped me in so many ways that have allowed me to achieve whatever I have been able to at this point in my life. I would like to make special mention of some of my teachers, Mr. Sarasi Mohan Das in Stewart School in Cuttack, Orissa, India, Professor Ganesh Prasad Dube, during my undergraduate career at Patna University, Patna, Bihar, India, Late Professor Meghnad Saha, the eminent physicis who contributed so much to the growth and organization of research in Physics in India and his son Late Professor Ajit Kumar Saha from Saha Institute of Nuclear Physics, Calcutta, India, both of whom were my research advisors during my graduate career, and finally, during my postdoctoral career, Professor Richard Bersohn from Columbia University and Professor Erwin L. Hahn and Professor Walter D. Knight of University of California, Berkeley. All of them helped me learn and appreciate the joys and spirit of Physics. I would like to thank all my students whose strong dedication to the advancement of scientific knowledge and highly infectious enthusiasm have been a source of great inspiration for me. Some of them I have been able to name in this talk and all of them have helped me to continue to learn both while they were (are) working with me and afterwards. I want to thank my parents and brother and sisters for their love and their strong encouragement to prepare myself for a career in education and research. I want to thank my wife, Basanta, and children, Puspa, Shankar and Surjya, for their continued love, affection and understanding which has contributed to the fullness of my life and for excusing me for the times I could not spend with them because of my involvement in Physics. The achievements that my children have had in their academic and professional careers have made me very proud. Lastly I would like to humbly express my respect for the Supreme Being for giving me the strength to be able to follow my dreams throughall the circumstances I have had to face in my life.

## REFERENCES


1. R. E. Watson, and A. J. Freeman, in *Hyperfine Interaction*, A. J. Freeman and R. B. Fraenkel. (Ed.), New York: Academic Press (1967); T. P. Das, *Hyperfine Interaction*, Vol. 34, 149 (1987).

2. T. P. Das, *Relativistic Theory of Electrons*, Chap. 7, New York: Harper and Row, (1973).

3. P. Jena, S. D. Mahanti and T. P. Das, *Phys. Rev.*, B1, 432 (1970); S. N. Ray and T. P. Das, *Phys. Rev.*, B16, 4794 (1977); T. P. Das and P. C. Schmidt, *Zeitschrift fur Naturforschung*, 41a, 47 (1986); P. C. Kelires, K. C. Mishra, and T. P. Das, *Hyperfine Interactions*, 34, 289 (1987); S. M. Mohapatra, B. N. Dev, T. M. Thundat, L. Luo, W. M. Gibson, K. C. Mishra, N. Sahoo and T. P. Das, *Reviews of Solid State Science*, World Scientific Publishing Company, Singapore, 4, 873 (1990); D. W. Mitchell, S. B. Sulaiman, N. Sahoo, T. P. Das, W. Potzel, and G. M. Kalvius, *Phys. Rev.*, B44, 6728 (1991).

4. *Mössbauer Isomer Shifts*, G. K. Shenoy, and F. E. Wagner, (Ed.), Amsterdam, North Holland, (1978).

5. W. Potzel, in *Mössbauer Spectroscopy Applied to Magnetism and Materials Science*, Vol. 1, F. J. Long, and F. Grandjean, (Ed.), New York: Plenum Press (1993); P. C. Kelires and T. P. Das, *Hyperfine Interactions*, 34, 285 (1987).

6. A. K. Saha, and T. P. Das, *Nuclear Induction*, Calcutta, Saha Institute of Nuclear Physics (1957); A. Abragam, and M. Goldman, *Nuclear Magnetism: Order and Disorder*, Oxford: Oxford University Press, (1982); C. P. Slichter, *Principles of Magnetic Resonance*, New York: Springer Verlag, (1990).

7. T. P. Das, and E. L. Hahn, *Nuclear Quadrupole Resonance Spectroscopy*, New York: Academic Press, (1957); E. A. C. Lucken, *Nuclear Quadrupole Coupling Constants*, New York: Academic Press, (1969).

8. *Mössbauer Spectroscopy Applied to Magnetism and Materials Science*, F. J. Long and F. Grandjean, (Ed.), New York: Plenum Press, (1993); G. K. Shenoy, and F. E. Wagner, Ref. 4.

9. E. N. Kaufman and R. J. Vianden, *Revs. of Modern Physics*, 51, 161 (1979); P. Singh, M. N. Nyayate, S. H. Devare, and H. G. Devare, *Phys. Rev.*, B39, 2308 (1989).

10. A. Schenck, *Muon Spin Rotation Spectroscopy*, Boston: Adam Hilger Publishers, (1985).

11. J. Stein, S. B. Sulaiman, N. Sahoo and T. P. Das, *Hyperfine Interactions*, 60, 849 (1990); H. H. Klauss, N. Sahoo, P. C. Kelires, W. Potzel, G. M. Kalvius, M. Frank, and W. Kreische, *Hyperfine Interactions*, 60, 853 (1990); D. W. Mitchell, T. P. Das, W. Potzel, G. M. Kalvius, H. Karzel, W. Schiessl, M. Steiner, and M. Kofferlein, *Phys. Rev.*, B48, 16449 (1993).

12. N. Sahoo, Sigrid Markert, T. P. Das, and K. Nagamine, *Phys. Rev.*, B41, 220 (1990); *Hyperfine Interactions*, 61, 1135 (1990).

13. S. B. Sulaiman, N. Sahoo, T. P. Das, O. Donzelli, E. Torikai, and K. Nagamine, *Phys. Rev.*, B44, 7028 (1991); S. B. Sulaiman, N. Sahoo, S. Markert, J. Stein, T. P. Das, and K. Nagamine, *Bulletin of Material Science (India)*, 14, 149 (1991).

14. S. B. Sulaiman, N. Sahoo, T. P. Das, and O. Donzelli, *Phys. Rev.*, B45, 7383 (1992).

15. S. B. Sulaiman, N. Sahoo, Sudha Srinivas, F. Hagelberg, T. P. Das, E. Torikai, and K. Nagamine, *Phys. Rev.*, B (in press); *Hyperfine Interactions* (in press).

16. S. B. Sulaiman, N. Sahoo, Sudha Srinivas, F. Hagelberg, T. P. Das, E. Torikai, and K. Nagamine, *Hyperfine Interactions*, 79, 901 (1993).

17. S. B. Sulaiman, Sudha Srinivas, N. Sahoo, T. P. Das, O. Donzelli, E. Torikai, and K. Nagamine, *Bull. Am. Phys. Soc.*, 38, 174 (1993); Under preparation for *Phys. Rev.* B.

18. Sudha Srinivas, S. B. Sulaiman, N. Sahoo, T. P. Das, E. Torikai, and K. Nagamine, *Bull. Am. Phys. Soc.*, 38, 176 (1993); Under preparation for *Phys. Rev.* B.

19. Sudha Srinivas, S. B. Sulaiman, N. Sahoo, T. P. Das, E. Torikai, and K. Nagamine, *Bull. Am. Phys. Soc.*, 39, 839 (1994); Under preparation for *Phys. Rev.* B.

20. M. Mali, D. Brinkmann, L. Pauli, J. Ross, and H. Zimmermann, *Phys. Lett.*, A124, 112 (1987); R. E. Walstedt, W. W. Warren, Jr., R. F. Bell, G. F. Brennert, G. P. Espinosa, J. P. Remeika, R. J. Cava, and E. A. Rietman, *Phys. Rev.*, B36, 5727 (1987); Y. Kitaoka, S. Hiramatsu, T. Kondo, and K. Asayama, *J. Phys. Soc. Japan*, 57, 30 (1988); C. H. Pennington, D. J. Durand, C. B. Zax, C. P. Slichter, J. P. Rice, and D. M. Ginsberg, *Phys. Rev.*, B37, 7944 (1988); T. Shimizu, H. Yasuoka, T. Imai, T. Tsuda, T. Takabatake, Y. Nakazawa, and M. Ishikawa, *J. Phys. Soc. Jpn.*, 57, 2494 (1988).

21. C. Coretsopoulous, H. C. Lee, E. Ramli, L. Reven, T. B. Rauchfuss, and E. Oldfield, *Phys. Rev.*, B39, 781 (1989).

22. M. Mali, I. Mangelschots, H. Zimmermann, and D. Brinkmann, *Physica*, C175, 581 (1991).

23. H. Lütgemeier, V. Florentiev, and A. Yakubowskii, in *Electronic Properties of High-$T_c$ Superconductors and Related Compounds*, Springer-Verlag, Berlin, Proceedings of the Conference at Kirchberg, (1990).

24. K. Ueda, T. Sugata, Y. Kohara, Y. Oda, M. Yamada, S. Kashiwai, and M. Motoyama, *Solid State Comm.*, 73, 49 (1990); K. Ishida, Y. Kitaoka, and K. Asayama, *J. Phys. Soc. Japan*, 58, 36 (1989).

25. K. Ishida, Y. Kitaoka, G. Zheng, and K. Asayama, *J. Phys. Soc. Japan*, 60, 3516 (1991).

26. H. Lütgemeier, and M. W. Pieper, *Solid State Comm.*, 64, 267 (1987).

27. J. Shore, S. Yang, J. Haase, D. Schwartz, and E. Oldfield, *Phys. Rev.*, B46, 595 (1992);
A. Yakubowskii, A. Egorov, and H. Lütgemeier, *Appl. Magn. Reson.*, 3, 665 (1992).

28. J. I. Budnick, A. Golnik, Ch. Niedermayer, E. Recknagel,, M. Rossmanith, A. Weidinger, B. Chamberland, M. Filipowski, and D. P. Yang, *Phys. Lett.*, A124, 103 (1987); B. Hitti, P. Birrer, K. Fischer, F. N. Gygax, E. Lippelt, H. Malletta, A. Schenck and M. Weber, *Hyperfine Interactions*, 63, 287 (1990); E. Torikai, H. Ishihara, K. Nagamine, H. Kitazawa, I. Tanaka, H. Kojima, S. B. Sulaiman, Sudha Srinivas, and T. P. Das, *Hyperfine Interactions*, 79, 915 (1993).

29. D. R. Hartree, *The Calculation of Atomic Structures*, New York: John Wiley and Sons Inc., (1957); C. C. J. Roothaan, *Revs. Mod. Phys.*, 23, 69 (1951).

30. K. A. Colbourn and J. Kendrick, in *Computer Simulation of Solids*, C. R. A. Catlow and W. C. Mackrodt, (Ed.), New York: Springer-Verlag, (1982); P. C. Kelires, Ph.D Thesis, State University of New York at Albany, (1987) (Unpublished); S. M. Mohapatra et. al., in Ref. 3; N. Sahoo, S. B. Sulaiman, K. C. Mishra, and T. P. Das, *Phys. Rev.*, B39, 13389 (1989); S. B. Sulaiman, N. Sahoo, K. C. Mishra, T. P. Das, and K. Bonde-Nielsen, *Hyperfine Interactions*, 34, 289 (1987); P. C. Kelires et. al., Ref. 3; H. H. Klauss et. al., Ref. 11; J. Stein et. al., Ref. 11; D. W. Mitchell et. al., Refs. 3 and 11.

31. N. Sahoo et. al., Ref. 30; S. B. Sulaiman et. al., Ref. 30; S. B. Sulaiman, Chapter 3, Ph.D Thesis, State University of New York at Albany, (1992) (Unpublished).

32. B. N. Dev, S. M. Mohapatra, N. Sahoo, K. C. Mishra, W. M. Gibson and T. P. Das, *Phys. Rev.*, B38, 13335 (1988); L. Luo, N. Sahoo, G. A. Smith, W. M. Gibson, and T. P. Das, *J. Phys. Soc. Japan*, 60, 3139 (1991); S. M. Mohapatra et. al., Ref. 3.

33. P. C. Kelires et. al., Ref. 3; H. H. Klauss, Ref. 11; J. Stein et. al., Ref. 11; D. W. Mitchell et. al., Ref. 3.

34. Ravindra Pandey and A. Barry Kunz, *Phys. Rev.*, B38, 10150 (1988); John M. Vail, R. Pandey and A. B. Kunz, *Revs. Solid State Science*, 5, 241 (1991); D. W. Mitchell, S. B. Sulaiman, N. Sahoo, T. P. Das, W. Potzel, G. M. Kalvius, W. Schiessl, H. Karzel, and M. Steiner, *Hyperfine Interactions*, 78, 403 (1993).

35. N. Sahoo, K. Ramani Lata and T. P. Das, *Theo. Chim. Acta.*, 82, 285 (1992).

36. *Gaussian Basis Sets forMolecular Calculations*, S. Huzinaga, (Ed.), Amsterdam: Elsevier (1984).

37. Gaussian 80, 86, 88, 90, and 92 Series of Programs, Gaussian Inc., Pittsburgh, Pennsylvania.

38. J. Stein et. al., Ref. 11.

39. D. W. Mitchell et. al., Ref. 3.

40. R. E. Watson and A. J. Freeman, Ref. 1; G. D. Gaspari, Wei-Mei Shyu, and T. P. Das, *Phys. Rev.*, 134, A852 (1964); J. A. Pople, and R. K. Nesbet, *J. Chem. Phys.*, 22, 571 (1964).

41. R. S. Mulliken, *J. Chem. Phys.*, 23, 1833 (1955).

42. S. B. Sulaiman et. al., Ref. 13.

43. H. M. Foley, R. M. Sternheimer, and D. Tycko, *Phys. Rev.*, 93, 734 (1954); S. N. Ray, T. Lee, and T. P. Das, *Phys. Rev.*, 49, 1108 (1974); R. M. Sternheimer, *Zeitschrift fur Naturforschung*, 41a, 24 (1985); T. P. Das and P. C. Schmidt, Ref. 3.

44. T. P. Das, and E. L. Hahn, Ref. 7, Chap. 1.

45. M. H. Cohen and F. Reif, in *Solid State Physics*, F. Seitz and D. Turnbull, (Ed.), New York: Academic Press, (1957), Vol. 5.

46. T. P. Das, Ref. 2, Chap. 8; N. Sahoo et. al., Ref. 30.

47. M. A. Beno, L. Sederholm, D. W. Capone II, D. G. Hinks, J. D. Jorgensen, J. D. Grace, I. K. Schuller, C. U. Segre, and K. Zhang, *Appl. Phys. Lett.*, 51, 57 (1987).

48. J. N. Roy, S. K. Mishra, K. C. Mishra, and T. P. Das, *J. Chem. Phys.*, 90, 7273 (1989).

49. R. W. Dougherty, S. N. Panigrahy, T. P. Das, and J. Andriessen, *Phys. Rev.*, A47, 2710 (1993).

50. S. N. Panigrahy, R. W. Dougherty, T. P. Das, and J. Andriessen. *Phys. Rev.*, A44, 121 (1991); S. N. Ray, Taesul Lee, and T. P. Das, *Phys. Rev.*, B8, 5291 (1973); J. E. Rodgers, R. Ray and T. P. Das, *Phys. Rev.*, A14, 543 (1976); S. N. Ray, and T. P. Das, *Phys. Rev.*, B16, 4790 (1977).

51. H. P. Schaefer III, R. A. Klemm, and F. E. Harris, *Phys.Rev.*, 181, 138 (1969).

52. J. S. Swinnea, and H. Steink, *Journal of Materials Research*, 2, 442 (1987).

53. H. Yasuoka, T. Shimizu, Y. Ueda, and K. Kosuge, *J. Phys. Soc. Japan*, 57, 2659 (1988).

54. Y. Yamada, K. Ishida, Y. Kitaoka, K. Asayama, H. Takagi, H. Iwabuchi, and S. Uchida, *J. Phys. Soc. Japan*, 57, 2663 (1988); P. Mendels and H. Alloul, *Physica*, C156, 355 (1988).

55. J. M. Tranquada, A. H. Moudden, M. S. Alvarez, A. J. Jacobson, J. T. Lewandowski, and J. M. Newsam, *Phys. Rev.*, B38, 2477 (1991).

56. T. A. Kaplan, S. D. Mahanti and Hyunju Chang, *Phys. Rev.*, B45, 2565 (1992); S. D. Mahanti, T. A. Kaplan, Hyunju Chang, and J. F. Harrison, *J. Appl. Phys.*, 73, 6105 (1993).

57. T. A. Kaplan and S. D. Mahanti, *J. Appl. Phys.*, 69, 5382 (1991).

58. F. Mila and T. M. Rice, *Physica*, C157, 561 (1989).

59. T. Tsuda, T. Shimizu, H. Yasuoka, K. Kishio, and K. Kitazawa, *J. Phys. Soc. Japan*, 57, 2908 (1988).

60. J. M. Longo and P. M. Raccah, *Journal of Solid State Chemistry*, 6, 526 (1973); C. P. Poole, Jr., T. Datta, and H. A. Frach, *Copper Oxide Superconductors*, New York: John wiley and Sons, (1988).

61. R. R. Sharma and T. P. Das, *J. Chem. Phys.*, 41, 3582 (1964).

62. P. Raghavan, *Atomic and Nuclear Data Tables*, 42, 248 (1989).

63. D. Vaknin, S. K. Sinha, D. E. Moncton, D. C. Johnston, J. M. Newsam, C. R. Safinya, and H. E. King, Jr., *Phys. Rev. Lett.*, 58, 2802 (1987); T. Frelhoft, J. E. Fischer, G. Shirane, D. E. Moncton, S. K. Moncton, S. K. Sinha, D. Vaknin, J. P. Remeika, A. S. Cooper, and D. Harshmann, *Phys. Rev.*, B36, 826 (1987); Bingxin X. Yang, S. Mitsuda, G. Shirane, Y. Yamaguchi, H. Yamauchi, and Y. Syono, *J. Phys. Soc. Japan*,56, 2283 (1987); Y. Yamaguchi, H. Yamauchi, M. Ohashi, H. Yamamoto, N. Shimoda, M. Kikuchi, and Y. Syono, *Japan J. Appl. Phys.*, 26, L447 (1987).

64. J. M. Longo and P. M. Raccah, Ref. 60; C. P. Poole, Jr. et. al., Ref.60. See also S. B. Sulaiman, Ref. 31, p.111.

65. T. Tsuda et. al., Ref. 59.

66. J. I. Budnick et. al., Ref. 28; B. Hitti et. al., Ref. 28.

67. L. P. Le, G. M. Luke, B. J. Sternlieb, Y. J. Uemura, J. H. Brewer, T. M. Riseman, D. C. Johnston, L. L. Miller, Y. Hidaka, and H. Murakami, *Hyperfine Interactions*, 63, 279 (1990); E. Torikai et. al., Ref. 28.

68. E. Torikai, K. Nagamine, H. Kitazawa, I. Tanaka, H. Kojima, S. B. Sulaiman, Sudha Srinivas, and T. P. Das, *Hyperfine Interactions*, 79, 921 (1993).

69. R. Saito, H. Kamimura, and K. Nagamine, *Physica*, C185-189, 1217 (1991).

70. T. McMullen, P. Jena, and S. N. Khanna, *International Journal of Modern Physics*, B5, 1579 (1991).

71. B. Hitti et. al., Ref. 28.

72. E. Torikai et. al., Ref. 28.

73. Tina Briere, O. Donzelli, and T. P. Das, *Indian Jour. Physics*, (in press).

74. O. Donzelli, Tina Briere, and T. P. Das, *Solid State Comm.*, 90, 663 (1994).

75. D. Coffey and S. A. Trugman, *Phys. Rev. Lett.*, 69, 176 (1992); P. Allemand, K. C. Khemani, A. Koch, F. Wudl, K. Holczer, S. Donovan, G. Grüner, and J. D. Thompson, *Science*, 253, 301 (1991).

76. S. Sayito and A. Oshiyama, *Phys. Rev. Lett.*, 66, 2637 (1991); S. C. Erwin and M. R. Pederson, *Phys. Rev. Lett.*, 67, 1610 (1991); S. C. Erwin and W. E. Pickett, *Science*, 254, 842 (1991).

77. A. F. Hebard, M. J. Rosseinsky, R. C. Haddon, D. W. Murphy, S. H. Glarum, T. T. M. Palstra, A. P. Ramirez, and A. R. Kortan, *Nature*, (London), 350, 600 (1991).

78. S. Ijima, *Nature*, 354, 56 (1991); R. E. Smalley, *Materials Science and Engineering*, B19, 1 (1993).

79. H. W. Kroto and M. Jura, *Astron. Astrophys.*, 261, 275 (1992).

80. "Fullerenes: Past and Present", David M. Cox, at this Conference.

81. "Orientational Order in Fullerenes", S. Moss, at this Conference.

82. "Electron State and Superconductivity in Alkali-doped $C_{60}$", S. Satpathy, at this Conference.

83. "The Physics and Chemistry of Fullerenes", Robert N. Compton, at this Conference.

84. "Electronic Excitations in Nanoscale Tubules with Helical Symmetry", P. J. Liu-Chung and A. K. Rajagopal, at this Conference.

85. E. J. Ansaldo, C. Niedermayer and C. E. Stronach, *Nature*, 353, 129 (1991); E. J. Ansaldo, J. J. Boyle, C. Niedermayer, G. D. Morris, J. H. Brewer, C. E. Stronach, and R. S. Carey, *Z. Phys.*, B86, 317 (1992).

86. R. F. Kiefl, J. W. Schneider, A. MacFarlane, K. Chow, T. L. Duty, T. L. Estle, B. Hitti, R. L. Lichti, E. J. Ansaldo, C. Schwab, P. W. Percival, G. Wei, S. Wlodek, K. Kojima, W. J. Romanow, J. P. MacCauley, Jr., N. Coustel, J. E. Fischer, A. B. Smith III, *Phys. Rev. Lett.*, 68, 1347 (1992).

87. R. F. Kiefl, T. L. Duty, J. W. Schneider, A. MacFarlane, K. Chow, J. W. Elzey, P. Mendels, G. D. Morris, J. H. Brewer, E. J. Ansaldo, C. Niedermayer, D. R. Noakes, C. E. Stronach, B. Hitti, and J. E. Fisher, *Phys. Rev. Lett.*, 69, 2005 (1992).

88. J. A. Howard, *Chem. Phys. Lett.*, 203, 540 (1993).

89. J. R. Morton, K. F. Preston, P. J. Krusic and L. B. Knight, Jr., *Chem. Phys. Lett.*, 204, 481 (1993).

90. E. Roduner, *The Positive Muon as a Probe in Free Radical Chemistry*, New York, (1988), p.3; A. Schenck, Ref. 10, Chap. 7.

91. Paul W. Percival and Stanislaw Wlodek, *Chem. Phys. Lett.*, 196, 317 (1992).

92. S. K. Estreicher, C. D. Latham, M. I. Heggie, R. Jones, and S. Oberg, *Chem. Phys. Lett.*, 196, 311 (1992).

93. M. J. S. Dewar and W. Thiel, *J. Am. Chem. Soc.*, 99, 4899 (1977).

94. T. A. Halgren and W. N. Lipscomb, *J. Chem. Phys.*, 58, 1569 (1973); D. S. Marynick and W. N. Lipscomb, *Proc. Natl. Acad. Sci. (USA)*, 79, 1341 (1982); L. Throckmorton and D. S. Marynick, *J. Comput. Chem.*, 6, 652 (1985).

95. N. Sahoo, S. K. Mishra, K. C. Mishra, A. Coker, C. K. Mitra, L. C. Snyder, T. P. Das, and A. Glodeanu, *Phys. Rev. Lett.*, 50, 913 (1983).

96. T. A. Claxton and S. F. J. Cox, *Chem. Phys. Lett.*, 207, 31 (1993); M. A. Boxwell, T. A. Claxton and S. F. J. Cox, *J. Chem. Soc. Faraday Trans.*, 89, 2957 (1993).

97. R. Tycko, G. Dabbagh, R. M. Fleming, R. C. Haddon, A. V. Makhija, and S. M. Zahurak, *Phys. Rev. Lett.*, 67, 1886 (1991).

98. Taesul Lee, and T. P. Das, *Phys. Rev.*, A6, 962 (1972); J. E. Rodgers, Taesul Lee, and T. P. Das, *Phys. Rev.*, A7, 51 (1973); J. E. Rodgers, and T. P. Das, *Phys. Rev.*, A8, 3675 (1973); J. E. Rodgers, Ph.D Thesis, University of California, Riverside, (1972) (Unpublished).

99. N. Paschedag, H. U. Suter, P. M. Maric, and P. F. Meier, *Phys. Rev. Lett.*, 70, 154 (1993).

# FULLERENES: PAST AND PRESENT

Donald M. Cox

*Corporate Research Laboratories*
*Exxon Research and Engineering Company*
*Annandale, New Jersey 07060*

The Characterization of fullerenes, newly synthesized fullerene compounds and carbon nanotubes is of critical importance in understanding the structure, composition, chemistry and physics of such materials. In this talk some of the important events which led to the discovery of fullerenes and nanotubes will first be described and then some recent results involving fullerenes, fullerene derivatives, and carbon nanotubes will be discussed.

First, interesting aspects involving the discovery, isolation and both molecular and solid state characterization of $C_{60}O$, the first fullerene epoxide will be discussed, e.g. the molecular rotational reorientation dynamics of solid $C_{60}O$ will be contrasted and compared to thos of solid $C_{60}$. Second, our mass spectrometric studies of highly fluorinated $C_{60}$ show the first direct experimental evidence for the existence of perfluorinated $C_{60,}$ $C_{60}F_{60}$. We also observe highly epoxided species, $C_{60}F_{2x}O_m$, for $M=0$ to 5, in high abundance. Lastly, some of our recent work in synthesizing and characterizing the carbon nanotubes will be discussed.

# SYNCHROTRON X-RAY STUDIES OF LOCAL ORIENTATIONAL ORDER IN SINGLE CRYSTAL $C_{60}$ ABOVE ITS ORDERING TRANSITION

S. C. Moss

*Physics Department*
*University of Houston*
*Houston, Texas*

We discuss two sets of synchrotron (NSLS) x-ray experiments on vapor-phase-grown single crystal $C_{60}$. The first utilizes Bragg peak intensities to derive the anisotropic number density on a single buckyball in the orientationally disordered phase at ~300K. This result represents the deviation from spherical symmetry of the $C_{60}$ molecule in the hindered rotation phase. From this result we can determine both the distribution of 2-, 3 and 5-fold axes with respect to the cubic coordinates and, via statistical mechanics, the cubic crystal field acting on a single buckyball. The second set of experiments explores the diffuse scattering which contains both single particle, or molecule, (self) scattering and the two particle (distinct) scattering which reflects the orientational pair correlations and which peaks, above the phase transition, at the fcc Brillouin zone boundaries. This distinct scattering carries information on the orientational pair interaction potential.

# PHASE SEPARATION IN HYDROGEN LOADED 123 COMPOUNDS OBSERVED BY CU NQR AND NMR

H. Lutgemeier, *IFF, Forschungszentrum Jülich,*

H. Schone, *IFF, Forschungszentrum Jülich and College of William & Mary, Williamsburg, VA*

and Y. Xu, *Inst fur Schicht und Ionentechnik, Forschungszentrum, Jülich*

The transition from superconducting to antiferromagnetic behavior in RE $Ba_2Cu_3O_7H_x$, RE = Y, Gd, has been studied by Cu NMR and NQR. In the Gd compounds the spectra of Cu(1) and Cu(2) sites could be separated by the strong relaxation induced by the magnetic moments of Gd on the neighboring Cu(2) sites. Already at small x of about 0.5 antiferromagnetic Cu(2) sites are revealed by an AF-NMR spectrum. At the same x we find a strong broadening of the "normal" NQR spectrum and a new sharp line at 27 MHz for $^{63}Cu$ indicating a new EFG of the Cu(1) sites in the AF phase. The transformation to the AF state is complete at about x=2.0. Annealing of the doped material at 350°C leads to the normal oxygen depleted 123 compounds proving that the cation lattice is not distorted even at high x. As in the case of pure YBCO no static magnetic moments exist at the Cu(1) sites of the hydrogen containing phase.

# THE INFLUENCE OF SPATIAL CORRELATIONS OF CHARGED IMPURITY CENTERS ON CARRIER SCATTERING IN SEMICONDUCTOR SYSTEMS

J. R. Anderson

*Department of Physics*
*University of Maryland*
*College Park, Maryland*

Studies of HgSe:Fe have shown that spatial correlations of the dominant scatterers, $Fe^{3+}$ ions, can lead to enhanced mobilities in this system when it contains both $Fe^{3+}$ and $Fe^{2+}$ ions. Such correlations occur when an impurity level becomes resonant with the Fermi level in a conducting band and the scatterers form a quasi-periodic structure. Pressure, temperature, impurity concentration, and magnetic field are parameters that can be used to examine the influence of these spatial correlations. There is also preliminary evidence that control of DX centers and doping in quantum wells and heterostructures of GaAs/AlGaAs can lead to reduced scattering due to quasi-periodic arrangements of the charged scattering centers. Experimental results will be compared with theory for systems including HgSe:Fe, $Hg_{1-x}Mn_xTe$, and GaAs/AlGaAs.

# HYPERFINE INTERACTIONS IN HIGH-TEMPERATURE SUPERCONDUCTORS

Detlef Brinkmann

*Physik Institut*
*University of Zurich*
*CH-8001 Zurich, Switzerland*

**Nuclear magnetic resonance (NMR) and nuclear quadrupole resonance (NQR) have made significant contributions to the understanding of the electronic properties of high-temperature superconductors. In this talk we will discuss studies performed in our laboratory on some typical representatives of the Y-Ba-Cu-O family such as $YBa_2Cu_3O_7$, $YBa_2Cu_4O_8$ and $Y_2Ba_4Cu_7O_{15}$.**

**Topics of the discussion will be electric field gradients, Knight shift and spin-lattice relaxation. These parameters are intimately connected with hyperfine fields and local electronic susceptibilities. We will show how studies of the temperature, pressure and orientational dependence of these parameters yield information concerning the superconductivity mechanism and other relevant features at the atomic scale.**

# SIZE-MISMATCH EFFECTS IN SEMICONDUCTOR ALLOYS

M. F. Thorpe

*Department of Physics and Astronomy*
*Michigan State University*
*East Lansing, MI USA*

**We show that the structural distortions that occur in crystalline semiconductor alloys can be understood in terms of the topological rigidity parameter. This single parameter allows all EXAFS results for nearest and next nearest neighbor mean distances to be described. This approach is very successful in III-V and II-VI semiconductors but fails in group IV alloys like Si/Ge. We propose a new structural model for group IV alloys (both crystalline and amorphous) that emphasizes the importance of H in tying off dangling bonds and reducing the topological rigidity of the networks.**

# LOCAL ORDER IN COLD CONDENSED LEAD CHLORIDE - THE NATURE OF THE PHASE TRANSFORMATION GLASS-CRYSTAL

Ralf Zuber and Konrad G. Weil

*Institut für Physikalische Chemie, Technische Hochschule Darmstadt*
*Petersenstrasse 20*
*6100 Darmstadt, Germany*

We condense lead chloride from the vapor onto lithium fluoride windows, kept at temperatures below 80 K. The films are characterized by UV-VIS - spectroscopy. From the shape of the exciton absorption band one can conclude that the films are highly disordered.

UV-Vis spectroscopy also enables us to monitor the disorder/order-transition at higher temperatures. From the kinetics of this transition it can be demonstrated that it is a first-order-phase-transition, initiated by a nucleation step.

The radial distribution function of the disordered state was determined by electron diffraction. It can be definitely demonstrated that the disordered state is non-crystalline and that its short range order is different from those in the crystalline and in the molten state.

A phenomenological model is presented which explains the metastability of the noncrystalline phase.

# PROBING THE STRUCTURE OF POROUS MEDIA USING NMR SPIN ECHOES

Pabitra N. Sen[1], Partha P. Mitra[1,2] and L. M. Schwartz[1]

[1] *Schlumberger-Doll Research, Old Quarry Road*
*Ridgefield, CT 06877-4108*
[2] *Department of Physics, Harvard University*
*Cambridge, MA 02138*

**The NMR pulsed field gradient spin echo (PFGSE) amplitude M(k,t) is a direct measure of the diffusion propagator of the molecules of a fluid diffusing in the pores of a porous medium. For small values of $k = \gamma g \delta$, where g is the gradient strength, $\delta$ is the duration of the gradient pulse, and $\gamma$ the gyromagnetic ratio, the PFGSE amplitude gives the diffusion coefficient D(t). The exact short-time diffusion coefficient $D(t)/D_o = 1 - (4\sqrt{D_o}tS)/(9V_p\sqrt{\pi}) - D_otHS/(12V_p) + \rho tS/6V_p + ...$, provides an important method for determining surface to volume ration $S/V_p$. Here the mean surface curvature $H = \langle 1/R_1 + 1/R_2 \rangle$. Combining early D(t) with the magnetization decay gives the surface relaxivity $\rho$. The long-time effective diffusion constant derived from PFGSE gives the information on tortuosity of the connected space. The diffusion coefficient measured by PFGSE equals that derived from electrical conductivity only when enhanced surface relaxation for magnetization is absent. Exact solutions with partially absorbing boundary conditions for a periodic structure and numerical simulations are used to illustrate the influence of $\rho$ on diffusion coefficient. M(k,t) can be well represented by a convolution of the connected structure factor of the pore space with an appropriate Gaussian propagator. This ansatz provides a model-independent way of obtaining the structure factor.**

# STRUCTURAL ORDER IN GLASSES*

W. J. Bresser and P. Boolchand

*Department of Electrical & Computer Engineering*
*University of Cincinnati, OH 45221-0030*

**Bulk glasses formed by supercooling melts display short- as well as medium-range order. In the past ten years substantial progress has emerged on the crucial issue of medium range order in glasses. Local probes like the Mössbauer effect, NQR, and NMR have played a significant role in this connection. Experiments on good glass formers ($As_2Se_3GeSe_2$) provide clear evidence of 'floppy molecular fragments' forming building blocks of these covalent networks. Specific examples of medium range order in semiconducting glasses will be discussed. Existence of medium range order apparently extends to oxide glasses as well, as revealed by recent neutron scattering experiments.**

*Supported by NSF grant 2DMR-92-07166

# NOVEL MAGNETISM TAILORED WITH NANOCRYSTALLINE STRUCTURES


K. V. Rao

*Department of Condensed Matter Physics*
*Royal Institute of Technology*
*Stockholm, Sweden*


It may be fair to claim that we are now entering a period of new materials and man-made structures inspired by a theme 'from clusters to crystals' not necessarily with translational periodicity. While maintaining some of the unique size-dependent properties even in the tailored bulk form, these materials are mostly metastable but with a rather long lifetime at ambient temperatures. These materials may be in the form of multilayers, or dispersed media where the unique particle size effect are preserved and can be exploited for applications.

In this presentation I will describe some examples of our recent studies of nanostructured materials with properties which cannot be attained otherwise:

1) a new class of tailored zero magnetostrictive materials in which nano-sized selected elemental entities are homogeneously dispersed in a glassy metallic media by controlled melt-spinning. These materials are specially suited for a new generation of read/write heads, -an emerging need of the future high density information storage devices.

2) A new class of extremely soft magnetic materials with energy loss characteristics better than that for any other material we know today. These are especially suited for high frequency transformer applications.

3) Novel giant-magnetoresistive materials for potential applications is the measurement of extremely weak magnetic fields with a precision almost approaching that of SQUIDS. This unusual giant-magnetoresistive property first discovered in tailored thin magnetic multilayers, and as yet unexplained from a theoretical point of view, has now also been observed in nano-structured heterogenous media. These materials are already at a developing stage of producing novel applications ranging from ABS systems in cars to high density information storage read/write devices.

In all the materials described above their unusual properties are derived from their unique size dependent magnetism further tailored by controlling the entities in terms of a balance between magnetic mean-free path verses their separation distance.

Finally, from a basic physics point of view we will demonstrate for the first time, that vanadium carries a magnetic moment as high as 0.4 Bohr magneton per atom, when tailored into a few nano-layered sandwiches with molybdenum.

Capabilities to produce nanostructured materials in various forms has no doubt opened new challenges to develop novel materials with unique properties for specialized applications.

# PARTICIPANTS


**R. ANDERSON**
Physics Department
University of Maryland
College Park, MD 20742

**S. N. BEHERA**
Institute of Physics
Sachivalaya Marg
Bhubaneswar, India

**S. M. BOSE**
Department of Physics
Drexel University
Philadelphia, PA 19104

**W. BRESSER**
Dept. of ECE, Mail location 30
University of Cincinnati
Cincinnati, Ohio 45221

**D. BRINKMAN**
Physik Institut
Universitat Zurich
CH-8001 Zurich
Switzerland

**M. CIFTAN**
U.S. Army Research Office
Physics Division
PO Box 12211
Research Triangle Park, NC 27709

**G. S. COLLINS**
Physics Department
Washington State University
Pullman, Washington 99164

**R. COMPTON**
Oak Ridge National Laboratory
Bldg. 4500-S, S114
Oak Ridge, TN 37831-6033

**D. COX**
Exxon Research Laboratory
Route 22 East
Clinton Township
Annandale, N.J. 08801

**T. P. DAS**
Physics Department
S.U.N.Y. at Albany
Albany, New York 12222

**P. ENTEL**
Dept. of Theoretical Physics
University GH Duisberg
Postfach 101629
D-4100 Duisberg 1, Germany

**F. HAGELBERG**
Physics Department
S.U.N.Y at Albany
Albany, New York 12222

**S. Hanna**
Department of Physics
Stanford University
Stanford, California 94305-4090

**P. Jena**
Physics Department
Virginia Commonwealth Univ.
Richmond, VA 23284

**E. A. Karlsson**
Department of Physics
University of Uppsala
Box 53- S-75121 Uppsala
Sweden

**S. N. Khanna**
Department of Physics
Virginia Commonwealth Univ.
Richmond, VA 23284

**R. Kirby**
4024 Oakley Green
Sarasota, FL 34235

**B. Lindgren**
Department of Physics
Uppsala University
Box 530
S-75121 Uppsala
Sweden

**S. D. Mahanti**
Physics Department
Michigan State University
East Lansing, MI 48824

**K. C. Mishra**
OSRAM-Sylvania Inc.
100 Endicott St.
Danvers, MA 01923

**S. C. Moss**
Physics Department
University of Houston
4800 Calhoun Road
Houston, TX 77204-5504

**K. B. Nielsen**
Institute of Physics &
Astronomy
Aarhus University
DK-8000 Aarhus C
Denmark

**A. K. Rajagopal**
Naval Research Laboratory
Code 6860.1
Washington, D.C. 20375-5000
Dr. Rajagopal

**B. K. Rao**
Physics Department
Virginia Commonwealth Univ.
Richmond, VA 23284

**K. V. Rao**
Dept. of Condensed Matter Physics
Royal Inst. of Tech.
Stockholm, Sweden S10044
Sweden

**S. S. Satpathy**
Department of Physics
University of Missouri
Columbia, MO 65211

**G. Schatz**
Facultat fur Physik
University of Konstanz
Universitat Strasse 10
Postfach 5560
D7750 Konstanz, Germany

**H. SCHONE**
Physics Department
College of William & Mary
Williamsburg, VA 23185

**P. N. SEN**
Schlumberger-Doll Research Center
PO Box 307
Old Quarry Road
Ridgefield, CT 06877

**K. R. SUBBASWAMY**
Department of Physics
University of Kentucky
Lexington, KY 40511-8433

**S. B. SULAIMAN**
Department of Physics
University of Sains Malaysia
Minden Penang, Malaysia

**M. F. THORPE**
Physics Department
Michigan State University
East Lansing, MI 48824

**L. TTERLIKKIS**
College of Pharmacy
Florida A&M University
Tallahassee, FL 32307

**J. VAIL**
Department of Physics
University of Manitoba
Winnipeg, Manitoba
Canada R3T 2N2

**K. G. WEIL**
Physikalische Chemie
Technische Hochschule Darmstadt
D-6100 Darmstadt
Petersenstrasse 20, Germany

# AUTHOR INDEX

# SUBJECT INDEX